Plant Genetic Engineering

PLANT BIOTECHNOLOGY SERIES

Edited by Don Grierson, B.Sc., Ph.D., C.Biol., F.I.Biol., Professor of Plant Physiology and Head of AFRC Research Group in Plant Gene Regulation, Department of Physiology and Environmental Science, University of Nottingham.

Most important phases of plant growth, development, and reproduction that affect food production and quality involve changes in plant gene expression. These include processes such as germination, flowering, ripening, seed development, formation of storage organs, senescence, and responses to alterations in the environment and to pathogens. Recent developments in plant physiology, biochemistry, and molecular biology mean that we are beginning to understand these processes in molecular terms. Most importantly, the availability of plant genetic engineering techniques means that we can do experiments *in planta.* This makes it possible to provide new scientific information about macro-molecular interactions and control mechanisms, and to alter the properties of plants in a directed and controlled way. We are now on the threshold of a new era, poised to exploit these techniques in manipulating the growth, development, and productivity of plants, making them more suitable for our needs.

This series reviews recent developments in plant biotechnology, shows how scientific understanding leads to commercial applications, and discusses opportunities and problems in this rapidly developing field of plant molecular breeding.

PLANT BIOTECHNOLOGY

Volume One

Plant Genetic Engineering

Edited by

DON GRIERSON, B.Sc., Ph.D., C.Biol., F.I.Biol.
Professor of Plant Physiology
and Head of AFRC Research Group in Plant Gene Regulation
Department of Physiology and Environmental Science
University of Nottingham

Blackie
Glasgow and London

Published in the USA by
Chapman and Hall
New York

Blackie & Son Limited.
Bishopbriggs, Glasgow G64 2NZ
and
7 Leicester Place, London WC2H 7BP

Published in the USA by
Chapman and Hall
a division of Routledge, Chapman and Hall, Inc.
29 West 35th Street, New York, NY 10001-2291

Softcover reprint of the hardcover 1st edition 1991

First published 1991

British Library Cataloguing in Publication Data

Plant genetic engineering.
1. Plants. Genetic engineering
I. Grierson, D. (Donald)
581.15

ISBN-13:978-94-010-6668-6

Library of Congress Cataloguing-in-Publication Data

Plant genetic engineering / edited by D. Grierson.
p. cm.—(Plant biotechnology: v. 1)
Includes bibliographical references and index.
ISBN-13:978-94-010-6668-6 e-ISBN-13:978-94-009-0403-3
DOI: 10.1007/978-94-009-0403-3

1. Crops—Genetic engineering. 2. Plant genetic engineering.
3. Transgenic plants. I. Grierson, Donald. II. Series.
SB123.57.P57 1990
631.5'23—dc20 90-2420
CIP

Phototypesetting by Thomson Press (India) Limited, New Delhi

Preface

Plant biotechnology offers important opportunities for agriculture, horticulture, and the food industry by generating new transgenic crop varieties with altered properties. This is likely to change farming practices, improve the quality of fresh and processed plant products, and reduce the impact of food production on the environment. The purpose of this series is to review the basic science that underpins plant biotechnology and to show how this knowledge is being used in directed plant breeding. It is intended for those involved in fundamental and applied research on transgenic plants in the academic and commercial sectors. The first volume deals with plant genes, how they work, and their transfer from one organism to another. Authors discuss the production and evaluation of the first generation of transgenic crops resistant to insects, viruses and herbicides, and consider aspects of gene regulation and targeting of their protein products to the correct cellular location. All the contributors are actively engaged in research in plant biotechnology and several are concerned directly with its commercial applications. Their chapters highlight the importance of a fundamental understanding of plant physiology, biochemistry, and cell and molecular biology for the successful genetic engineering of plants. This interdisciplinary approach, which focuses research from traditionally separate areas, is the key to further developments which are considered in subsequent volumes.

Don Grierson

Contributors

Alan B. Bennett Mann Laboratory, Department of Vegetable Crops, University of California, Davis, CA 95616

John W.S. Brown Department of Biological Sciences, University of Dundee DD1 4HN

Sherri M. Brown Monsanto Co., 700 Chesterfield Village Parkway, St. Louis, MO 63198

Kenneth W. Buck Department of Biology, Imperial College of Science, Technology and Medicine, London SW7

John Draper Department of Botany, University of Leicester, University Road, Leicester LE1 7RH

Charles S. Gasser Department of Biochemistry and Biophysics, University of California, Davis, CA 95616

Angharad M.R. Gatehouse Plant Molecular Biology Group, Department of Biological Sciences, University of Durham, South Road, Durham DH1 3LE

John A. Gatehouse Plant Molecular Biology Group, Department of Biological Sciences, University of Durham, South Road, Durham DH1 3LE

Vaughan A. Hilder Plant Molecular Biology Group, Department of Biological Sciences, University of Durham, South Road, Durham DH1 3LE

Katherine W. Osteryoung Mann Laboratory, Department of Vegetable Crops, University of California, Davis, CA 95616

Colin Robinson Department of Biological Sciences, University of Warwick, Coventry CV4 7AL

Rod Scott Department of Botany, University of Leicester, University Road, Leicester LE1 7RH

Alan G. Smith Department of Horticultural Science, University of Minnesota, St. Paul, MN 55108

David Stalker Calgene, Inc., 1920 Fifth Street, Davis, CA 95616

Robbie Waugh Scottish Crop Research Institute, Invergowrie, Dundee DD2 5DA

Contents

1 Plant gene structure and expression

R. WAUGH and J.W.S. BROWN

1.1 Introduction

The cloning of the first plant nuclear protein-encoding gene almost a decade ago signified the beginning of plant molecular biology as a discipline in its own right. Since then the number of plant genes isolated and studied by molecular techniques has increased exponentially. Up to 1986 the majority of cloned plant genes belonged to relatively few gene families— storage protein genes, rubisco small subunit genes and chlorophyll a/b-binding protein genes— which reflected both the ease of isolation of their respective mRNAs and a widespread interest in chloroplasts. More recently, improved techniques and expertise and the potential for plant genetic manipulation have stimulated the detailed investigation of a much broader range of plant genes. The genes encoding many enzymes and structural proteins involved in fundamental cellular processes and specific developmental processes such as ripening, flowering and embryogenesis have now been isolated. As a result more information on the biochemical and physiological processes in which the gene products are involved is becoming available.

Previous analyses of plant gene structure (Messing *et al.*, 1983; Heidecker and Messing, 1986; Casey and Domoney, 1987; Shotwell and Larkins, 1989) were based on DNA sequence data of a limited range of plant genes and on comparisons with already well-characterised animal and yeast genes. Although a broader database now exists, these earlier analyses are still extremely valuable because the molecular biology of these particular genes is currently the most advanced. In this review, we will aim, firstly, to summarise the basic structure of plant genes as formulated in the previous analyses, as the general consensus is still valid. Secondly, we will highlight some of the newer, more functional information which is becoming available on both protein- and RNA-coding genes. Much of this information, particularly in the latter section, has been established from studies on animals and yeast. We have, therefore, tried to relate this information to the often purely structural information available in plants. Finally, we have tried to restrict the references to more general review articles, although for the most recent work this has not been possible. Thus, by necessity, our coverage has been selective, and in many instances other examples which are equally valid have been omitted.

1.2 Protein-coding genes

1.2.1 *Structure of protein-coding genes*

Protein-coding genes in eukaryotes are transcribed by RNA polymerase II. Comparisons of plant and other eukaryotic gene sequences have identified, at a general level, conservation of regulatory or structural sequences in the 5′, 3′ and coding regions which are required for accurate and coordinated gene expression. The sequences 5′ of the transcription start site are generally referred to as the promoter region and are involved in transcriptional initiation and regulation, whereas those 3′ of the coding sequence are essential for transcription termination and polyadenylation. In the 5′ region, two sequence elements in particular have been identified: the 'TATA' and the 'CAAT' boxes (Figure 1.1). The TATA box, located 25–40 nucleotides (nt) upstream of the transcription start site, is required for the accurate initiation of transcription by RNA polymerase II. In plants, the TATA box has the

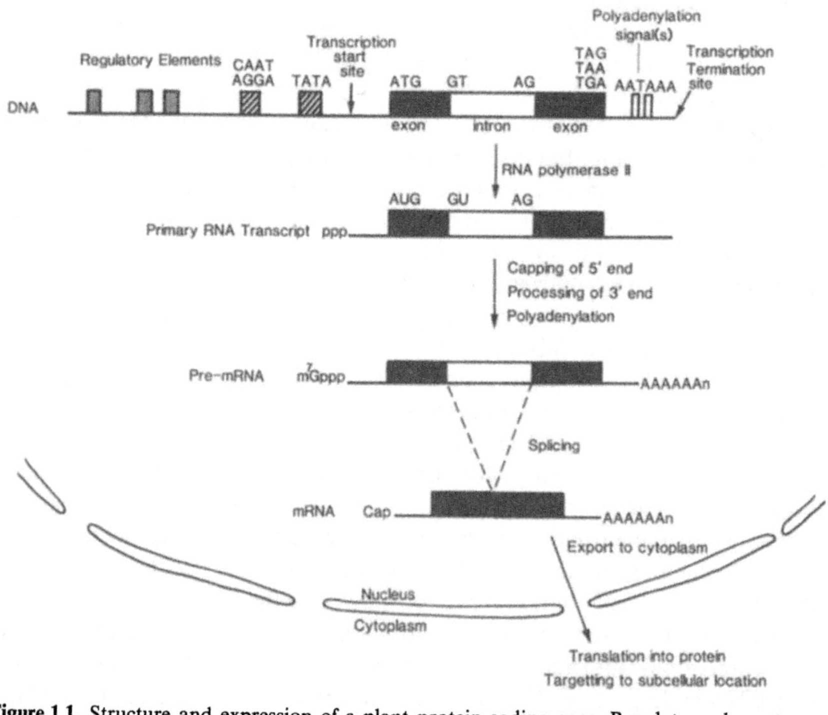

Figure 1.1. Structure and expression of a plant protein-coding gene. Regulatory elements are shown by hatched boxes in the 5′ region of the gene. The coding region is shown by filled boxes (exons) interrupted by open boxes (introns) and containing translation, initiation and termination codons. The 3′ region contains the signals for polyadenylation and termination. The primary mRNA transcript is processed by capping, 3′ end processing, polyadenylation and splicing (intron removal) involving the conserved intron sequences -GU and AG-, before being exported into the cytoplasm for translation.

consensus sequence TCACTATATATAG (Joshi, 1987a). Further upstream, the CAAT box is considered important in regulating the transcriptional activity of the promoter. The importance of these sequences to the expression of a range of plant genes has been confirmed by deleting the regions of the 5′ flanking sequence containing them and examining the effect on expression in transformed plant cells. In some plant genes alternative element, the 'AGGA' box, may replace the CAAT box function. Other 5′ DNA regions that govern developmental regulation and tissue-specific expression are discussed later. Transcription begins 40–80 nt upstream of the start of the coding region of the gene, usually at an adenosine nucleotide (Joshi, 1987a). The primary mRNA transcript (pre-mRNA) therefore contains an untranslated 5′ leader sequence.

A more complex 5′ gene structure involving two promoters (P1 and P2) each with more than one transcription start site, has been described for some of the zein storage protein genes from maize (Langridge and Feix, 1983). One of the two transcription start sites in the P2 promotor region, 50 base pairs (bp) upstream of the coding sequence, has been shown to be a functional 3′ splice site instead of a transcription start site (Brown and Feix, 1990), raising the possibility of an intron in the 5′ untranslated region of the otherwise intronless zein mRNAs. A further example of a gene containing more than one promoter is the nuclear-encoded chloroplast glutamine synthetase gene of bean (J.M. Cock and J.V. Cullimore, unpublished results).

The coding region of the mRNA transcript begins with the translation initiation codon, AUG, followed by an open reading frame ending in one of three stop codons, UAG, UAA or UGA (Figure 1.1). The translation initiation codon, AUG, is usually found within a conserved sequence, AACAAUGGCT (Joshi, 1987a), which is selected by the 40S ribosomal subunit scanning the mRNA sequence from the 5′ end (Rogers *et al.*, 1985; Kozak, 1987). The open reading frame of the coding sequence of the pre-mRNA may be interrupted by one or more non-coding intervening sequences (introns) which all have highly conserved -GT and AG-dinucleotides at their 5′ and 3′ ends respectively (Breathnach and Chambon, 1981).

Finally, the 3′ region contains a conserved sequence element, AATAAA (Joshi, 1987b), presumably specifying the site of addition of a stretch of adenosine nucleotides (polyadenylation), as in animal cells, and less well-defined sequences promoting dissociation of the polymerase from the template and the termination of transcription (Figure 1.1). The primary transcript (pre-mRNA) therefore consists of the coding sequence with intron sequences (if present in the gene) and 5′ and 3′ untranslated regions which contain all the information required for its processing to a mature mRNA.

1.3 Regulation of plant gene expression

The regulation of gene expression is governed by the rates of transcription of the gene, processing of the pre-mRNA, export of the mRNA into the cytoplasm, degradation of the mRNA in both the nucleus and cytoplasm and translation. Although gene expression is mainly regulated at the transcriptional level, reflected in the abundance of mRNA transcripts of a gene, the various post-transcriptional processes are also important, particularly because of their effect on mRNA stability. The regulation of gene expression at the transcriptional level in plants is likely to be extremely complex, involving dynamic interactions between regions of the DNA, structural proteins and other specialised proteins (transcription factors) in general and gene-specific ways. The flanking regions of protein-coding genes contain an array of sequence elements which can bind proteins, and the interactions between these modular DNA–protein complexes lead to a refined regulation of rates of transcription (reviewed by Dynan, 1989). The study of the regulation of plant gene expression has proceeded at several different levels: transcriptional regulation involving the identification of conserved sequence elements, expression analysis in transformed plant tissue, DNA–protein binding analyses, DNA methylation, and post-transcriptional regulation involving mRNA processing and stability.

1.3.1 *Transcriptional regulation*

The analysis of transcriptional regulation of plant gene expression is possible as a result of the isolation of genomic clones and the development of stable plant transformation and transient gene expression technology (see Chapter 2). Transcriptional regulation studies involve the identification of conserved sequence elements in the flanking regions of genes, the analysis of expression of mutated gene sequences in transformed plant tissue and the examination of interactions between specific sequence elements (*cis*-acting sequences) and transcription factors (*trans*-acting factors). After introducing the principles behind such investigations, some of the most advanced studies of the identification of *cis*-acting sequences and *trans*-acting factors and their effect on gene expression will be described.

1.3.1.1 *Conserved sequence elements.* In addition to the basic TATA box promoter element, required for correct transcription initiation, and common upstream elements such as the CAAT or AGGA boxes, other *cis*-acting sequence elements regulate or modulate gene transcription. Generally, sequences which stimulate transcription of genes are called enhancers, and sequences which reduce transcription levels are called silencers. Enhancer elements have been extensively studied in viral systems and classically exert their effect independently of orientation and over a large distance. Other sequence elements which regulate the transcription of specific classes of genes have also been widely reported. Unlike enhancers, these elements are generally located at relatively fixed positions in the flanking regions of their respective genes.

The initial stage in an analysis of the 5′ flanking region of a gene is the localisation of the transcription start site by S1 mapping or primer extension analyses. Having located the transcription start site, the TATA and CAAT elements can usually be identified upstream in the 5′ DNA sequence. Other potential *cis*-acting DNA sequence elements which may be involved in transcriptional regulation can be identified by a comparison of 5′ flanking sequences of genes belonging to the same multigene family or genes transcribed under similar conditions, in the same stage of development or in the same tissue or organ. Examples of such conserved sequence motifs have been described for a number of seed storage protein gene families, in genes expressed during nodulation of leguminous plants by *Rhizobium* spp., or in response to heat shock, anaerobic conditions and light (Tables 1.1 and 1.2). The value of such comparisons is limited, however, and sequences identified in this way must be shown to be of functional significance, by both expression analyses *in vivo* and *in vitro* assays involving interaction with proteins.

1.3.1.2. *Analysis of gene expression.* Analysis of gene expression in animal *in vitro* transcription systems has allowed the effect of different mutations on normal transcription to be readily assessed. The lack of a plant nuclear *in vitro* transcription system and the lack of a routine stable transformation system for monocotyledonous plants have forced investigators to develop an array of alternative methods to analyse gene expression (reviewed in Weising *et al.*, 1988). One disadvantage in expression analyses has been the use of non-homologous systems (e.g. *Acetabularia* or yeast), particularly when analysing

Table 1.1. Conserved sequence elements in plant genes.

Element	Sequence	Reference
Prolamin box	ACAtgTGTAAAGGTGAA$^{t}_{g}$NAGATGA$^{gt}_{tg}$CATGTAT	Kreis *et al.* (1986)
Legumin box	TCCATAGCCATGCATGCTGAAGAATGTC	Bäumlein *et al.* (1986)
Vicillin box	GCCACCTCaattt	Gatehouse *et al.* (1986)
Zein	CACATGTGTAAAGGT	Maier *et al.* (1987)
Soybean	CATGCAT and $\substack{a\\t\\c}$AACACA$^{aa}_{c\,\substack{t\\c}}$	Goldberg (1986); Dickenson *et al.* (1988)
Nodulin box	ATCCAAACAATCAATTTTACCAATC	Stougaard *et al.* (1987)
Heat shock[a]	CTNGAANNTTCNAG	Gurley *et al.* (1986)
Anaerobic response	CGGTTT——TGGTTT	Walker *et al.* (1987)
Wound response	T$^{g}_{t}$GTTGAAATA$^{a}_{t}$A——TAGT$^{a}_{t}$AAAT$^{aa}_{gt}$TATGA	Siebertz *et al.* (1989)

[a]*Drosophilia* heat shock consensus with which plant heat shock sequences are compared.

Table 1.2. Protein-binding sequence elements in light-regulated genes.

DNA-binding factor	Recognition box	Sequence	Genes	References
GBF	G-box	5′ TCTTACACGTGGCA 3′	*cabE*, *rbcS*, CHS	Giuliano *et al.* (1988) Schulze-Lefert *et al.* (1989) Schindler and Cashmore (1989)
GC-1	GC-rich element	12 bp sequence	*cabE*	Schindler and Cashmore (1989)
GA-1	GATA box	5′ GTGTGG–TA–TA–G 3′	*cabE*, *rbcS1A*	Schindler and Cashmore (1989)
GT-1	GT box	5′ GTGTGGTTAATATG 3′	*cabE*, *rbcS3A*	Schindler and Cashmore (1989) Green *et al.* (1988)
AT-1	AT-1 box	5′ AATATTTTTATT 3′	*cabE*, *rbcS*	Schindler and Cashmore (1989) Vatta and Cashmore (1989)

cab, cytochrome a/b-binding protein genes; *rbcS*, small subunit of rubisco genes; CHS, chalcone synthase.

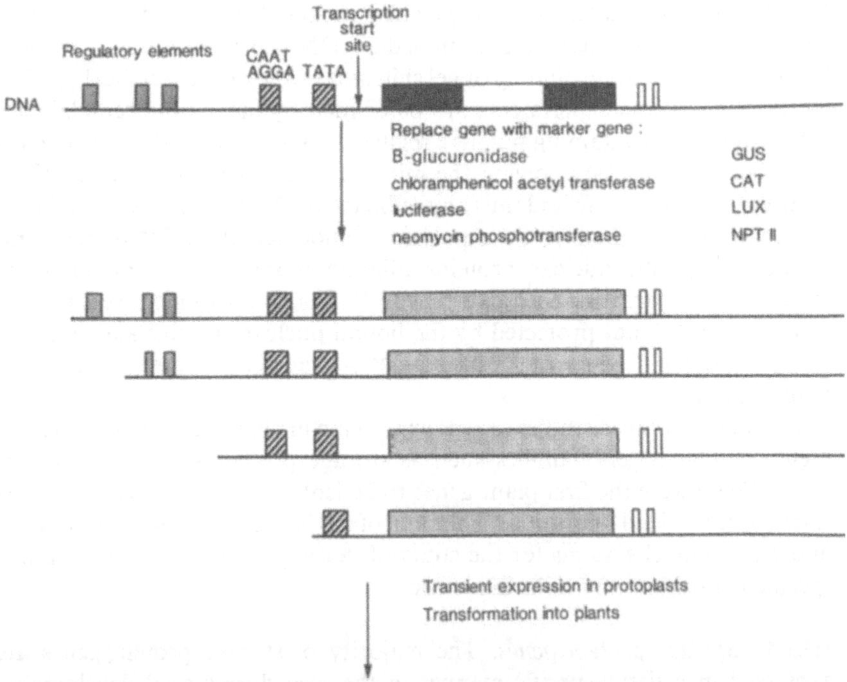

Figure 1.2. 5′deletion expression analysis. The coding region of a cloned gene can be replaced with a reporter gene and successive deletions removing various regulatory sequences are produced. Following transient or stable transformation, the effect of loss of these sequences on expression under particular conditions can be assayed.

monocotyledonous genes. However, as techniques constantly improve, especially with stable transformation of monocotyledonous tissues with the particle gun (Klein *et al.*, 1987), it is becoming possible to transform homologous cells and tissue, thus increasing the relevance of results on gene expression.

The standard procedure for investigating the function of different regions of the promoter is usually to replace the coding sequence with that of a reporter gene, e.g. chloramphenicol acetyltransferase (CAT), neomycin phosphotransferase (NPT II), β-glucuronidase (GUS) or luciferase (LUX) (see Weising *et al.*, 1988), and to construct a series of 5′ deletions which successively remove portions of the 5′ flanking sequence (Figure 1.2). The 5′ deletion constructs are then reintroduced into plant cells where their effect on the levels of expression and response to external stimuli are measured by assaying expression of the reporter gene. Although deletion analyses provide some delimitation of essential sequences in plant promoters, deletions are often large and furnish only limited information about the regions essential for transcriptional regulation. Internal deletions of the 5′ flanking sequence and site-directed mutagenesis of specific nucleotide sequences are used in the same way and can provide more refined information on gene expression.

1.3.1.3 Trans-*acting factors.* The precise location of the important sequences within the 5′ region may be determined by DNA–protein binding analyses based on nitrocellulose binding or gel shift assays and, more specifically, DNA footprinting. These experiments are performed by incubating radiolabelled DNA fragments containing putative regulatory sequences with crude nuclear extracts and either (1) applying the mixture to a nitrocellulose filter which tightly binds the protein and thus any radioactive DNA which has bound to it, (2) separating the mixture on a polyacrylamide gel where DNA fragments bound by specific nuclear proteins migrate more slowly than unbound fragments or (3) treating the mixture with DNAse I or exonuclease III which digests the DNA not protected by the bound nuclear proteins and creates a gap or 'footprint' when the resulting DNA fragments are separated on a sequencing gel.

The most advanced studies of gene expression in plants have centred around highly regulated gene families such as storage protein and light-regulated genes. These were the first plant genes to be isolated, are transcribed at high levels, which aided the initial isolation of cDNA clones, and are of great interest as model systems for the study of developmental regulation, tissue-specific expression and light induction.

1.3.1.4 *Storage protein genes.* The majority of storage protein genes are expressed in a tissue-specific manner in the seed during seed development (notable exceptions are the tuber-specific proteins of potato). They are expressed at high levels, often in a highly coordinated fashion. Many seed storage protein genes have been isolated from a number of the major cereal and legume crop plants. A decade of intensive study has led to a detailed understanding of their gene structure and to some extent expression (Messing *et al.*, 1983; Heidecker and Messing, 1986; Casey and Domoney, 1987; Goldberg *et al.*, 1989; Shotwell and Larkins, 1989).

The signals by which the developing seed regulates the tissue-specific and coordinated expression of storage proteins are currently being investigated by the procedures described above. DNA sequence motifs conserved in the 5′ region of storage protein genes from a wide range of crop species have been identified (Table 1.1), and several have been confirmed as being important for the correct expression of their respective genes by deletion analysis and expression studies *in vivo*. One of the first motifs identified was the prolamin or −300 box found in wheat, barley and maize storage protein genes (Forde *et al.*, 1985). Deletion analyses of the 5′ flanking region of a wheat glutenin gene in transformed tobacco have shown that a region of the 5′ sequence containing the prolamin box confers tissue-specific gene expression (Colot *et al.*, 1987; Robert *et al.*, 1989). A 15 bp region about 260 nt upstream of the transcription start site which is conserved in maize zein storage protein genes (Table 1.1) and is similar to the prolamin box was the first potential regulatory sequence to be defined by DNA–protein binding procedures (Maier *et al.*,

1987). This sequence was shown to bind proteins in an endosperm extract (zein genes are only expressed in the maize endosperm) and not in an extract from maize coleoptiles, pointing to possible tissue-specific binding (Maier *et al.*, 1988). The interacting DNA sequence in this case has the ability to form a hairpin loop structure (Figure 1.3). Such DNA secondary structures may be one signal by which *trans*-acting factors recognise and bind to individual regulatory motifs. They are considered important in the assembly of transcription factor complexes on the promoters of a number of genes (e.g. rRNA genes, see below). However, this conserved sequence was found to be unimportant for expression in transformed tobacco, which, given the generally poor level of expression and regulation of zein genes in dicotyledonous tissue, presumably reflects differences in *trans*-acting factors or recognition in the non-homologous tissue (Schernthaner *et al.*, 1988).

The glycinin and β-conglycinin storage protein genes and the Kunitz trypsin inhibitor genes are specifically expressed in developing soybean embryos. Deletion analyses of members of both glycinin and β-conglycinin gene families have shown that each contains two regions essential for correct expression. The -1 to -159 region in the β-conglycinin and the -1 to -65 region in the glycinin genes are required for embryo-specific expression (Chen *et al.*, 1986, 1988; and see Nielsen *et al.*, 1989), while sequences further

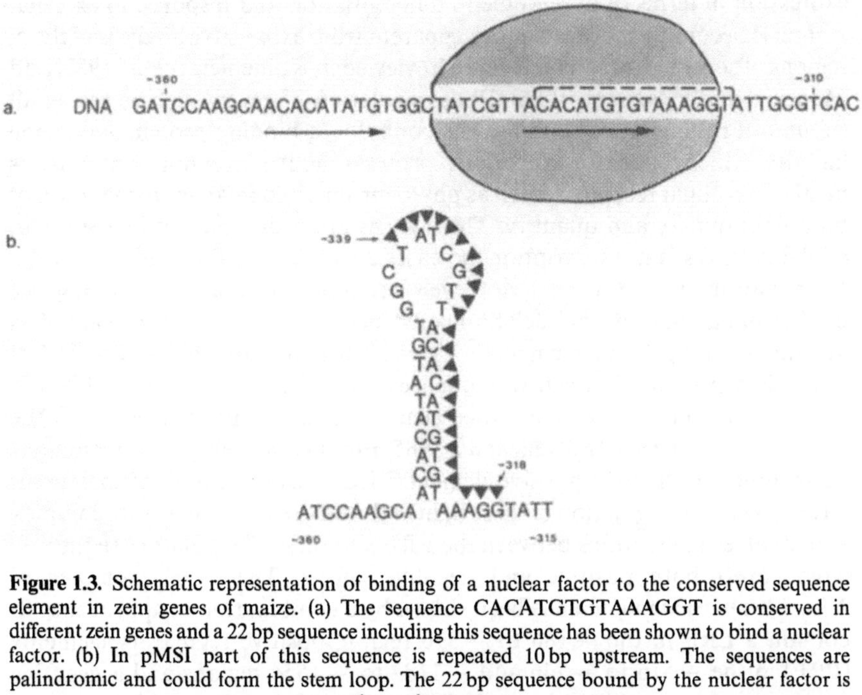

Figure 1.3. Schematic representation of binding of a nuclear factor to the conserved sequence element in zein genes of maize. (a) The sequence CACATGTGTAAAGGT is conserved in different zein genes and a 22 bp sequence including this sequence has been shown to bind a nuclear factor. (b) In pMSI part of this sequence is repeated 10 bp upstream. The sequences are palindromic and could form the stem loop. The 22 bp sequence bound by the nuclear factor is shown by arrows.

upstream are essential for quantitative regulation. The latter regions contain several CATGCATG and CAACACA sequence motifs thought to be important for the binding of transcription factors required for the expression of legume seed protein genes (Harada *et al.*, 1989). The glycinin 5′ region also contains a legumin box found in many legume storage protein genes (Bäumlein *et al.*, 1986) (Table 1.1). The −69 to −257 region of a different β-conglycinin gene has also been shown to be important for enhanced seed-specific gene expression. Joining this region to a CaMV 35S promoter enhanced the expression of a marker gene in maturing seeds. This occurred independently of orientation and indicated that the sequences between −69 and −257 can function as a seed-specific transcriptional enhancer (Chen *et al.*, 1988). More recently, two nuclear DNA-binding factors have been identified which interact with specific sequences in this region (Allen *et al.*, 1989). In addition to the above examples, other seed storage protein gene families are also under investigation by expression and DNA–protein binding analyses. Sequences involved in seed-specific gene expression have been identified for legumin genes of pea (Shirsat *et al.*, 1989), hordein genes of barley (Marris *et al.*, 1988), helianthin genes of sunflower (Jordano *et al.*, 1989) and the phaseolin (Bustos *et al.*, 1989) and lectin (Riggs *et al.*, 1989) genes of bean.

1.3.1.5 *Light-regulated genes.* The complexity of the regulation of plant gene expression in terms of development, differentiation and response to external stimuli is becoming more and more apparent from exhaustive analysis of the 5′ regions of some light-regulated genes (reviewed in Kuhlemeier *et al.*, 1987, and Manzara and Gruissem, 1988). Photoregulation of genes, such as the small subunit of rubisco (*rbcS*) and the chlorophyll a/b-binding protein (*cab*) gene families which code for abundant proteins active in photosynthesis, is mediated by light receptors, such as phytochrome, and involves recognition of both light quality and quantity. These genes are transcribed at low levels in etiolated leaves, but transcription increases 20- to 50-fold after light induction. Examination of mRNA transcript levels and reporter gene activity of a series of 5′ terminal and internal deletion constructions in transformed plants has identified two upstream elements which confer light inducibility. The first of these elements was confined to sequences around the TATA box, while the second was present between this element and 410 nt upstream of the transcription start site (Morelli *et al.*, 1985; Fluhr *et al.*, 1986). This region has been termed the light-responsive element (LRE). However, subsequent studies have shown that regulation of these light-induced genes is much more complex and involves interactions between the LRE and other 5′ regulatory sequences which exert both positive and negative effects. The *cabE* promoter of *Nicotiana plumbaginifolia* has been shown by a combination of approaches to contain a LRE in the −186 to −396 region, positive regulatory elements (PRE) in the −516 to −746 and −1182 to −1554 regions and a negative regulatory element (NRE) in the −746 to −1182 region in addition to the sequences surrounding the TATA box (Castresana *et al.*, 1988). These regions have been extensively analysed by DNA-protein binding assays, and at least five different *trans*-acting nuclear proteins bind to these regions (Table 1.2). The proteins, GT-1, GBF, GA-1, GC-1 and AT-1, are all present in nuclear extracts from light-grown leaf material and bind to multiple copies of specific sequences in the *cis*-acting regulatory elements described above. Similar binding sequences have been determined in other light-responsive genes, for example the chalcone synthase genes of parsley (Schulze-Lefert *et al.*, 1989) and a rice phytochrome gene (Kay *et al.*, 1989).

In the case of AT-1, which interacts with a negative regulatory element in the *cabE* gene, binding appears to be regulated by reversible phosphorylation (Datta and Cashmore, 1989). This finding, in conjunction with more extensive information from animals and yeast, suggests that phosphorylation of DNA-binding proteins is important as a means of controlling their activation and deactivation and thereby regulating gene expression. Indeed, the GT-1 factor is present in both dark- and light-grown tissue but only binds in the latter, suggesting that it is modified, perhaps by phosphorylation, or is in some way prevented from binding to its target site in the dark (Green *et al.*, 1987).

The specific DNA sequences with which some of these factors interact have also been identified in the 5′ regions of unrelated genes, suggesting that they may play a general rather than specific role in regulating gene expression. For example, the consensus sequence for the binding of AT-1 has been identified in the 5′ region of a soybean leghaemoglobin gene (Jensen *et al.*, 1988), a bean β-phaseolin gene (Bustos *et al.*, 1989) and a sunflower helianthinin gene (Jordano *et al.*, 1989). Similarly, GC-1- and GA-1-binding motifs have been found in the 5′ regions of plant viral promoters (Mullineaux *et al.*, 1984; Benfey *et al.*, 1989). The overall transcriptional modulation of gene expression, therefore, must involve a balance of general DNA–protein interactions and specific interactions brought about in response to other signals.

Conserved sequences present in the 5′ regions of other plant genes which are coordinately regulated have also been identified. A sequence contained within − 139 to − 102 bp upstream of the transcription start site of plant genes induced by *Rhizobium* has been shown by deletion analysis to be required for expression during nodulation (Stougaard *et al.*, 1987). DNA–protein binding sites have also been identified for genes involved in tomato fruit ripening (Cordes *et al.*, 1989). The *cis*-acting sequences required for wound induction are contained in the − 571 to − 1022 region of the 5′ flanking sequence of *wun-1* (Siebertz *et al.*, 1989). Currently, in order to examine these interactions in more detail, clones of DNA-binding protein genes are being isolated by screening cDNA expression libraries with labelled DNA fragments containing the target sequence. Although not isolated by this procedure, the *opaque-2* gene of maize, which has for many years been known to regulate zein gene expression, has recently been cloned (Hartings *et al.*, 1989). The protein sequence encoded by this gene contains many of the structural motifs described for animal DNA-binding proteins. In the near future the detailed characterization of the genes and proteins of other plant transcription factors will provide further information on the mechanism of DNA–protein interactions and the overall regulation of transcription.

1.3.2 *DNA methylation*

For transcription factors to interact with specific sequence elements the DNA must satisfy both structural and conformational criteria. One structural feature which has long been implicated in determining the accessibility of DNA sequence to the transcriptional machinery is DNA methylation. Although it is well established that plant DNA contains nucleotides which are modified post-replicatively by DNA methylation, the precise function of the modifications is not clear. The principal methylated nucleotide is 5-methylcytosine (5-mC), which occurs in CpG dimers or the sequence CpNpG. Plant DNA is more extensively methylated than animal DNA, 5-mc occurring, for example, in up to 32% of all the cytosine residues in maize DNA. However, the overall pattern of C-methylation is not random but depends on both sequence and genome location, occurring mostly in repeated or heterochromatic regions. Furthermore, the pattern of methylation is likely to be dynamic, with increasing evidence indicating a correlation between low levels of methylation (hypomethylation) and gene expression.

The main approach to investigating the effect of methylation on gene expression in plants involves digestion of the DNA with restriction enzymes which are resistant or sensitive to 5-mC in their recognition sequences and by subsequent detection of the DNA restriction fragments under investigation by Southern blotting. Direct evidence for the role of methylation has come from studies on the deactivation of foreign DNA integrated at random sites in the plant genome (Hepburn *et al.*, 1983). By sequentially transforming tobacco plants with two different selective reporter gene constructs, Matzke *et al.* (1989) found that in a high proportion of the secondary transformants expression of the first reporter gene was lost as a result of DNA methylation. However, when these plants were cross- or self-pollinated, the expression of the original reporter gene was restored in progeny which no longer contained the second introduced reporter gene construct. The introduction of a second foreign gene was therefore demonstrated to suppress the expression of an unlinked foreign gene via directed DNA methylation. Similar results have also been observed for endogenous plant DNA sequences, particularly rRNA genes (Kaufman *et al.*, 1987) and the transposable elements *Ac* and *Mu* (Chomet *et al.*, 1987). The activation of genes by incubation of plant cells in the presence of 5-azacytosine, which becomes incorporated into the genome and inhibits the maintenance methylase, has also been demonstrated. For example, the demethylation of the promoter region of maize zein storage protein genes in embryo cells due to incorporation of 5-azocytosine correlated with increased transcriptional activity (Bianchi and Viotti, 1988).

The implications of the relationship between hypomethylation of at least the promoter region and gene expression are that *in vivo* methylation either prevents the binding of specific transcription factors that are essential for promoter activity or promotes the organisation of the DNA into chromatin structures that are inaccessible to the transcriptional machinery. Whatever the model, methylation and demethylation of specific sequences is clearly an important and well-regulated process.

1.3.3 *Post-transcriptional regulation*

In the nucleus immediately after transcription, the primary mRNA (pre-mRNA) transcripts are bound with a number of specific proteins, forming complexes called heterogeneous ribonucleoprotein (hnRNP) particles. The pre-mRNAs undergo a series of processing events prior to their export into the cytoplasm where they can be translated into protein. The pre-mRNAs are modified at both their 5′ and 3′ ends, and intron sequences are removed. The efficiency and extent to which particular mRNAs are processed, and their relative stabilities, may play a critical role in the regulation of expression.

Whether a gene is under transcriptional or post-transcriptional control is determined from *in vitro* nuclear run-on or run-off analyses. Nuclei are isolated and previously initiated mRNA transcripts allowed to elongate in the presence of a radioactive nucleotide. The labelled mRNAs are hybridised to specific gene sequences, and the level of hybridisation is used as a measure of the relative *in vivo* transcription rates of the genes under study. If these rates correlate with the relative steady-state levels of mRNAs *in vivo*, the genes can be assumed to be under direct transcriptional control, as has been shown for a number of plant genes, particularly storage protein genes (Walling *et al*., 1986; Goldberg *et al*., 1989; Nielsen *et al*., 1989). Discrepancies between the estimated transcription rates and steady-state levels provide an indication of post-transcriptional regulation whereby the turnover or stability of different mRNAs varies. Post-transcriptional regulation has been demonstrated for the autoregulation of phytochrome (Quail *et al*., 1986), legumin gene expression in response to sulphur deficiency (Beach *et al*., 1985) and soybean genes expressed during embryogenesis (Goldberg *et al*., 1989). However, the exact nature of such post-transcriptional regulation has not yet been determined but must involve one or more of the aspects of mRNA processing which can affect mRNA stability.

1.3.4 *mRNA processing*

1.3.4.1 *Capping*. At the 5′ end of the transcript, the addition of a guanosine residue through a 5′–5′ triphosphate linkage is followed by its rapid methylation at the N7 position to form the characteristic 5′–7-methylguanosine residue termed the ‘cap’. The specificity of this process appears to be mediated solely by the triphosphate group at the 5′ end of the mRNA. The cap structure itself helps to protect the pre-mRNA from exonucleolytic degradation and appears to be required, at least *in vitro*, for pre-mRNA splicing. In animal systems there is evidence that the binding of cap-binding proteins to the 5′ end of the mRNA may be involved in the initiation of translation.

1.3.4.2 *Polyadenylation.* The processing of the 3′ end of pre-mRNAs occurs in the nucleus prior to other processing events such as splicing. Following endonucleolytic cleavage of the primary transcript, a stretch of adenosine nucleotides (the poly-A tail) is added. In contrast to the promoter regions, far less information is available about the 3′ end sequences responsible for termination and processing. The modifications which occur at the 3′ end are largely specified by sequences present in the original DNA template. In animals, two sequence elements have been identified which are essential for correct 3′ end formation: the highly conserved AATAAA box situated 10–30 nt upstream of the poly-A addition site and the less highly conserved YGTGTTYY motif (where Y is a pyrimidine) immediately 3′ to the cleavage site (reviewed by Platt, 1986). A compilation of plant nuclear gene 3′ sequences has indicated that, although often imperfect, similar sequence motifs are also present in plant genes (Joshi, 1987b). Of the plant genes surveyed, only 39% showed an unaltered AATAAA motif, and 54% exhibited one base substitution. Selection of the cleavage and poly-A addition sites appears to involve recognition of the upstream and downstream *cis*-acting signals by the processing machinery. Although the similarities in sequences between plants and animals imply conservation of function, there are several examples of animal 3′ processing signals being poorly recognised in transformed plants. When chimeric genes containing poly-A signals from the human growth hormone and animal viral genes were introduced into tobacco, utilisation of the 3′ signals was both aberrant and inefficient. In all three cases mRNAs ending at an authentic plant poly-A signal could be detected (Hunt *et al.*, 1987). It appeared that the animal AATAAA poly-A signal was recognised but that the plants chose to process at a different site. Such observations suggest that either the sequence requirements for 3′ end processing and polyadenylation or their utilisation differ significantly between plants and animals.

The effect of removing the 3′ end region on the expression of a chimeric reporter gene has recently been investigated in both single cells and transformed plants (Ingelbrecht *et al.*, 1989). By removing the AATAAA- and YGTGTTYY-like boxes by deletion, longer transcripts were produced, but at only 8–10% of the level of the normal mRNAs, suggesting that the longer mRNAs were intrinsically unstable. Polyadenylation appeared to have occurred at cryptic sites in the plasmid vector sequence which resembled authentic plant 3′ signals. When the 3′ sequences in the original construct were replaced with 3′ regions from a selection of plant genes (e.g. *rbcS*, extensin, 2S seed protein and chalcone synthase) which are normally expressed at different levels in different tissues, the authentic 3′ signals were utilised, but the level of reporter gene expression varied by as much as 60-fold. The observed differences were suggested to be the result of altered mRNA stability because the respective mRNA levels paralleled the natural levels of expression of these genes. In plant cells, the length of the poly-A tail has recently been shown to have a marked effect on mRNA stability (Gallie *et al.*, 1989), and therefore factors influencing polyadenylation may have profound effects on gene expression. The integrity of the AATAAA box may represent one level at which polyadenylation efficiency, and hence differential mRNA stability, may occur. Many plant genes contain several potential AATAAA motifs in their 3′ ends (Joshi, 1987b). Although very little is known about the mechanism of poly-A site selection, if as suggested 3′ end structure influences mRNA stability, then the choice of poly-A signals represents another level at which regulation can occur.

In animal systems the poly-A tail is believed to protect the mRNA from ribonuclease digestion and to have a role in mRNA transport to the cytoplasm and translation (Brawerman, 1987, 1989). As with the cap structure, it appears that specific proteins called poly-A-binding proteins (PABP) recognise and bind to the poly-A tail, contributing to mRNA stability. In plants a PABP has been isolated from germinating pea seeds (Sieliwanowicz, 1987), and although it seems to have a role in translation initiation *in vitro* its precise role *in vivo* is unclear. Thus, although many of these features remain to be tested experimentally, whatever their functional significance, the origin, selection and structure of the 3′ end clearly play a fundamental role in plant gene expression.

1.3.4.3 *Pre-mRNA splicing.* The majority of plant nuclear protein-coding genes contain intervening sequences (introns) which, in order to generate an open reading frame for translation, must be removed from the pre-mRNA by splicing. Pre-mRNA splicing is a complex process involving interactions between intron sequences, U-type small nuclear ribonucleoprotein particles (UsnRNPs) and other less well-defined proteins. Various UsnRNPs (U1, U2, U4, U5 and U6) have been shown to be essential for the splicing of pre-mRNAs in animals and yeast (reviewed in Krainer and Maniatis, 1988). The mechanism of splicing has been extensively studied due to the availability of *in vitro* splicing systems. The initial event in splicing is the assembly of a ribonucleoprotein complex on the pre-mRNA, called the spliceosome. Splicing then proceeds by a two-step pathway. Firstly, the pre-mRNA is cleaved at the 5′ splice site and a 5′–2′ phosphodiester bond is formed between the phosphate at the 5′ end of the intron and the 2′-OH of an adenosine residue (branch point) lying within the intron near the 3′ splice site. Secondly, the pre-mRNA is cleaved at the 3′ splice site, releasing the intron as a lariat RNA, and the two exons are ligated together (Figure 1.4). Splicing therefore requires the accurate and efficient recognition of the exon–intron borders in the pre-mRNA molecules. One level of this recognition process is the interactions between the conserved sequences at the 5′ and 3′ splice sites and the branch point of the

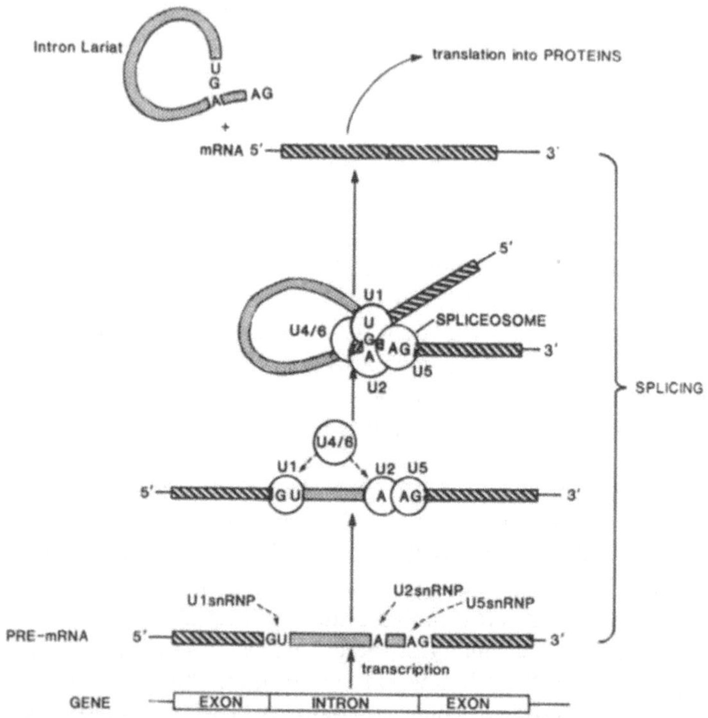

Figure 1.4. Pre-mRNA splicing. A protein-coding gene is transcribed into a pre-mRNA containing 5′ and 3′ non-translated regions and exon (hatched boxes) and intron (dotted boxes) sequences. The intron contains the conserved -GU and AG- dinucleotides at its 5′ and 3′ ends respectively and an internal branch point sequence (A). UsnRNPs associate with the pre-mRNA to form the spliceosome in which the intron sequence is excised as a lariat RNA and the exon sequences joined to produce a mature mRNA which can be exported to the cytoplasm for translation into a protein.

intron with U1, U5 and U2 snRNPs. In the case of the U1 and U2 snRNAs, base pairing between the UsnRNA and the intron sequences is involved in these interactions.

Due to the absence of a plant *in vitro* splicing system, very little is known about splicing in plants except that plant introns, like animal introns, exhibit great sequence variation. The 5′ and 3′ splice junctions are similar between plants and animals, with the exception that plant 3′ splice junctions have a much less emphasised polypyrimidine stretch which is required for splicing in animals. Indeed, functional analysis of plant pre-mRNA splicing using synthetic intron constructs in transformed protoplasts has suggested that the high AU content of plant introns may be one of the more important determinants of the splicing process in plants, replacing the requirement for a polypyrimidine stretch at the 3′ splice site (Goodall and Filipowicz, 1989). Attempts to splice plant introns in animal *in vitro* and *in vivo* systems have met with mixed success, and as yet no animal intron has been successfully spliced in plants (Barta *et al.*, 1986; Brown *et al.*, 1986a; Wiebauer *et al.*, 1988). Although branch point-like sequences have been observed in plant introns (Brown, 1986), functional branch points have not been demonstrated in plant pre-mRNA splicing, and a synthetic intron lacking a branch point sequence is still spliced in protoplasts (Goodall and Filipowicz, 1989), suggesting that although the basic process is similar differences do exist.

A difference in splicing between the two plant families, monocotyledons and dicotyledons, has been proposed on the basis of two monocot. introns being inefficiently spliced when transferred to dicot. plants (Keith and Chua, 1986). Subtle differences have been suggested to exist between the 5′ and 3′ splice junctions of monocots. and dicots. (Hanley and Schuler, 1987) to explain this difference in splicing. However, due to the overall variation in splice junction and intron sequences, and given that some monocot. genes are processed in dicots., such differences may only be valid on a gene-to-gene or intron-to-intron basis.

The role of some intron sequences in influencing levels of gene expression has been demonstrated for maize *Adh-1* and *bronze-1* introns (Callis *et al.*, 1987; Oard *et al.*, 1989). The mode of action by which this occurs is not well understood but would appear to be an effect on mRNA transport or stability, possibly involving the efficiency with which particular pre-mRNAs are spliced. Particular mRNAs may require specific factors for both hnRNP formation and for splicing to occur, implying a regulation of expression at the level of pre-mRNA splicing. Different pre-mRNAs may have different interactions with snRNPs, and variations in UsnRNA sequence or protein components may affect the affinity of different UsnRNPs for particular introns.

Differential or alternative splicing of pre-mRNAs has been demonstrated in a number of animal gene systems (reviewed in Leff *et al.*, 1986; Breitbart *et al.*, 1987). Alternative splicing produces more than one mRNA transcript from a single pre-mRNA by the selection of different 5′ and 3′ splice sites in different combinations. In most cases the selection of particular splice sites is involved in the tissue-specific and developmental expression of these genes. In plants the only documented example of alternative splicing of a nuclear-encoded gene is that of rubisco activase from spinach and *Arabidopsis thaliana* (Werneke *et al.*, 1989). The alternative splicing in the rubisco activase pre-mRNAs utilises two 3′ splice sites separated by 22 nt. Pre-mRNA transcripts retaining the 22 nt of the intron have a reading frame shift and are translated into a smaller protein with a different C-terminus. The functional relevance of this processing has not yet been determined but, along with the examples of genes with multiple transcription start sites or multiple promoters, the prospect of differential or alternative splicing playing a role in plant gene regulation can be considered.

1.3.4.4 *mRNA editing.* mRNA editing is the process whereby individual bases are substituted post-transcriptionally to alter specifically the nucleotide sequence of a mRNA and thus alter the amino acid sequence of the protein product. Editing has recently been shown to occur in wheat mitochondrial cytochrome C oxidase where certain cytidine residues are converted to uridine (Covello and Gray, 1989). The generality of this process and whether it occurs in the nucleus remains to be demonstrated.

1.4 Translational control

In the ribosome-scanning model proposed for translation in animals, the cap structure and other factors are recognised by the 40S ribosomal subunit which scans along the RNA until the first AUG is encountered (Kozak, 1983, 1987). The 60S subunit then associates with the 40S subunit to form the functional 80S ribosome. The model predicts that the first AUG codon encountered will be utilised as the initiation codon. Ribosome scanning has also been demonstrated in plants (Rogers *et al.*, 1985). However, some plant genes, for example the lectin genes of *Phaseolus vulgaris*, contain AUG codons in their 5′ untranslated leader sequences. Other genes, for example a soybean and maize actin gene and a maize glutathione-*S*-transferase gene, contain short reading frames in their 5′ untranslated regions (reviewed in Joshi, 1987a). The sequences surrounding these upstream AUG codons are often poorly matched with the consensus and are presumably read over by the ribosome. Where short reading frames exist the possibilities of translational control, as found in yeast, (reviewed in Hinnebusch, 1988) or reinitiation of translation (Peabody and Berg, 1986) must be considered. The sequences surrounding the initiation codon from −3 to +4 have also been shown to be important for optimum levels of translation. In plants these sequences form the consensus AACAAUGGCT (Joshi, 1987a), which differs from those found in animals and yeast (Lüctke *et al.*, 1987). Most importantly, there is a marked preference for an adenosine residue at position −3.

In the cytoplasm, a separate set of proteins from those making up the nuclear hnRNPs associates with the mRNAs to form cytoplasmic or polysomal messenger ribonucleoprotein particles (cmRNPs and pmRNPs respectively). The pmRNPs represent actively translated mRNAs, while non-translated mRNAs are sequestered in cmRNPs. The mechanism of transport of the mRNAs from nucleus to cytoplasm and the transition from nuclear to cytoplasmic RNPs, and from cmRNPs to pmRNPs, are unknown areas at present. In animals, the complexing of the mRNAs into mRNPs has been suggested to play an important role in mRNA stability. Whether this occurs via sequestration of the mRNA into complexes where it is protected from nuclease attack is unknown, but certain mRNAs do exist for long periods of time in specialised tissues (e.g. seeds) in a form which is resistant to degradation. In *Drosophila*, for example, ribosomal protein cmRNAs are translated at specific stages during development and this regulation has been proposed to occur via interactions of the cmRNAs with small cytoplasmic ribonucleoprotein particles (scRNPs) (Schmid *et al.*, 1984).

Another factor which regulates translational efficiency is the abundance of amino-acylated tRNAs. Due to the degeneracy of the genetic code, several different codons in the RNA sequence can specify the incorporation of the same amino acid into the polypeptide chain. In different species, however, all possible codons are not present randomly, leading to the over-representation of certain preferred codons (codon usage bias). Codon usage bias is usually compensated for by a similar bias in the abundance of amino-acylated tRNAs. Thus, if the mRNA to be translated contains a high proportion of rare codons, its translation will be impaired in comparison with a mRNA containing only preferred codons as a result of the availability of the required charged tRNAs.

mRNA secondary structure also influences translation *in vitro*, where the formation of stable secondary structures in the 5′ region of a mRNA has been shown to prevent efficient recognition of the initiation codon by the ribosome. Stable secondary structures formed between the 5′ and 3′ ends of a zein gene transcript, which contained an imperfect inverted repeat in the 5′ and 3′ untranslated regions, drastically reduced translation. Deletion of the inverted repeats resulted in a 50-fold increase in the rate of translation (Spena *et al.*, 1985).

1.5 Differential expression

1.5.1 *Multigene families*

A major consideration of plant gene expression is gene organisation and, in particular, the organisation and expression of individual gene members of multigene families. Many of the plant genes so far isolated are members of multigene families (e.g. storage protein, small subunit of rubisco (*rbcS*), leghaemoglobin, nodulin and chalcone synthase genes). Multigene families have evolved from single ancestral genes by duplication and amplification events through unequal crossing over and chromosomal mutations. Plants, by virtue of the vegetative growth phase employed by many species, may be tolerant to the accumulation of mutations in genes, particularly when multiple copies are present. In addition, the high incidence of polyploidy leads to natural increases in gene or allele number and to introgression of genetic material from one species to another. In many cases members of a gene family may be clustered while other members may be found as single entities. The gene clusters and the dispersed genes may be physically linked on the same chromosome or reside on different chromosomes.

A natural consequence of multigene families and their genomic organisation and mode of evolution may underlie the phenomenon of differential gene expression, where different members of a family are expressed in the same tissue to different degrees, or in different tissues. This can be explained on a number of different levels. Firstly, individual genes accumulate mutations in their 5′ and 3′ flanking regions which affect specific *cis*-acting sequences required for normal regulation of expression. The degree of homology of flanking regions varies at the sequence level, both in the number and size of point mutations and small insertions and deletions and in the length of the regions showing homology. For example, such variation is seen in the 5′ and 3′ regions of maize zein genes (Brown *et al.*, 1986b; Kriz *et al.*, 1987). Secondly, dispersed genes, by virtue of how they have arisen (translocation of chromosome segments, introgressions from other species, integration as pseudogenes), may be located in areas of the genome with varying activity where expression is expedited or depressed from normal levels. For example, two zein genes have been shown by their distal flanking sequences to be located near highly repetitive sequences which could affect their ability to be expressed (Kriz *et al.*, 1987). Finally, little is known of the nature of the mode of action, stability or availability of *trans*-acting factors in the nucleus. Classical genetic studies have identified a number of loci which regulate storage protein gene expression (e.g. the *opaque* and *floury* mutants of maize) on chromosomes other than those on which are located the zein structural genes which they affect. If the products of these regulatory loci are not available to the promoters of the structural genes on an equal basis, because of the chromosomal organisation within the nucleus or uneven diffusion or localisation in the nuclear matrix, different structural gene loci may be expressed at different levels.

Very careful investigations into the expression of individual petunia *rbcS* genes which are expressed differentially have, for the first time, demonstrated that sequences 3′ to the translation start site, in addition to 5′ sequences, are important in regulating levels of expression. The role of introns was also investigated, especially as the most highly expressed gene contained three introns while the rest of the gene family contained only two. Although effects were observed, no clear explanation of their role could be formulated (Dean *et al.*, 1988, 1989a, b). Differential expression has also been described in other *rbcS* gene families (reviewed in Manzara and Gruissem, 1988) and, for example, in the Kunitz trypsin inhibitor family of soybean (Jofuku and Goldberg, 1989).

1.5.2 *Pseudogenes*

A further consequence of the gene duplication and mutation processes, which have increased gene numbers, is the accumulation of mutated or truncated genes which, although non-functional, are observable as hybridising sequences by Southern analysis. When more than one gene is present, mutations (e.g. small insertions, deletions and base substitutions) can be tolerated at the phenotypic level. Many examples of mutated genes have been observed in storage protein multigene families. In one maize zein gene, a base substitution changes the translational initiation codon ATG to CTG so that the mRNA is presumably not translated (Spena *et al.*, 1983). Certain zein genes contain stop codons within their coding region and, given an intact promoter, may be transcribed normally but translated as truncated polypeptides (Kridl *et al.*, 1984). A legumin gene from pea, in addition to stop codons and small deletions in the open reading frame, has a large deletion in the second intron which removes the 3′ splice site (Bown *et al.*, 1985). Presumably, this gene will be transcribed but produce a truncated or mutated protein. However, in the one example where this has been examined, there was no accumulation of mRNA from a soybean Kunitz trypsin inhibitor gene which contained a frame shift mutation resulting in a premature stop codon (Jofuku *et al.*, 1989). Translocation events may separate coding regions from promoter sequences, truncate genes or bring about novel combinations of promoters and structural genes. An isolated 780 bp sequence containing a zein promoter region has been found 3.5 kb downstream from an intact zein gene but with no zein coding region within the 2 kb downstream of the truncated promoter (Quayle *et al.*, 1989). The coding sequence of a mutant glycinin gene has been split in half, probably by a chromosomal inversion (Cho *et al.*, 1989). Other mutations can be caused by the insertion of transposable elements. Transposable elements are discrete, mobile pieces of DNA which can excise from the DNA and insert at another location. Genes may be inactivated by transposons interrupting the coding sequence or separating a gene from its promoter by insertion in the 5′ transcribed region (Döring and Starlinger, 1986).

Finally, processed pseudogenes, in which reverse transcripts of mRNA are incorporated into the genome, have been observed for a potato actin gene (Drouin and Dover, 1987) and a tomato U1snRNA gene (Kiss *et al.*, 1989). Processed pseudogenes, therefore, lack promoter regions and introns and usually contain a stretch of T's at the 3′ end, corresponding to part of the poly-A tail. In contrast to some of the mutated genes, pseudogenes are unlikely to be expressed.

1.6 RNA-coding genes

Although proteins are responsible for many of the features which characterise specific cell types or responses to various stimuli, gene expression involves a complex series of interactions between the DNA and mRNAs with a range of protein and RNA molecules. The importance of RNA molecules in gene expression is perhaps reflected in the fact that only a small percentage of the total RNA synthesised in the nucleus is destined to enter the cytoplasm for translation into proteins. In contrast to protein-coding genes, which are transcribed by RNA polymerase II, processed and exported to the cytoplasm, genes coding for RNA products utilise different transcriptional components and follow alternative processing pathways.

The RNA-coding genes can be conveniently categorised according to the RNA polymerase responsible for their synthesis. Three RNA polymerase activities have been identified in the cells of higher eukaryotes based on their sensitivity to the fungal toxin α-amanitin. All have a molecular weight of greater than 500 000 daltons but differ in both the number and identity of their peptide components.

1.7 RNA genes transcribed by RNA polymerase I

RNA polymerase I is resistant to α-amanitin and is composed of 6–10 peptide subunits of molecular mass ranging from 8 to 185 kDa. It is located exclusively in the nucleolus and is responsible for transcription of the 18S, 5.8S and 25S rRNA genes which, along with 80 or so different polypeptides, form the basic components of the ribosome (reviewed in Grierson and Covey, 1988). The rRNA gene units are transcribed as a single polycistronic precursor RNA, which is subject to a range of processing steps including nucleolytic cleavage and base modifications. They are methylated at approximately 100 sites, mainly on the ribose moiety, and processed at internal, 5′ and 3′ sites prior to their association with ribosomal proteins which combine to make up the ribosomes. The 18S rRNA is destined to form part of the 40S ribosomal subunit, while the 5.8S and 25S rRNAs combine with a 5S rRNA and ribosomal proteins to form the 60S ribosomal subunit. The individual subunits are assembled in the nucleus prior to export into the cytoplasm. It has been estimated that a typical plant cell contains as many as 10 million nuclear-encoded ribosomes (80S ribosomes) both free in the cytoplasm and bound to endoplasmic reticulum. To synthesise and assemble this vast number of ribosomes, which are required to satisfy the translational demands of abundant and diverse protein-coding mRNAs, the synthesis of ribosomal components must be efficiently and coordinately maintained. A series of mechanisms have evolved which combine to sustain the required high levels of transcription.

Structural analysis of rDNA from a range of organisms including a number of plants has indicated that similar organisational arrangements have evolved in higher eukaryotes. rDNA in plants is organised in units containing 18S, 5.8S and 25S rRNA genes together with external and internal spacer regions (Figure 1.5a). The rRNA genes are highly reiterated in tandem arrays with the units varing in both length and number, even between closely related species. In maize the rRNA gene units occur at approximately 1650–11 500 copies per haploid genome with all copies located on a single chromosome. In wheat rDNA repeat units are also variable in number but rDNA loci are located on several different chromosomes (Smith and Flavell, 1974).

rRNA synthesis occurs in the nucleolus, where the heterochromatin is decondensed (reviewed in Grierson, 1982). However, even during active transcription some rDNA remains condensed in heterochromatin, indicating that the plant genome contains more rRNA gene sequences than are actually required (Flavell *et al.*, 1986). Information from chromosome manipulations and *in situ* hybridisations in wheat (especially those involving alien chromosomes) indicated that rRNA transcription at different nucleolus organiser

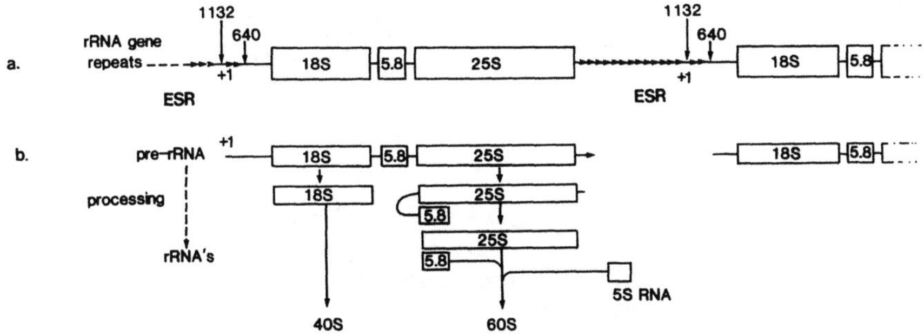

Figure 1.5. Structure and processing of rRNA gene units. (a) Structure of rRNA gene units showing the 18S, 5.8S and 25S rRNA coding regions, the external spacer region (ESR), the position of the ESR repeat units (arrows) and S1 signals 1132 and 640 nt upstream of the start of the 18S coding region. (b) Processing of the primary pre-rRNA transcript. The primary transcript is processed to produce the 18S rRNA for ribosomal 40S subunit production and the 5.8S and 25S rRNAs which are associated by base pairing and, together with a 5S RNA, produce the ribosomal 60S subunit.

regions could be induced or repressed in different nuclear backgrounds, implying differential expression of rRNA loci (Flavell *et al.*, 1986). The demand for rRNA is therefore partly satisfied by increasing the number of copies of rRNA transcriptional units at a limited number of chromosomal locations which, during periods of active transcription, become partly or fully decondensed and thus available for interaction with RNA polymerase I.

Another response to the transcriptional demands is a highly specialised promoter structure in the external spacer region (ESR) of the rRNA gene unit (Figure 1.6a). The ESR is important because it contains all the DNA signals required for the accurate and efficient transcription of rDNA. The best-characterised plant ESRs are those from maize (McMullen *et al.*, 1986; Toloczyki and Feix, 1986) and wheat (Barker *et al.*, 1988). In both cases the ESRs contain an array of direct repeats which are also characteristic of rDNA ESRs from a variety of other sources.

ESRs have been reported to contain sequences required for transcription initiation (Haltiner-Jones *et al.*, 1988), transcription termination (Grummt *et al.*, 1986; Labhart and Reeder, 1987), enhancer-like sequences (Moss, 1983) and hotspots for recombination (Voelkel-Meiman *et al.*, 1987). In both maize and wheat the transcription start sites have been located by S1 mapping and primer extension. The wheat ESR contains four sets of distinctive repeating units, all of which exhibit some homology with the putative promoter sequence.

A detailed analysis of wheat rRNA transcripts by northern blotting has been performed (Vincentz and Flavell, 1989) in an attempt to characterise sequences responsible for transcription initiation and termination and to explore whether transcriptional read-through from one gene unit to the next occurs, as has been demonstrated in *Xenopus* and *Drosophila* (Labhart and Reeder, 1986;

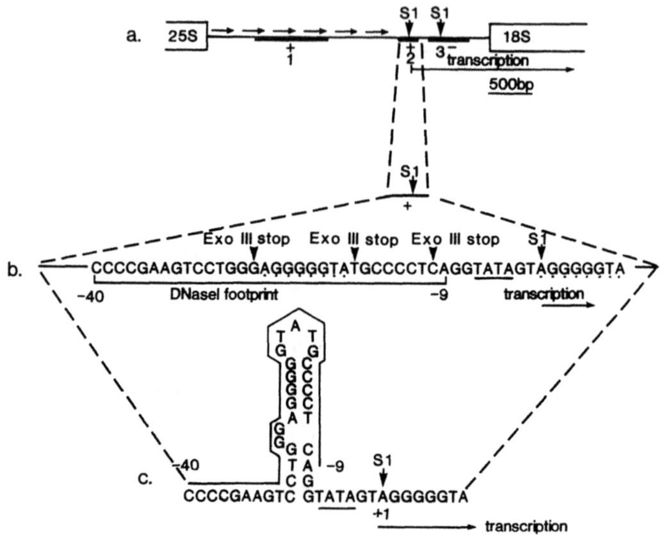

Figure 1.6. Maize rRNA gene promoter structure. (a) The external spacer region between maize rRNA gene units showing the positions of the homologous repeats (horizontal arrows) and the transcription start site and a processing site as determined by S1 mapping. The location of three DNA fragments used in DNA–protein binding assays are shown by thick black lines and + and – reflect positive and negative binding results respectively. (b) Sequence of the rRNA gene promoter region showing fine mapping of the transcription start site (+ 1) and the position of signals obtained by exonuclease III and the footprint obtained with DNAse I in DNA footprinting experiments. An 8 bp direct repeat in this region is highlighted (dots). (c) Possible secondary structure formed by the maize rRNA promoter. The start of transcription (+ 1) and S1 mapping signal and the region protected from digestion by DNAse I are indicated.

Tautz and Dover, 1986). At the 5′ end, two major S1 signals were detected 1132 nt and 640 nt upstream of the 18S rRNA gene (Figure 1.5a). These were proposed to represent the *bona fide* transcription start and a putative RNA-processing site. As the RNA species beginning at the truncated 640 nt S1 signal was present in greater abundance, it was suggested that the processing step occurred fairly rapidly after transcription. Using probe DNAs specific to the ESR and to the 3′ end of the 25S sequence, the 3′ ends of the rRNA transcripts were also located. Again, two major 3′ ends were detected, one corresponding precisely to the 3′ end of the 25S gene and another 10 nt longer. At much lower concentrations, transcripts extending more than 750 nt past the 3′ end were also found. It was therefore suggested that transcription continues through into the ESR and that the major detected 3′ ends represent the products of the rapid processing of larger precursor RNAs.

Information from several organisms has suggested that transcription by RNA polymerase I requires the formation of a preinitiation complex on the DNA template because purified RNA polymerase I fails to bind to target promoter sequences (see Roeder, 1976). RNA polymerase I can, however, bind efficiently to preformed transcription factor complexes bound to the promoter region (Cavanaugh and Thompson, 1986, Kownin *et al.*, 1987). Schmitz *et al.*, (1989) performed a series of experiments to examine protein binding to the maize rRNA gene ESR which suggested that a complex must also form on the maize ribosomal DNA prior to the binding of RNA polymerase I.

DNAse footprints of the promoter identified a region −9 to −40 nt upstream of the transcription start site which was specifically protected from digestion by DNAse I (Figure 1.6b). This position correlated well with the DNA–protein binding sites on rDNA promoters reported in animal systems which are located close to their respective transcription initiation sites (Kownin *et al.*, 1987; Reeder *et al.*, 1987). However, the boundaries of sequences protected from digestion by DNAse I do not necessarily coincide with the sequences recognised by DNA-binding proteins (Borgmeyer *et al.*, 1984). Footprints obtained using exonuclease III, which is the preferred method for multifactorial protein–DNA binding studies, revealed three distinct signals corresponding to positions −10, −17 and −26 nt, indicating the participation of more than one protein in the binding reaction (Figure 1.6b). Several RNA polymerase I promoters have been demonstrated to have the potential to form hairpin loop structures (Rogers and Bendich, 1987). The maize rRNA promoter region also has the potential to form such a structure, the stem-loop region corresponding to the area protected from DNAse I and exonuclease III digestion (Figure 1.6c). The ability of several proteins to bind to the maize rRNA promoter region was further demonstrated by south-western and gel retardation experiments. Three promoter-binding proteins were detected in maize endosperm nuclear extracts. However, as only crude nuclear extracts were used in these experiments, the nature of these proteins remains unclear.

These lines of evidence have allowed a model for rRNA gene expression in maize to be proposed (Figure 1.7). The model proposes that a protein which

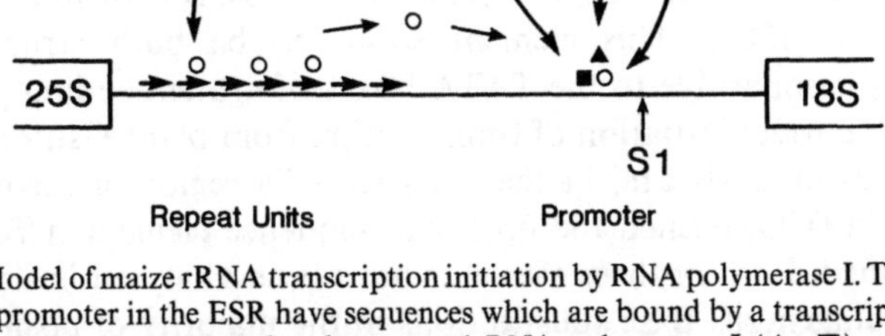

Figure 1.7. Model of maize rRNA transcription initiation by RNA polymerase I. The repeat units (arrows) and promoter in the ESR have sequences which are bound by a transcription factor (o). Once bound to the promoter, other factors and RNA polymerase I (▲, ■) can bind to the promoter/factor complex, allowing transcription to be initiated. The repeat units, by binding the initial factor, ensure a supply of factor for the promoter and thus RNA polymerase I.

recognises both the promoter and the repeats in the ESR must bind to the promoter region as a prerequisite for the association of other protein factors. The repeat units in the ESRs therefore serve to sequester the primary binding protein to ensure its continued supply at the promoter (which may have a distinct stem-loop secondary structure) where its binding promotes formation of the pre-initiation complex which is essential for the transcription by RNA polymerase I.

1.8 RNA genes transcribed by RNA polymerase II

Distinguishable from RNA polymerase I by its acute sensitivity to α-amanitin, an RNA polymerase II-like activity is responsible for the synthesis of a range of small RNA molecules. The most notable of these are the major UsnRNAs, which are integral components of the spliceosome complex responsible for the removal of introns from pre-mRNAs. Present in high concentrations in the cell nucleus, the UsnRNAs show none of the elaborate promoter sequence modifications exhibited by rRNA genes. In plants the UsnRNAs are coded for by multigene families and show heterogeneity in both size and sequence (Egeland *et al.*, 1989). Detailed analyses of *Arabidopsis* UsnRNA genes have indicated that approximately 10–15 U2snRNA and eight or nine U5snRNA genes are present in the *Arabidopsis* genome (Vankan and Filipowicz, 1988; Vankan *et al.*, 1989). In maize and potato there are between 25 and 40 U2snRNA hybridising sequences (Brown and Waugh, 1989, and unpublished observations). Recently, eight tomato U1snRNA genes have been isolated on four independent lambda clones (Abel *et al.*, 1989). On two clones, three functional U1 genes were found, indicating that at least local clustering on the genome can occur. A processed pseudogene for U1snRNA has also been characterised from tomato (Kiss *et al.*, 1989) and has probably arisen from RNA-mediated events, as has been proposed for human UsnRNA pseudogenes (Bernstein *et al.*, 1983).

Two conserved promoter elements in the 5′ upstream region which are essential for plant UsnRNA gene expression have been identified. Firstly, a TATA box, about 30 nt upstream of the transcription start site, is present in all of the functional UsnRNA genes which have been isolated from plants (Vankan and Filipowicz, 1988, 1989; Abel *et al.*, 1989; Brown and Waugh, 1989; Vankan *et al.*, 1989). This element seems to be both structurally and functionally comparable to the TATA boxes of protein-coding genes and is required for correct initiation of transcription from plant UsnRNA genes. A further 33–34 nt upstream, in the −68 to −78 region, a second element, RTCCCACATCG, termed the upstream sequence element (USE), has also been identified. Both of these elements have been analysed in detail by site-directed mutagenesis. Base substitutions in the majority of positions significantly decreased the level of expression in transformed protoplasts (Vankan and Filipowicz, 1989). Deletion and insertion mutagenesis have further indicated that the DNA sequences both between the two boxes and between the TATA box and the transcription start site are not essential for promoter activity. However, the spacing between the USE and TATA box, which in a fully functioning promoter represents three turns of the DNA helix, can critically determine the level of expression. These regulatory sequence elements are substantially different from those found in animal UsnRNA genes, which do not have a TATA sequence and contain proximal and distal sequence elements in positions -50 to -60 and -200 to -250 from the transcription start site (Dahlberg and Lund, 1988). Adjacent to the 3′ end of the plant UsnRNA-coding regions is a conserved sequence, CAN_{4-8} AGTN(A/T)AA, which may be involved in 3′ end processing or termination (Vankan *et al.*, 1989).

Processing of the major UsnRNAs follows a different pathway from that of mRNAs. Although little is known about the processing and transport of UsnRNAs and the formation of UsnRNPs in plants, this area has been well studied in animal systems (Mattaj, 1988). The initial transcripts are synthesised close to the nuclear membrane and transported immediately into the cytoplasm where processing events take place. The exact reason for this is not known, but is probably to prevent inactivation by avoiding polyadenylation, nuclear processing events or by association with hnRNP proteins (Mattaj, 1988). The primary UsnRNA transcripts contain a 7-methylguanosine cap structure like mRNAs and additional nucleotides at the 3′ end. Once in the cytoplasm, the cap structure is modified to 2,2,5-trimethylguanosine, internal bases are modified and the additional 3′ end nucleotides are cleaved to yield the mature UsnRNA. While still in the cytoplasm, snRNP proteins associate with the UsnRNA, and the assembled individual UsnRNP particles are reimported into the nucleus where they can perform their role in pre-mRNA splicing.

Efficient transcription from UsnRNA promoters also appears to require coupling to *bona fide* UsnRNA 3′ end sequences, and recognition of the 3′ end signals for correct termination requires that transcription proceed from UsnRNA promoters (Hernandez and Weiner, 1986). Transcripts initiating at normal RNA polymerase II promoters do not terminate at UsnRNA 3′ end signals, and the resulting transcripts are cleaved and polyadenylated. Substitution of mRNA promoters with those from UsnRNAs is generally not successful either, as transcription is often arrested at cryptic UsnRNA 3′ end sequences. Although this suggests that transcription takes place in a transcriptional complex, the precise nature of the interaction between the 5′ and 3′ ends has not been characterised.

In vitro competition experiments between UsnRNA promoters and other promoters have shown that normal mRNA promoters do not efficiently interfere with transcription of the UsnRNA genes. However, some strong RNA polymerase III promoters do compete with UsnRNA transcription (Dahlberg and Lund, 1988). Taken together, these experiments suggest that the polymerase II-like activity responsible for the expression of these small RNAs may be different from the classical RNA polymerase II activity described for protein-coding genes and that it may share a common transcription factor with RNA polymerase III (see below). Such a relationship is also suggested by the fact that U6snRNA genes are transcribed by RNA polymerase III in animal systems and by the similarities between some of the subunits of the three polymerase enzymes, which have presumably arisen from a single evolutionary ancestor.

1.9 RNA genes transcribed by RNA polymerase III

RNA polymerase III is responsible for the transcription of a variety of small nuclear and cytoplasmic RNA molecules, including tRNAs, U6snRNA, 4.5S and 5S RNA genes. In contrast to both RNA polymerase I and II, which recognise DNA sequences or preinitiation complexes 5′ to the coding regions of their target genes, RNA polymerase III recognises intragenic transcription signals.

Many 5S RNA-coding genes have been isolated from plants, and their structure is well established (Gerlach and Dyer, 1980, Goldsborough *et al.*, 1982). In pea for example, the genes for 5S RNAs are present in multiple copies (4000 ± 1000) arranged in three major tandem arrays of repeat units consisting of the coding region (120 nt) and a non-transcribed spacer (Ellis *et al.*, 1988). Like the ribosomal 25S, 18S and 5.8S RNAs, the 5S RNA-coding sequences exhibit a low level of sequence and length microheterogeneities, with the spacer regions being much more variable. They are generally not genetically linked to the former, but in wheat the rDNA genes are found on groups 1,5 and 6 chromosomes (Smith and Flavell, 1974); those for 5S RNAs are located at different loci on groups 1 and 5 chromosomes (McIntyre *et al.*, 1988).

tRNA genes have also been isolated from a range of plant species where they are represented in single and multiple copies, the latter often in clusters on the plant genome (G.A. Green *et al.*, 1986, 1987). However, some confusion exists with regard to the origin of several of the cloned nuclear tRNA genes, as recent evidence suggests that they may be of chloroplast or mitochondrial origin (Joyce and Gray, 1989; Szweykowska-Kulinska *et al.*, 1989). U6snRNAs and 4.5S-like RNAs have also been shown to be present in plant nuclei (Kiss *et al.*, 1987).

Although information is therefore available on 5S and tRNA gene structure and organisation in plants, almost all of the information regarding expression must be inferred from animals and yeast. In general terms the genes transcribed by RNA polymerase III can be divided into two groups based on the organisation of the DNA sequence elements that control their transcription. In both of these classes (type 1 and type 2), intragenic DNA sequences termed internal control regions (ICRs) have been demonstrated to play a major role. Type 1 genes are represented by the 5S RNA genes and type 2 genes by the tRNA genes. The type 1 promoter ICR consists of two functional domains: box A and box C. Site-directed mutagenesis and deletion analysis of *Xenopus* and *Drosophila* 5S RNA genes have revealed that the ICR is a discontinuous structure extending over a region of the gene from about 50 to 100 nt past the transcription start site. In the *Xenopus* gene, three different elements determine promoter strength. Box A (+ 50 to + 64) is separated from box C (+ 80 to + 97) by an intermediate element (+ 67 to + 72) (Peiler *et al.*, 1987). Between these boxes, bases 65–66 and 73–79 are spacer elements in which sequence alterations do not influence transcription. Alterations in the spacing between the elements does however have serious effects, severely reducing or destroying transcriptional activity. 5S genes isolated from a wide range of plant species are highly homologous with those from *Xenopus* and *Drosophila* (Hemleben and Werts, 1988). Although the functionality of their ICR elements remains to be tested, the homology suggests that they perform a similar role (Erdmann and Wolters, 1986) (Figure 1.8).

The type 2 promoter is also composed of two domains: box A and box B. In *Xenopus* tRNA genes, box A is situated at positions + 8 to + 19 and is structurally and functionally similar to the type 1 box A. Some variability in the distance separating the A and B boxes is tolerated in the type 2 ICR, but, if the spacing is either too large or too small, adventitious, weakly functioning substitutes (pseudo A or B boxes) can be utilised. Both boxes contain nucleotides which are highly conserved in the majority of tRNAs. In the tRNA genes isolated from plants, ICRs similar to those in animal tRNA genes have been observed. Both type 1 and type 2 genes contain a simple consensus sequence consisting of a cluster of four or more T residues surrounded by GC-rich sequences on the non-coding strand which acts as a termination signal.

These sequence elements themselves are not sufficient for promoter activity, and ancillary factors are required for gene expression. Three separate transcription factors determining promoter specificity of RNA polymerase III

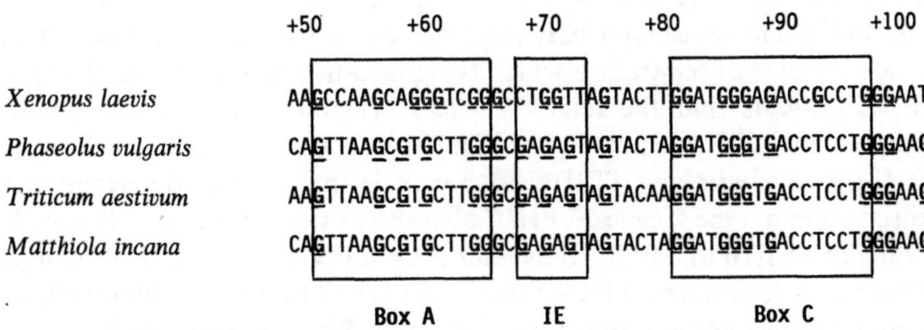

Figure 1.8. Homology between the *Xenopus* 5S RNA internal control region and those from bean, wheat and *Matthiola*. Sequences were obtained from Hemleben and Werts (1989) and Erdmann and Wolters (1986). Box A, Box B and the intermediate element, IE, were positioned according to Peiler *et al.* (1987) as outlined in the text. The G-residues, which are potential sites of interaction with TFIIIA, are highlighted.

have been identified in animals and yeast (reviewed in Geiduschek and Tocchini-Valentini, 1988). Two of these, transcription factors B and C (TFIIIB and C), are required for the transcription of both types of genes. The third, TFIIIA, which is specifically required for 5S RNA synthesis, is one of the most extensively studied of all eukaryotic transcription factor proteins (Miller *et al.*, 1985; Rhodes and Klug, 1986).

A model for RNA polymerase III transcription has been proposed for the *Xenopus* 5S RNA genes. Firstly, TFIIIA binds to the ICR over an extended binding site of almost 50 bp, with the strongest association to box C and the intermediate element. TFIIIC then binds to, and helps stabilise, the ICR–TFIIIA complex. Finally, TFIIIB is thought to recognise and bind to the TFIIIA/C structure, forming a stable preinitiation transcription factor complex on the ICR which is recognised by RNA polymerase III.

The TFIIIA protein from *Xenopus* is a 38.5 kDa protein containing a 30 kDa internal domain organised into nine repeat structures. Each repeat consists of a 30 amino acid motif containing two cysteine and two histidine residues at precisely repeated positions which tetrahedrally complex one Zn^{2+} ion (Diakun *et al.*, 1986). In addition, aromatic and hydrophobic amino acids are clustered at characteristic positions in each repeat. The resulting molecule has an extended structure containing nine substructures known as 'zinc fingers'. Detailed analysis of TFIIIA DNA footprints on the 5S ICR has shown that there are nine zones of interaction located roughly 5 bp apart along the ICR, which coincide with the occurrence of G residues either singly, in pairs or in triplets at a periodicity of 5.5 nucleotides in the major groove of the DNA (Glover, 1989). Although, no RNA polymerase III transcription factors have been isolated from plants, zinc finger-binding domains have been identified in proteins coded for by transposable elements (Vodkin and Vodkin, 1989).

As with the 5S RNA genes, putative regulatory sequences of plant tRNA ICRs have not been functionally delineated. However, as some plant tRNAs are expressed and processed in Hela cell extracts, much about their transcription can again be inferred from studies on yeast and animals (van Tol *et al.*, 1987). The conserved ICR sequence elements, box A and box B, can tolerate a range of separation which is significant because, unlike the type 1 ICR, it suggests that the relative helical orientation of the two boxes is unimportant. TFIIIA is not required for tRNA transcription. Instead, in yeast tRNA genes, binding of TFIIIC to box B is the primary determinant for transcription of type 2 genes. TFIIIC also shows weak binding to box A. The binding of TFIIIB to TFIIIC may alter the affinity of the TFIIIB–C complex for the box A region, and it is the formation of the TFIIIB–C complex which is essential for recognition and transcription by RNA polymerase III.

Although the ICRs obviously constitute essential promoter elements, the 5′ flanking sequences in the majority of, if not all, genes transcribed by RNA polymerase III perform at least a modulating function. Comparative sequence analysis has revealed conserved sequences within individual classes of genes in different genera and that the effects of these 5′ sequences are often only seen in homologous systems (Rubacha *et al.*, 1984). In many tRNA genes, a TATA box sequence is found in the 5′ region, but its significance has yet to be established. Recent evidence has indicated that, in *Xenopus* at least, a transcription factor distinct from those already described binds to the 5′ region and stimulates transcription from a somatic 5S RNA gene (see Murphy *et al.*, 1989).

The importance of 5′ sequences in RNA polymerase III-mediated transcription is much more apparent in the expression of U6snRNAs. Many features of the 5′ upstream sequences of U6 are similar to those of the other major UsnRNAs which are transcribed by RNA polymerase II. Although a box A-like sequence is present within the coding region, and a row of T's (which act as a simple transcription termination signal for RNA polymerase III) at the 3′ end, no sequences homologous to box B or C elements of the ICR are present. Transcription appears therefore to be directed by a combination of upstream promoter and enhancer elements resembling RNA polymerase II rather than RNA polymerase III signals in both sequence and location. As these elements are highly conserved in U6 it is feasible that the same transcription factor interacts with both U6 and U1–U5 snRNA genes. Thus RNA polymerase II (or at least the RNA polymerase II-like activity) and RNA polymerase III may share at least one ancillary transcription factor (Murphy *et al.*, 1989).

All of the RNAs are subject to a range of post-transcriptional processing events which are essential to their function. These include specific 5′ and 3′ end modifications, intron removal (in the case of some tRNAs), extensive base modifications and the formation of distinct secondary structures, and all have been described previously (Grierson, 1982, 1988). It must however be stressed that the expression of a functional gene product involves a combination of all these processes and that variations in any one step represent, at least potentially, a separate stage at which regulation can occur.

References

Abel, S., Kiss, T. and Solymosy, F. (1989) Molecular analysis of eight U1 RNA gene candidates from tomato that could potentially be transcribed into U1 RNA sequence variants differing from each other in similar regions of secondary structure. *Nucl. Acids Res.* **17**: 6319.

Allen, R.D., Bernier, F., Lessar, P.A. and Beachy, R.N. (1989) Nuclear factors interact with a soybean β-conglycinin enhancer. *Plant Cell* **1**: 623.

Barker, R.F., Harberd, N.P., Jarvis, M. and Flavell, R.B. (1988) The structure and evolution of the intergenic region in a ribosomal DNA repeat of wheat. *J. Mol. Biol.* **201**: 1.

Barta, A., Sommergruber, K., Thompson, D., Hartmuth, K., Matzke, M.A. and Matzke, A.J.M. (1986) The expression of a nopaline synthase growth hormone chimaeric gene in transformed tobacco and sunflower callus tissue. *Plant Mol. Biol.* **6**: 347.

Bäumlein, H., Wobus, U., Pustell, J. and Kafatos, F.C. (1986) The legumin gene family: structure of a B type gene of *Vicia faba* and a possible legumin gene specific element. *Nucl. Acids Res.* **14**: 2707.

Beach, L.R., Spencer, D., Randall, P.J. and Higgins, T.J.V. (1985) Transcriptional and post-transcriptional regulation of storage protein gene expression in sulphur-deficient pea seeds. *Nucl. Acids Res.* **13**: 999.

Benfey, P.N. and Chua, N.-H. (1989) Regulated genes in transgenic plants. *Science* **244**: 174.

Bernstein, L.B., Mount, S.M. and Weiner, A.M. (1983) Pseudogenes for human small nuclear RNA U3 appear to arise by integration of self-primed reverse transcripts of the RNA into new chromosomal sites. *Cell* **32**: 461.

Bianchi, M.W. and Viotti, A. (1988) DNA methylation and tissue-specific transcription of the storage protein genes of maize. *Plant Mol. Biol.* **11**: 203.

Birnstiel, M.L. (1988) *Structure and Function of Major and Minor Small Nuclear Ribonucleoprotein Particles.* Springer Verlag, Berlin.

Borgmeyer, U., Nowock, J. and Sippel, A.E. (1984) The TGGCA-binding protein: a eukaryotic nuclear protein recognising a symmetrical sequence on double-stranded linear DNA. *Nucl. Acids Res.* **12**: 4295.

Bown, D., Levasseur, M., Croy, R.R.D., Boulter, D. and Gatehouse, J.A. (1985) Sequence of a pseudo gene in the legumin gene family of pea (*Pisum sativum* L.) *Nucl. Acids Res.* **12**: 4527.

Brawerman, G. (1987) Determinants of mRNA stability. *Cell* **48**: 5.

Brawerman, G. (1989) mRNA decay: finding the right target. *Cell* **57**: 9.

Breathnach, R. and Chambon, P. (1981) Organisation and expression of eukaryotic split genes coding for proteins. *Ann. Rev. Biochem.* **50**: 349.

Breitbart, R.E., Andreadis, A. and Nadal-Ginard, B. (1987) Alternative splicing: a ubiquitous mechanism for the generation of multiple protein isoforms from single genes. *Ann. Rev. Biochem.* **56**: 467.

Brown, J.W.S. (1986) A catalogue of splice junction and putative branchpoint sequences from plant introns. *Nucl. Acids Res.* **14**: 9549.

Brown, J.W.S. and Feix, G. (1990) A functional splice site in the 5′ untranslated region of a zein gene. *Nucl. Acids Res.* **18**: 111.

Brown, J.W.S and Waugh, R. (1989) Maize U2snRNAs: gene sequence and expression. *Nucl. Acids Res.* **17**: 8991.

Brown, J.W.S., Frendewey, D. and Feix, G. (1986a) Accurate *in vitro* splicing of two pre-mRNA plant introns in a HeLa cell nuclear extract. *EMBO J.* **5**: 2749.

Brown, J.W.S., Wandelt, C., Feix, G., Neuhaus, G. and Schweiger, H.G. (1986b) The upstream regions of zein genes: sequence analysis and expression in the unicellular green alga *Acetabularia. Eur. J. Cell Biol.* **42**: 161.

Bustos, M.M., Guiltannan, M.J., Jordano, J., Begum, D., Kalkan, F.A. and Hall, T.C. (1989) Regulation of β-glucuronidase expression in transgenic tobacco plants by an A/T-rich *cis*-acting sequence found upstream of a french bean β-phaseolin gene. *Plant Cell* **1**: 839.

Callis, J., Fromm, M. and Walbot, V. (1987) Introns increase gene expression in cultured maize cells. *Genes and Development* **1**: 1183.

Casey, R. and Domoney, C. (1987) The structure of plant storage protein genes. *Plant Mol. Biol. Reporter* **5**: 261.

Castresana, C., Garcia-Luque, I., Alonso, E., Malik, V.S. and Cashmore, A.R. (1988) Both positive and negative elements mediate expression of a photoregulated *cab* gene from *Nicotiana plumbaginifolia. EMBO J.* **7**: 1929.

Cavanaugh, A.H. and Thompson, E.A. (1986) Hormonal regulation of transcription of rDNA. *J. Biol. Chem.* **261**: 12738.

Chen, Z.-L., Schuler, M.A. and Beachy, R.N. (1986) Functional analysis of regulatory elements in a plant embryo specific gene. *Proc. Natl Acad. Sci. USA* **83**: 8560.

Chen, Z.-L., Pan, N.S. and Beachy, R.N. (1988) A DNA sequence element that confers seed specific enhancement to a constitutive promoter. *EMBO J* **7**: 297.

Cho, T.-J., Davies, C.S., Fischer, R.L., Turner, N.E., Goldberg, R.B. and Nielsen, N.C. (1989) Molecular characterisation of an aberrant allele for the Gy_3 glycinin gene: A chromosomal rearrangement. *Plant Cell* **1**: 339.

Chomet, P.S., Wessler, S. and Dellaporta, S.L. (1987) Inactivation of the maize transposable element Activator (*Ac*) is associated with its DNA modification. *EMBO J* **6**: 295.

Colot, V., Robert, L.S., Kavanagh, T.A., Bevan, M.W. and Thompson, R.D. (1987) Localisation of sequences in wheat endosperm protein genes which confer tissue-specific expression in tobacco. *EMBO J.* **6**: 3559.

Cordes, S., Deikman, J., Margossian, L.J. and Fischer, R.L. (1989) Interaction of a developmentally regulated DNA-binding factor with sites flanking two different fruit-ripening genes from tomato. *Plant Cell* **1**: 1025.

Covello, P.S. and Gray, M.W. (1989) RNA editing in plant mitochondria. *Nature* **341**: 662.
Dahlberg, J.E. and Lund, E. (1988) The genes and transcription of the major small nuclear RNAs. In *Structure and Function of Major and Minor Small Nuclear Ribonucleoprotein Particles* ed. Birnstiel, M. Springer Verlag, Berlin, 38.
Datta, N. and Cashmore, A.R. (1989) Binding of a pea nuclear protein to promoters of certain photoregulated genes is modulated by phosphorylation. *Plant Cell* **1**: 1069.
Dean, C., Jones, J., Favreau, M., Dunsmuir, P. and Bedbrook, J. (1988) Influence of flanking sequences on variability in expression levels of an introduced gene in transgenic tobacco plants. *Nucl. Acids Res.* **16**: 9267.
Dean, C., Favreau, M., Bond-Nutter, D., Bedbrook, J. and Dunsmuir, P. (1989*a*) Sequences downstream of translation start regulate quantitative expression of two petunia *rbcS* genes. *Plant Cell* **1**: 201.
Dean, C., Favreau, M., Bedbrook, J. and Dunsmuir, P. (1989*b*) Sequences 5′ to translation start regulate expression of petunia *rbcS* genes. *Plant Cell* **1**: 209.
Diakun, G.P., Fairall, L. and Klug, A. (1986) EXAFS study of the zinc-binding in the protein transcription factor IIIA. *Nature* **324**: 689.
Dickenson, C.D., Evans, R.P. and Nielsen, N.C. (1988) RY repeats are conserved in the 5′ flanking regions of legume seed protein genes. *Nucl. Acids Res.* **16**: 371.
Döring, H.-P. and Starlinger, P. (1986) Molecular genetics of transposable elements in plants. *Annu. Rev. Genet.* **20**: 175.
Drouin, G. and Dover, G.A. (1987) A plant processed pseudogene. *Nature* **328**: 557.
Dynan, W.S. (1989) Modularity in promoters and enhancers. *Cell* **58**: 1.
Egeland, D.B., Sturtevant, A.P. and Schuler, M.A. (1989) Molecular analysis of dicot and monocot small nuclear RNA populations. *Plant Cell* **1**: 633.
Ellis, T.H.N., Lee, D., Thomas, C.M., Simpson, P.R., Cleary, W.G., Newman, M.-A. and Burcham, K.W.G. (1988) 5S RNA genes in *Pisum*: Sequence, long range and chromosomal organisation. *Mol. Gen. Genet.* **214**: 333.
Erdmann, V.A. and Wolters, J. (1986) Collection of published 5S, 5.8S and 4.5S RNA sequences. *Nucl. Acids Res.* **14**: r1.
Flavell, R.B., O'Dell, M., Thompson, W.F., Vincentz, M., Sardana, R. and Barker, R.F. (1986) The differential expression of ribosomal RNA genes. *Phil. Trans. R. Soc. Lond. B* **314**: 385.
Fluhr, R., Kuhlemeier, C., Nagy, F. and Chua, N.-H. (1986) Organ-specific and light-induced expression of plant genes. *Science* **232**: 1106.
Forde, B.G., Heyworth, A., Pywell, J. and Kreis, M. (1985) Nucleotide sequence of a B1 hordein gene and the identification of possible upstream regulatory elements in endosperm storage protein genes from barley, wheat and maize. *Nucl. Acids Res.* **13**: 7327.
Gallie, D.R., Lucas, W.J. and Walbot, V. (1989) Visualizing mRNA expression in plant protoplasts: factors influencing efficient uptake and translation. *Plant Cell* **1**: 301.
Gatehouse, J.A., Evans, I.M., Croy, R.R.D. and Boulter, D. (1986) Differential expression of genes during legume seed development. In Differential gene expression and plant development. *Phils. Trans. Roy. Soc. Lond. B* **314**: 25.
Geiduschek, E.P. and Tocchini-Valentini, G.P. (1988) Transcription by RNA polymerase III. *Ann. Rev. Biochem.* **57**: 873.
Gerlach, W.L. and Dyer, T.A. (1980) Sequence organisation of the repeating units in the nucleus of wheat which contain 5S rRNA genes. *Nucl. Acids Res.* **8**: 4851.
Giuliano, G., Pichersky, E., Malik, V.S., Timko, M.P., Scolnik, P.A. and Cashmore, A.R. (1988) An evolutionary conserved protein binding sequence upstream of a plant light-regulated gene. *Proc. Natl. Acad. Sci. USA* **85**: 7089.
Glover, C.V.C. (1989) Sequence-specific protein–DNA recognition by transcriptional regulatory proteins. *Plant Mol. Biol. Reporter* **7**: 183.
Goldberg, R.B. (1986) Regulation of plant gene expression. In Differential gene expression and plant development. *Phil. Trans. Roy. Soc. Lond. B* **314**: 1.
Goldberg, R.B., Barker, S.J. and Perez-Grau, L. (1989) Regulation of gene expression during plant embryogenesis. *Cell* **56**: 149.
Goldsborough, P.B., Ellis, T.H.N. and Lomonossoff, G.P. (1982) Sequence variation and methylation of the flax 5S RNA genes. *Nucl. Acids Res.* **10**: 4501.
Goodall, G.J. and Filipowicz, W. (1989) The AU-rich sequences present in the introns of plant nuclear pre-mRNAs are required for splicing. *Cell* **58**: 473.

Green, G.A., Weil, J.H. and Steinmetz, A. (1986) The sequences of two nuclear genes and a pseudogene for tRNA (pro) from the higher plant *Phaseolus vulgaris*. *Plant Mol. Biol.* **7**: 207.

Green, G.A., Marechal, L., Weil, J.H. and Guillemaut, P. (1987) A *P. vulgaris* mitochondrial tRNA (leu) is identical to its cytoplasmic counterpart. Sequencing and *in vivo* transcription of the gene corresponding to the cytoplasmic tRNA (leu). *Plant Mol. Biol.*

Green, P.J., Kay, S.A. and Chua, N.-H. (1987) Sequence specific interactions of a pea nuclear factor with light-responsive elements upstream of the *rbcS-3A* gene. *EMBO J.* **6**: 2543.

Grierson, D. (1982) RNA processing and other post transcriptional modifications. In *Nucleic Acids and Proteins in Plants. II. Structure, Biochemistry and Physiology of Nucleic Acids*, eds. B. Parthier and D. Boulter. Springer Verlag, Berlin, 192.

Grierson, D. and Covey, S.N. (1988) *Plant Molecular Biology*, 2nd ed. Blackie, Glasgow.

Grummt, I., Kuhn, A., Barstch, I. and Rosenbauer, H. (1986) A transcription terminator located upstream of the mouse rDNA initiation site affects rRNA synthesis. *Cell* **47**: 901.

Gurley, W.B., Czarnecka, E., Nagao, R.T. and Key, J.L. (1986) Upstream sequences required for efficient expression of a soybean heat shock gene. *Mol. Cell Biol.* **6**: 559.

Haltiner-Jones, M., Learned, R.M. and Tjian, R. (1988) Analysis of clustered point mutations in the human ribosomal RNA gene promoter by transient expression *in vivo*. *Proc. Natl Acad. Sci. USA* **85**: 669.

Hanley, B.A. and Schuler, M.A. (1987) Plant intron sequences: evidence for distinct groups of introns. *Nucl. Acids Res.* **16**: 7159.

Harada, J.J., Barker, S.J. and Goldberg, R.B. (1989) Soybean β-conglycinin genes are clustered in several DNA regions and are regulated by transcriptional and post-transcriptional processes. *Plant Cell* **1**: 415.

Hartings, H., Maddaloni, M., Lazzaroni, N., Di Fonzo, N., Motto, M., Salamini, F and Thompson, R. (1989) The O2 gene which regulates zein deposition in maize endosperm encodes a protein with structural homologies to transcriptional activators. *EMBO J.* **8**: 2795.

Heidecker, G. and Messing, J. (1986) Structural analysis of plant genes. *Ann. Rev. Plant Physiol.* **37**: 439.

Hemleben, V. and Werts, D. (1988) Sequence organisation and putative regulatory elements in the 5S rRNA genes of 2 higher plants (*Vigna radiata* and *Matthiola incana*). *Gene* **62**: 165.

Hepburn, A.G., Clarke, L.E., Pearson, L. and White, J. (1983) The role of cytosine methylation in the control of nopaline synthase gene expression in a plant tumour. *J. Mol. Appl. Genet.* **2**: 315.

Hernandez, N. and Weiner, A.M. (1986) Formation of the 3′ end of U1 snRNA requires compatible snRNA promoter elements. *Cell* **47**: 249.

Hinnebusch, A.G. (1988) Novel mechanisms of translational control in *Saccharomyces cerevisiae*. *Trends in Genet.* **4**: 169.

Hunt, A.G., Chu, N.M., Odell, J.T., Nagy, F. and Chua, N.-H. (1987) Plant cells do not properly recognise animal gene polyadenylation signals. *Plant Mol. Biol.* **8**: 23.

Ingelbrecht, I.L.W., Lieve, M.F., Dekeyser, R.A., Van Montagu, M.C. and Depicker, A.G. (1989) Different 3′ end regions strongly influence the level of gene expression in plants. *Plant Cell* **1**: 671.

Jensen, O.E., Marcker, K.A., Schell, J. and de Bruijn, F.J. (1988) Interaction of a nodule-specific, *trans*-acting factor with distinct DNA elements in the soybean leghaemoglobin *lbc*$_3$ 5′ upstream region. *EMBO J.* **7**, 1265.

Jofuku, K.D. and Goldberg, R.B. (1989) Kunitz trypsin inhibitor genes are differentially expressed during the soybean life cycle and in transformed tobacco plants. *Plant Cell* **1**: 1079.

Jofuku, K.D., Schipper, R.D. and Goldberg, R.B. (1989) A frameshift mutation prevents Kunitz trypsin inhibitor mRNA accumulation in soybean embryos. *Plant Cell* **1**: 427.

Jordano, J., Almoguera, C. and Thomas, T.L. (1989) A sunflower helianthin gene upstream sequence ensemble contains an enhancer and sites of nuclear protein interaction. *Plant Cell* **1**: 855.

Joshi, C.P. (1987*a*) An inspection of the domain between putative TATA box and translation start site in 79 plant genes. *Nucl. Acids Res.* **15**: 6643.

Joshi, C.P. (1987*b*) Putative polyadenylation signals in nuclear genes of higher plants: a compilation and analysis. *Nucl. Acids. Res.* **15**, 9627..

Joyce, P.B.M. and Gray, M.W. (1989) Aspartate and asparagine tRNA genes in wheat mitochondrial DNA: a cautionary note on the isolation of tRNA genes from plants. *Nucl. Acids Res.* **17**: 7865.

Kaufman, L.S., Watson, J.C. and Thompson, W.F. (1987) Light regulated changes in DNase I hypersensitivity sites in the rRNA genes of *Pisum sativum*. *Proc. Natl Acad. Sci. USA* **84**: 1550.

Kay, S.A., Keith, K., Shinozaki, K., Chye, M.-L. and Chua, N.-H. (1989) The rice phytochrome gene: structure, autoregulated expression, and binding of GT-1 to a conserved site in the 5′ upstream region. *Plant Cell* **1**: 351.

Keith, B. and Chua, N.-H. (1986) Monocot. and dicot. pre-mRNAs are processed with different efficiencies in transgenic tobacco. *EMBO J.* **5**: 2419.

Kiss, T., Antal, M. and Solymosy, F. (1987) Plant small nuclear RNAs. II. U6 RNA and a 4.5S-like RNA are present in plant nuclei. *Nucl. Acids Res.* **15**: 543.

Kiss, T., Abel, S. and Solymosy, F. (1989) A plant pseudogene for U1 RNA. *Plant Mol. Biol.* **12**: 709.

Klein, T.M., Wolf, E.D., Wu, R. and Sanford, J.G. (1987) High velocity microprojectiles for delivery of nucleic acids into living cells. *Nature* **327**: 70.

Kownin, P., Bateman, E. and Paule, M.R. (1987) Eukaryotic RNA polymerase I binding is directed by protein contacts with transcription initiation factors and is DNA sequence independent. *Cell* **50**: 693.

Kozak, M. (1983) Translation of insulin-related polypeptides from messenger RNAs with tandemly reiterated copies of the ribosome binding site. *Cell* **34**: 971.

Kozak, M. (1987) An analysis of 5′ non-coding sequences from 699 vertebrate messenger RNAs. *Nucl. Acids Res.* **15**, 8125.

Krainer, A.R. and Maniatis, T. (1988) In *Transcription and Splicing*, eds. Hames, B.D. and Glover, D.M. IRL Press, Oxford 131.

Kreis, M., Williamson, M.S., Forde, J., Schmutz, D., Clark, J., Buxton, B., Pywell, J., Marris, C., Henderson, J., Harris, N., Shewry, P.R., Forde, B.G. and Miflin, B.J. (1986) Differential gene expression in the developing barley endosperm. In Differential gene expression and plant development. *Phil. Trans. Roy. Soc. Lond. B* **314**: 13.

Kridl, J.C., Viera, J., Rubenstein, I. and Messing, J. (1984) Nucleotide sequence analysis of a zein genomic clone with a short open reading frame. *Gene* **28**: 113.

Kriz, A.L., Boston, R.S. and Larkins, B.A. (1987) Structural and transcriptional analysis of DNA sequences flanking genes that encode 19kDa zeins. *Mol. Gen. Genet.* **207**: 290.

Kuhlemeier, C., Green, P.J. and Chua, N.-H. (1987) Regulation of gene expression in higher plants. *Ann. Rev. Plant Physiol.* **38**: 221.

Kuhlemeier, C., Cuozzo, M., Green, P.J., Goyvaerts, E., Ward, K. and Chua, N.-H. (1988) Localisation and conditional redundancy of regulatory elements in *rbcS-3A*, a gene encoding the small subunit of ribulose-bisphosphate carboxylase. *Proc. Natl Acad. Sci. USA* **85**: 4662.

Labhart, P. and Reeder, R.H. (1986) Characterization of three sites of RNA 3′ end formation in the *Xenopus* ribosomal gene spacer. *Cell* **43**: 431.

Labhart, P. and Reeder, R.H. (1987) A 12-base pair sequence is an essential element of the ribosomal gene terminator in *Xenopus laevis*. *Mol. Cell Biol.* **7**: 1900.

Langridge, P. and Feix, G. (1983) A zein gene of maize is transcribed from two widely separated promoter regions. *Cell.* **34**: 1015.

Leff, S.E., Rosenfeld, M.G. and Evans, R.M. (1986) Complex transcriptional units: diversity in gene expression by alternative RNA processing. *Ann. Rev. Biochem.* **55**: 1091.

Lüctke, H.A., Chow, K.G., Mickel, F.S., Moss, K.A., Kern, H.F. and Scheele, G.A. (1987) Selection of AUG codons differs in plants and animals. *EMBO J.* **6**: 43.

McIntyre, C.L., Reddy, P. and Appels (1988) The 5S DNA loci in the Triticeae. *Seventh International Wheat Genetics Symposium.* Vol. 1: 769.

McMullen, M.D., Hunter, B., Phillips, R.L. and Rubenstein, I. (1986) The structure of the maize ribosomal DNA spacer region. *Nucl. Acids Res* **14**: 4953.

Maier, U.-G., Brown, J.W.S., Toloczyki, C. and Feix, G. (1987) Binding of a nuclear factor to a consensus sequence in the 5′ flanking region of the zein genes from maize. *EMBO J.* **6**: 17.

Maier, U.-G., Brown, J.W.S., Schmitz, L.M., Schwall, M., Dietrich, G. and Feix, G. (1988) Mapping of tissue-dependent and independent protein binding sites to the 5′ upstream region of a zein gene. *Mol. Gen. Genet.* **212**: 241.

Manzara, T. and Gruissem, W. (1988) Organization and expression of the genes encoding ribulose-1,5-bisphosphate carboxylase in higher plants. *Photo. Res.* **16**: 117.

Marris, C., Gallois, P., Copley, J. and Kreis, M. (1988) The 5′-flanking region of a barley B hordein gene controls tissue-specific and developmental specific CAT expression in tobacco plants. *Plant Mol. Biol.* **10**: 359.

Mattaj, I.W., (1988) UsnRNP assembly and transport. In *Structure and Function of Major and Minor Small Nuclear Ribonucleoprotein Particles*, ed. Birnsteil, M.L., Springer Verlag, Berlin, 100.

Matzke, M.A., Primig, M., Trnovsky, J. and Matzke, A.J.M. (1989) Reversible methylation and inactivation of marker genes in sequentially transformed tobacco plants. *EMBO J.* **8**: 643.

Messing, J., Geraghty, D., Heidecker, G., Hu, N.-T., Kridl, J. and Rubenstein, I. (1983) Plant gene structure. In *Genetic Engineering of Plants*, eds. Kosuge, T., Meredith, C.P. and Hollaender, A., Plenum Press, New York, 211.

Miller, J., McLachlan, A.D. and Klug, A. (1985) Repetitive zinc-binding domains in the protein transcription factor IIIA from *Xenopus* oocytes. *EMBO J.* **4**: 1609.

Morelli, G., Nagy, F., Fraley, R.T., Rogers, S.G. and Chua, N.-H. (1985) A short conserved sequence is involved in the light inducibility of a gene encoding the ribulose 1-5-bisphosphate carboxylase small subunit of pea. *Nature* **315**: 200.

Moss, T. (1983) A transcriptional function for the repetitive ribosomal spacer in *Xenopus laevis*. *Nature* **302**: 223.

Mullineaux, P.M., Donson, J., Morris-Krsinich, B.A.M., Boulton, M.I., and Davies, J.W. (1984) The nucleotide sequence of maize streak virus DNA. *EMBO J.* **3**: 3063.

Murphy, S., Moorefield, B. and Pieler, T. (1989) Common mechanisms of promoter recognition by RNA polymerases II and III. *Trends Genet.* **5**: 122.

Nielsen, N.C., Dickinson, C.D., Cho, T.-J., Thanh, V.H., Scallon, B.J., Fischer, R.L., Sims, T.L., Drews, G.N. and Goldberg, R.B. (1989) Characterisation of the glycinin gene family in soybean. *Plant Cell* **1**: 313.

Oard, J.H., Paige, D. and Dvorak, J.L. (1989) Chimeric gene expression using maize introns in cultured cells of bread wheat. *Plant Cell Reports* **8**: 156.

Peabody, D.S. and Berg, P. (1986) Termination–reinitiation occurs in the translation of mammalian cell mRNAs. *Mol. Cell Biol.* **6**: 2704.

Peiler, T., Hamm, J. and Roeder, R.G. (1987) The 5S gene internal control region is composed of three distinct sequence elements organised as two functional domains with variable spacing. *Cell* **48**: 91.

Platt, T. (1986) Transcription termination and the regulation of gene expression. *Ann. Rev. Biochem.* **55**: 339.

Quayle, T.J.A., Brown, J.W.S. and Feix, G. (1989) Analysis of distal flanking regions of maize 19 kDa zein genes. *Gene* **80**: 249.

Reeder, R.H., Pennock, D., McStay, B., Roan, J., Tolentino, E. and Walker, P. (1987) Linker scanner mutagenesis of the *Xenopus laevis* ribosomal gene promoter. *Nucl. Acids. Res.* **15**: 7429.

Rhodes, D. and Klug, A. (1986) An underlying repeat in some transcriptional control sequences corresponding to half a double helical turn of DNA. *Cell* **46**: 123.

Riggs, C.D., Voelker, T.A. and Chrispeells, M.J. (1989) Cotyledon nuclear proteins bind to DNA fragments harbouring regulatory elements of phytohaemagglutinin genes. *Plant Cell* **1**: 609.

Robert, L.S., Thompson, R.D. and Flavell, R.B. (1989) Tissue-specific expression of a wheat high molecular weight glutenin gene in transgenic tobacco. *Plant Cell* **1**: 569.

Roeder, R.G. (1976) In *RNA Polymerase*, eds. Losick, R. and Chamberlin, M.J. Cold Spring Harbor Laboratory, Cold Spring Harbor, New York, 285.

Rogers, S.O. and Bendich, A.J. (1987) *Plant Mol. Biol.* **9**: 509.

Rogers, S.G., Fraley, R.T., Horsch, R.B., Levine, A.D., Flick, J.S., Brand, L.A., Fink, C.L., Mozer, T., O'Connell, K. and Sanders, P.J. (1985) Evidence for ribosome scanning during translation initiation of mRNAs in transformed plant cells. *Plant Mol. Biol. Reporter.* **3**: 111.

Rubacha, A., Sumner III, W. and Beckingham, K. (1984) Conserved 5′ flank homologies in dipteran 5S RNA genes that would function on A form DNA. *Nucl. Acids Res.* **12**: 8193.

Schernthaner, J.P., Matzke, M.A. and Matzke, A.J.M. (1988) Endosperm-specific activity of a zein gene promoter in transgenic tobacco plants. *EMBO J.* **7**: 1249.

Schindler, U. and Cashmore, A.J. (1989) At least five different nuclear proteins with distinct DNA recognition sites interact with the *Cab-E* promoter of *N. plumbaginifolia*. In *Molecular Communication in Higher Plants.* EMBO Symposium, Heidelberg, 186.

Schmid, H.P., Akhayat, O., Martins de Sa, C., Puvion, F., Koehler, K. and Scherrer, K. (1984) The prosome: a ubiquitous morphologically distinct RNP particle associated with repressed RNPs and containing specific scRNA and a characteristic set of proteins. *EMBO J.* **3**: 29.

Schmitz, L.M., Maier, U.-G., Brown, J.W.S. and Feix, G. (1989) Specific binding of nuclear proteins to the promoter region of a maize nuclear rRNA gene unit. *J. Biol. Chem.* **264**: 1467.

Schulze-Lefert, P., Dangl, J.L., Becker-Andre, M., Hahlbrock, O and Schulz, W. (1989) Inducible *in vivo* DNA footprints define sequences necessary for UV light activation of the parsley chalcone synthase gene. *EMBO J.* **8**: 651.

Shirsat, A., Wilford, N., Croy, R. and Boulter, D. (1989) Sequences responsible for the tissue specific promoter activity of a pea legumin gene in tobacco. *Mol. Gen. Genet.* **215**: 326.

Siebertz, B., Logemann, J., Willmitzer, L. and Schell, J. (1989) *Cis*-analysis of the wound-inducible promoter *wun1* in transgenic tobacco plants and histochemical localisation of its expression. *Plant Cell* **1**: 961.

Sieliwanowicz, B. (1987) The influence of poly(A)-binding proteins on translation of poly(A)$^+$ RNA in a cell free system from embryo axes of dry pea seeds. *Biochim. Biophys. Acta* **908**: 54.

Shotwell, M.A. and Larkins, B.A. (1989) The biochemistry and molecular biology of seed storage proteins. In *The Biochemistry of Plants: A Comprehensive Treatise*, Vol. XV., New York, Academic Press, in press.

Smith, D.B. and Flavell, R.B. (1974) Variation in nucleolar organiser rRNA gene multiplicity in wheat and rye. *Chromosoma* **47**: 327.

Spena, A., Viotti, A. and Pirotta V. (1983) Two adjacent genomic zein sequences: structure, organisation and tissue-specific restriction pattern. *J. Mol. Biol.* **169**: 799.

Spena, A., Krarse, E. and Dobberstein, B. (1985) Translation efficiency of zein mRNA is reduced by hybrid formation between the 5′ and 3′ untranslated region. *EMBO J.* **4**: 2153.

Stougaard, J., Sandal, N.N., Gron, A., Kühle, A. and Marcker, K.A. (1987) 5′ analysis of the soybean leghaemoglobin *lbc*$_3$ gene: regulatory elements required for promoter activity and organ specificity. *EMBO J.* **6**: 3565.

Szweykowsky-Kulinska *et al.* 1989. A nuclear tRNA (ser) UGA gene from wheat *Triticum vulgare* var. Aria. *Gene* **77**: 163.

Tautz, D. and Dover, G.A. (1986) Transcription of the tandem array of ribosomal DNA in *Drosphila melangoster* does not terminate at any fixed point. *EMBO J.* **5**: 1267.

Toloczyki, C. and Feix, G. (1986).Occurrence of 9 homologous repeat units in the external spacer region of a nuclear maize rRNA gene unit. *Nucl. Acids Res.* **14**: 4969.

van Tol, H., Stange, N., Gross, H.J. and Beier, H. (1987) A human and a plant intron-containing tRNA (tyr) gene are both transcribed in a Hela cell extract but spliced along different pathways. *EMBO J.* **6**: 35.

Vankan, P. and Filipowicz, W. (1988) Structure of U2 snRNA genes of *Arabidopsis thaliana* and their expression in electroporated protoplasts. *EMBO J.* **7**: 791.

Vankan, P. and Filipowicz, W. (1989) A U-snRNA gene-specific upstream element and a −30 'TATA box' are required for transcription of the U2 snRNA gene of *Arabidopsis thaliana. EMBO J.* **8**: 3875.

Vankan, P., Edoh, D. and Filipowicz, W. (1989). Structure and expression of the U5 snRNA gene of *Arabidopsis thaliana.* Conserved upstream sequence elements in plant U-RNA genes. *Nucl. Acids Res.* **16**: 10425.

Vincentz, M. and Flavell, R.B. (1989) Mapping of ribosomal RNA transcripts in wheat. *Plant Cell* **1**: 579.

Vodkin, M.H. and Vodkin, L.O. (1989) A conserved zinc finger domain in higher plants. *Plant Mol. Biol.* **12**: 593.

Voelkel-Meiman, K., Keil, R.L. and Roeder, G.S. (1987) Recombination-stimulating sequences in yeast ribosomal DNA correspond to sequences regulating transcription by RNA polymerase II. *Cell* **48**: 1071.

Walling, L., Drews, G.N. and Goldberg, R.B. (1986) Transcriptional and post-transcriptional regulation of soybean seed protein mRNA levels. *Proc. Natl Acad. Sci. USA* **83**: 2123.

Walker, J.C., Howard, E.A., Dennis, E.S. and Peacock, W.J. (1987) DNA sequences required for anaerobic expression of the maize alcohol dehydrogenase 1 gene. *Proc. Natl. Acad. Sci. USA* **84**: 6624.

Witbauer, K., Herrero, J.-J. and Filipowicz, W. (1988) Nuclear pre-mRNA processing in plants: distinct modes of 3′-splice site selection in plants and animals. *Mol. Cell. Biol.* **8**: 2042.

Weising, K., Schell, J. and Kahl, G. (1988) Foreign genes in plants: transfer, structure, expression and application. *Ann. Rev. Plant Physiol.* **22**: 421.

Werneke, J.M., Chatfield, J.M. and Ogren, W.L. (1989) Alternative mRNA splicing generates the two ribulosebisphosphate carboxylase/oxygenase activase polypeptides in spinach and *Arabidopsis. Plant Cell* **1**: 815.

2 Gene transfer to plants

J. DRAPER and R. SCOTT

2.1 Introduction

2.1.1 *General concepts*

Gene transfer technology underpins both crop plant genetic engineering, where it is an obvious prerequisite, and the plant sciences generally, where it is becoming an essential tool for tackling fundamental problems in physiology, biochemistry, developmental biology, molecular biology and genetics. In particular, the use of transgenic plant technology will undoubtedly lead to the development of genetic-based strategies for gene cloning, enabling the linkage of genotype to phenotype, and therefore will provide revolutionary new approaches for identifying genes useful in pure and applied research.

Much of the impetus for the development of transgenic plant technology has arisen from innovation in mammalian and yeast gene transfer systems. Likewise, the success of mammalian transformation technology owes much to the very great advances made in fertilisation biology, in which ovules fertilised *in vitro* and implanted in a receptive uterus develop into perfectly normal embryos. To create transgenic animals, transforming DNA is simply micro-injected into the nuclei of the embryonic cells at a very early stage in embryo development (Hammer *et al.*, 1985). Although the original animal is usually chimeric for the transferred gene, where germline tissues arise from the transformed cell lineages, wholly transformed individuals can be isolated in the next generation.

In contrast to research on transgenic mammals, where established gene transfer techniques have proved applicable to the relatively few species involved, the enormous diversity of plants and the very large number of different plant species under study have complicated the development of equivalent techniques. Specialised *in vitro* fertilisation technology allowing the manipulation of the ovule, zygote or early embryo is not yet routinely available for plants. In addition, the genetic stocks representative of individual species (plant varieties) and the plant growth conditions used in different laboratories are so variable that it is unlikely that a few standard organisms or techniques will ever predominate in the future.

Transformation of plants involves the stable introduction of DNA sequences, usually into the nuclear genome of the species under investigation. One obvious application of such technology is the addition of specific new, or modified, genetic traits to existing plant varieties. Such genetic engineering approaches potentially have a number of advantages over conventional breeding strategies. The normal tool of the plant breeder, sexual hybridisation, imposes a definite limit to the scope of gene transfer, confining the breeder to the variability present in the normal gene pool. The artificial gene transfer techniques go far beyond the simple removal of this barrier, since, in theory, they will allow the creation of varieties containing completely man-made genes. More modestly, circumventing the sexual process should speed up the whole process of plant improvement, since the transfer of short pieces of DNA containing one or a few genes to an already top-performing variety will remove the need for lengthy backcrossing procedures that currently have to be used. The development of efficient gene transfer procedures for crop species is obviously important. It should be stressed, however, that a major research effort is also required in the search for, and design of, suitable genes for transfer that might improve a crop plant.

This chapter aims to provide information on the current techniques for gene transfer in plants and will be mainly concerned with the production of stably transformed, non-chimeric, plants for use in applied and fundamental research. In particular it is hoped to stress the utility and limitations of each approach and to identify future avenues of research that might prove fruitful. For a detailed description of transformation protocols readers are advised to consult specialist practical texts (Draper *et al.*, 1988; Gelvin and Schilperoort, 1988). In addition, we will also consider the application of transgenic plant technology in novel approaches for gene cloning. For detailed discussions on the use of gene transfer techniques in plant genetic engineering and fundamental cellular and molecular biology the reader should consult other chapters in this book and the companion volume.

2.1.2 *Target plant cells for transformation*

The main aim of any gene transfer procedure is stably to insert DNA into the nuclear genome of cells capable of giving rise to a whole transformed plant. Transformation without regeneration, and regeneration without transformation, are of limited value. In most species it is often difficult to identify such target cell types; several possible transformation routes exist which are outlined in Table 2.1.

In an ideal world, gene transfer would be possible without perturbing normal developmental processes. Gene transfer into pollen (or possibly egg cells) might give rise to genetically transformed gametes which could be used to generate directly wholly transformed plants via normal fertilisation procedures. Pollen, especially, would appear to offer an ideal cellular target for transformation; indeed, the exploration of pollen transformation techniques

Table 2.1. Theoretical routes for recovery of transformed plants.

Target cell type	Method of obtaining transformed plant
(1) Cultured cells	Organogenesis or embryogenesis via a callus phase
(2) Immature embryo or organ meristem cells	*In vitro* plant regeneration from transformed cell lineages following continued development of the embryo or organ
(3) Cells in immature embryos, shoot and flower meristems	Continued development of embryo, shoot or flower to give rise to a chimeric plant. Pollen derived from transformed cell lineages used to produce transformed seed via normal fertilisation
(4) Pollen	Direct production of transformed plants via fertilisation with DNA-treated developing, mature or germinating pollen
(5) Zygote	Direct development of transformed plant

has been under way since the early 1970s but has yet to be developed as a routine technology. Although technically more difficult, the insertion of DNA sequences into zygotes *in vivo*, or *in vitro* followed by embryo rescue, might also be used to produce transformed individuals. However, neither of these approaches has been demonstrated conclusively in any plant species. An alternative strategy is to transform single cells in complex tissues, such as embryos or meristems, which would then continue their normal development. In these cases the resulting plants would be chimeric for the transformed genotype, and, just as in the mammalian system described earlier, one would hope that some transformed cell lineages would eventually give rise to transformed gametes, or tissues capable of exhibiting totipotency (see below). Significantly, both of these approaches have recently met with success, but they have yet to be widely evaluated.

It has been recognised for several decades that totipotent plant cells can be stimulated to regenerate into whole plants *in vitro*, via organogenesis (shoot formation) or embryogenesis, provided they are given the correct hormonal and nutritional conditions (Skoog and Miller, 1957). Adventitious shoots or embryos are thought to arise from single activated cells, and thus, providing totipotent cells can be identified which are both competent and accessible for gene transfer, genetic transformation is a possibility and will give rise directly to non-chimeric transformed plants. Indeed transformation techniques reliant on plant regeneration from *in vitro* cultured tissue have been described for many species (see Draper *et al.*, 1988, and Lindsey and Jones, 1989, for lists of species). However, *in vitro* plant regeneration techniques often impose a degree of 'genome stress', especially if plants are regenerated via a callus phase. This can often result in the appearance of chromosomal and/or genetic abnormalities in the regenerated plants, a phenomenon commonly referred to as 'somaclonal variation' (Evans and Sharp, 1986). The appearance of such random genetic changes is usually undesirable and could have disastrous effects on the genetic make-up of normally high-performing varieties. In contrast, transformed plants generated using techniques which do not perturb normal development should not suffer from such deleterious side-effects. It remains to be seen if a marketable plant variety can be developed from transformed plants derived via organogenesis in tissue culture.

2.1.3 *Transformation vector considerations*

Vectors used for the genetic transformation of eukaryotic cells have several features in common (Figure 2.1A) and plant transformation vectors are no exception. Most vectors carry 'marker' genes which allow the recognition of transformed cells, by either selection or screening. These genes are dominant, usually of microbial origin, and placed under the control of strong, constitutive, eukaryotic promoters, often of viral origin (Figure 2.1B). Table 2.2 lists the most popular selectable marker genes used in plant transformation vectors; these include constructs affording resistance to antibiotics such as kanamycin or methotrexate, and genes which allow growth in the presence of herbicides such as glyphosate or bialaphos. For successful selection, the target plant cells must be susceptible to relatively low concentrations of the antibiotic or herbicide in a non-leaky manner. Screenable markers ('reporter genes') tend to be developed from bacterial genes coding for easily assayed enzymes, such as chloramphenicol acetyl transferase, β-glucuronidase, luciferase, nopaline synthase and octopine synthase (see Section 2.2.4.5). Screenable marker genes are obviously only useful if enzymes with comparable activities are not present

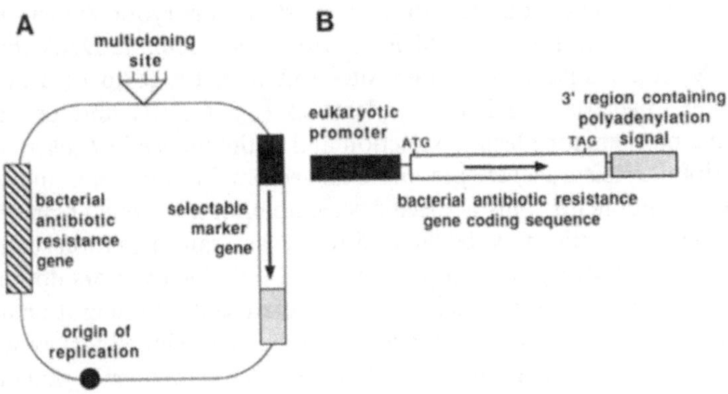

Figure 2.1. (A) Generalised vector for the transformation of eukaryotic cells. (B) Detail of the selectable marker gene.

Table 2.2. Selectable marker genes.

Gene	Enzyme encoded	Selective agent(s)	Reference
Antibiotics			
ble	Enzymic activity not known!	Bleomycin	Hille *et al.* (1986)
dhfr	Dihydrofolate reductase	Methotrexate Trimethoprim	Herrera-Estrella *et al.* (1983)
hpt	Hygromycin phosphotransferase	Hygromycin B	Van den Elzen *et al.* (1985)
npt II	Neomycin phosphotransferase	G418 Kanamycin Neomycin	Bevan *et al.* (1983) Herrera-Estrella *et al.* (1983)
Herbicides			
als	Mutant forms of acetolactate synthase	Chlorsulfuron Imidazolinones	Haughn *et al.* (1988)
aro A	5-Enolpyruvylshikimate-3-phosphate synthase	Glyphosate (Roundup)	Comai *et al.* (1985) Shah *et al.* (1986)
bar	Phosphinothricin acetyltransferase	Phosphinothricin (Bialaphos)	De Block *et al.* (1987)

in non-transformed plant tissues. Table 2.3 presents a list of common reporter genes and some of their properties. The utility of any particular gene construct as a transformation marker varies depending on the plant species and explant involved (e.g. Hauptman *et al.*, 1988); to date kanamycin resistance is probably the most widely used selectable marker phenotype, and β- glucuronidase is finding favour as a versatile reporter gene.

Other components common to most modern eukaryotic transformation vectors include features required for various recombinant DNA manipulations. Multiple unique restriction sites (often in the form of a synthetic polylinker), bacterial origins of replication (e.g. ColE1) and prokaryotic selectable markers for plasmid selection and maintenance in *Escherichia coli* (e.g. antibiotic resistance) are present in all vectors. Plasmids containing these common components, plus a specific selectable marker gene engineered for expression in plants, may be used directly as transformation vectors in physical DNA delivery strategies (see Section 2.3). Such vectors do not have any features which facilitate their transfer to plant cells or integration into the plant nuclear genome. Some of the vector systems developed for use with the natural gene transfer method based on *Agrobacterium* virulence genes and T-DNA border sequences (see Section 2.2) have wide host range replication and transfer functions to allow conjugation from *E. coli* to *Agrobacterium* and plasmid maintenance in both bacterial hosts.

Table 2.3. Screenable marker ('reporter') genes.

Gene	Enzyme encoded	Substrate(s) and assays	Reference
CAT	Chloramphenicol acetyl transferase	[^{14}C]chloramphenicol and acetyl CoA. TLC separation of acetylated [^{14}C]chloramphenicol—detection by autoradiography	Herrera-Estrella *et al.* (1983)
lac Z	β-Galactosidase	As β-glucuronidase. Problems with background activity in some species	Helmer *et al.* (1984)
GUS	β-Glucuronidase	Range of substrates depending on assay. Colourimetric, fluorometric, and histochemical techniques available	Jefferson *et al.* (1987)
lux	Luciferase:		
	bacterial	Decanal and $FMNH_2$.	Koncz *et al.* (1987)
	insect	ATP and O_2 and luciferin Bioluminescent assays: Quantitative tests on extracts or *in situ* tissue assays with activity detected by exposure of X-ray film	Ow *et al.* (1986)
npt-II	Neomycin phosphoryl-transferase	Kanamycin and [^{32}P]ATP. *In situ* assay on enzyme fractionated by non-denaturing PAGE; enzyme detected by autoradiography.	Reiss *et al.* (1984)
		Quantitative dot-binding assay on reaction products	McDonnell *et al.* (1987)

2.2 Vectors based on the *Agrobacterium* Ti plasmid

Since the turn of the century the soil bacterium *Agrobacterium tumefaciens* has been implicated in gall formation at the wound sites of many dicotyledonous plants. This tumour-inducing capability is due to the presence of a large Ti (tumour-inducing) plasmid in virulent strains of *Agrobacterium.* Likewise, Ri (root-inducing) megaplasmids are found in virulent strains of *Agrobacterium rhizogenes,* the causative agent of 'hairy root' disease. Ti and Ri plasmids and the molecular biology of crown gall and hairy root induction have been studied in great detail over the past decade (see the following for reviews: Lichtenstein and Fuller, 1987; Klee *et al.*, 1987a; Zambryski *et al.*, 1989). The present section will consider only those features of Ti plasmid biology relevant to the design of plant transformation vectors and the development of gene transfer systems.

2.2.1 *The Ti plasmid as a natural plant transformation vehicle*

Early DNA hybridisation studies revealed that most Ti plasmids had four regions in common (Figure 2.2A). Deletion, transposon mutagenesis and complementation studies revealed that two of these regions (B and C) were involved in plasmid conjugation and replication. *Agrobacterium* strains carrying Ti plasmids mutated in region A exhibited attenuated oncogenicity

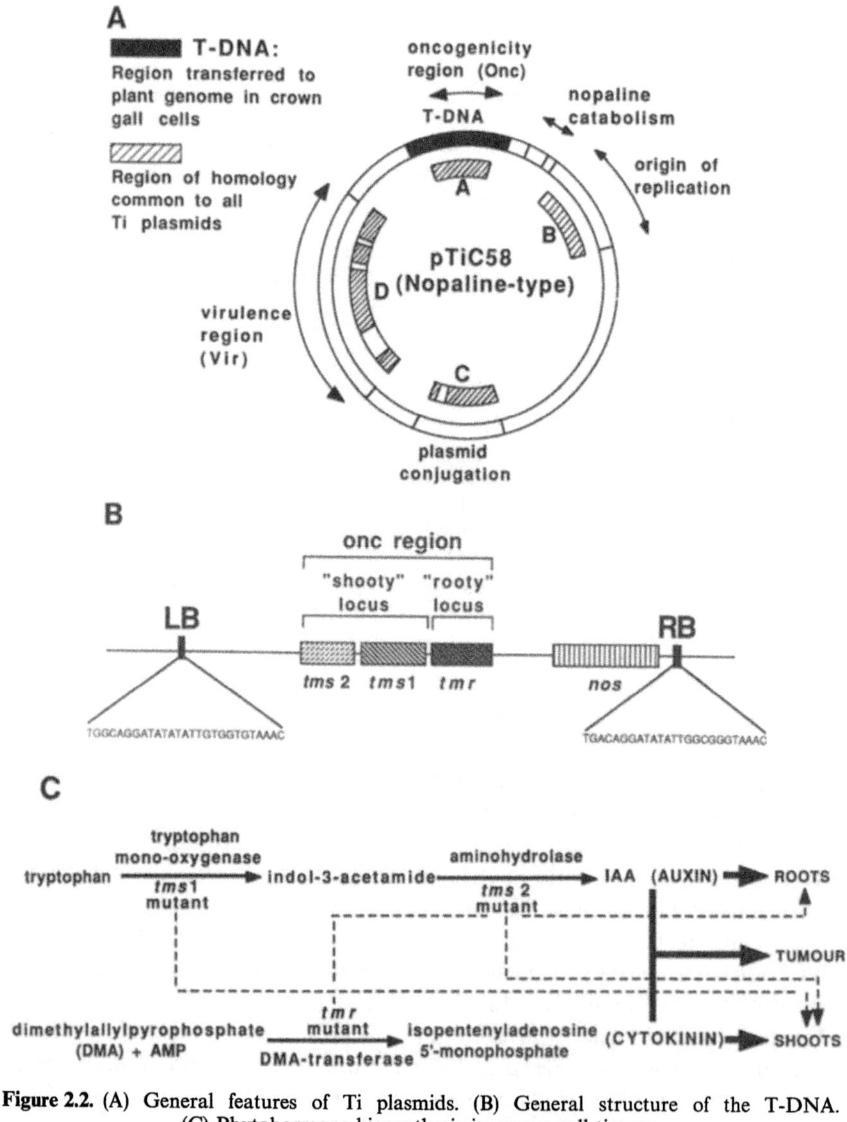

Figure 2.2. (A) General features of Ti plasmids. (B) General structure of the T-DNA. (C) Phytohormone biosynthesis in crown gall tissues.

and/or induced tumours with an altered morphology ('shooty' or 'rooty' mutant galls). Sequences homologous to region A were always transferred to the plant nuclear genome during transformation, and so this region is usually referred to as the T-DNA (transferred DNA; Figure 2.2B). In tumour tissue, genes carried on the transferred DNA are expressed from their own promoters and are directly responsible for oncogenicity (tumour formation). At least three of these genes direct the constitutive synthesis of the phytohormones auxin and cytokinin (Figure 2.2C), and are therefore responsible for the induction and maintenance of cell division in the neoplastic outgrowths. Crown galls may be excised from host plants and grown axenically in tissue culture, provided that the growth of *Agrobacterium* is inhibited using antibiotics such as cefotaxime and carbenicillin, which are non-toxic to plant cells. The T-DNA also codes for one or more enzymes that direct the synthesis of unusual amino acid or sugar derivatives, collectively referred to as 'opines'; two of the most widely studied enzymes are octopine synthase and nopaline synthase. In addition, each Ti plasmid carries genes outside the T-DNA region which code for enzymes that catabolise the opine(s) specific on the transferred DNA; thus each *Agrobacterium* strain is able to utilise at least one opine as a unique source of carbon and nitrogen.

Mutations in region D of the Ti plasmid often abolish virulence; this large section of the Ti plasmid (35–40 kb) is usually referred to as the virulence (*vir*) region and plays a crucial role in the transfer of the T-DNA into the plant nuclear genome (see Section 2.2.2). From these early studies it became clear that the Ti plasmid is capable of transferring genes directly from *Agrobacterium* to the plant cell nuclear genome. More recently, a great deal of effort has gone into elucidating the mechanism of the T-DNA transfer process and into developing efficient plant transformation vectors from components of Ti and Ri plasmids.

2.2.2 *The process of T-DNA transfer*

2.2.2.1 *T-DNA and 25 bp repeat sequences.* The mechanism of T-DNA transfer is not fully understood, but several excellent reviews discuss this remarkable interkingdom gene transfer phenomenon (Lichtenstein and Fuller, 1987; Zambryski *et al.*, 1989). Amongst the early discoveries was that none of the T-DNA-encoded genes are required for T-DNA transfer; this suggested that this region of the plasmid plays a passive role in the transfer process. It also appears that the T-DNA does not influence the site of insertion since T-DNA inserts were found to be at random locations in the genome and present at a range of copy numbers (averaging 2–3) within individual transformed cell lines. However, the T-DNA inserts in plant tumours were almost always of a constant size, and normally colinear with the T-DNA region on the Ti plasmid; this fact has very important practical implications as it suggests that each T-DNA is a discrete unit which is inserted into the plant genome without modification. A discovery of great importance was that the T-DNA regions on all Ti and Ri plasmids are flanked by almost perfect 25 bp direct repeat sequences (reviewed by Gheysen *et al.*, 1985). Elegant experiments proved that these 25 bp repeat sequences, particularly those on the right border of the T-DNA, are absolutely required for T-DNA transfer and that they function in a *cis*-acting and polar fashion (Wang *et al.*, 1984). It was soon shown that any DNA sequence could be transferred to plant cells as long as it was flanked by the 25 bp repeat sequences in the correct orientation (Rubin, 1986). The T-DNA junctions with plant genomic DNA in transformed cells ended within or close (internal) to the 25 bp repeat sequences.

2.2.2.2 *Virulence functions.* Any mutations in the virulence region of a Ti plasmid greatly attenuated or abolished tumorigenesis, and so it was suggested that this region was somehow responsible for T-DNA transfer from *Agrobacterium* to plant cells. However, physical separation of the T-DNA and virulence regions onto different plasmids did not affect T-DNA transfer provided both plasmids were present in the same *Agrobacterium* cell; this indicated that the virulence functions were *trans*-acting, a factor which proved to be important in future transformation vector design.

The virulence region is organised into six operons (Figure 2.3); *virA, B, D* and *G* are absolutely required for virulence, whilst strains mutated in *virC* and *virE* are strongly attenuated for tumour formation. With the exception of *virG* and *virA*, all the operons are polycistronic. The *virA* locus is expressed constitutively under all conditions. The *virG* locus is expressed at low levels in vegetative cells but is rapidly induced to higher expression levels by extracts from wounded plant tissue. All other *vir* loci are exclusively plant-inducible, and it has been shown that small phenolic compounds, such as acetosyringone (Figure 2.3), present in plant exudates are responsible for *vir* gene induction

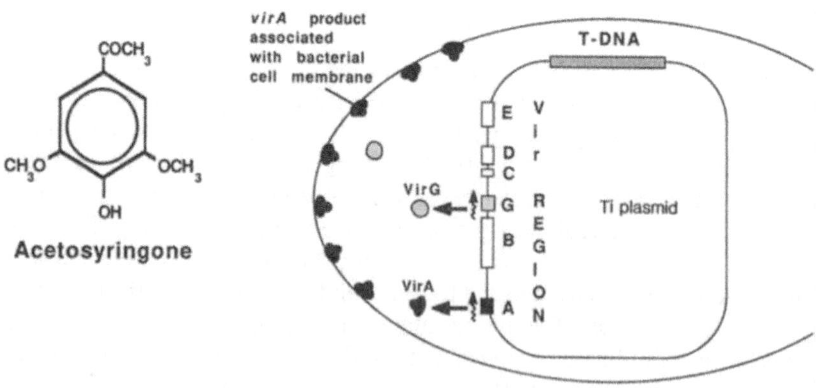

Figure 2.3. Ti plasmid gene expression in vegetative *Agrobacterium* prior to *vir* gene induction by the plant phenolic compound acetosyringone.

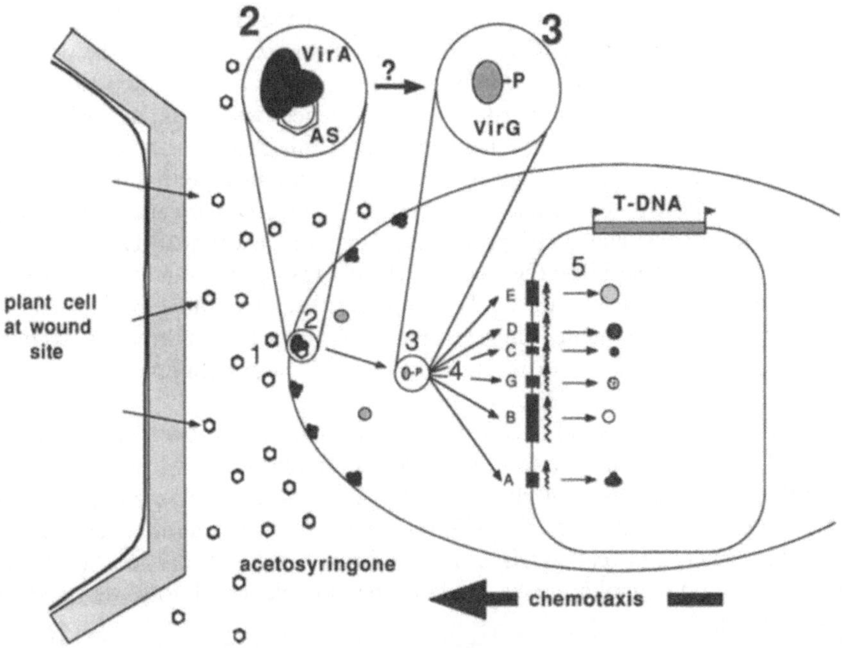

Figure 2.4. Key events during the induction of the Ti plasmid *vir* genes by acetosyringone and other wound-related compounds: 1, acetosyringone released by plant cells at the wound site; 2, acetosyringone 'activates' VirA protein located in the bacterial membrane; 3, activated VirA modifies VirG protein, possibly by phosphorylation; 4, modified VirG binds to the operator **sequences of the *vir* genes resulting in their activation or up-regulation, 5.**

(Stachel *et al.*, 1986a). It is now known that the *virA* and *virG* gene products regulate the expression of the other *vir* loci (Figure 2.4). The *virA* product (VirA) is located on the inner membrane of *Agrobacterium* cells and is probably a chemoreceptor which senses the presence of acetosyringone (AS) in the environment. Signal transduction proceeds via activation (possibly phosphorylation) of VirG, which then stimulates transcription by binding to operator sequences in the promoters of the plant-inducible *vir* genes (Stachel and Zambryski, 1986; Leroux *et al.*, 1987; Zambryski *et al.*, 1989). It is interesting to note that *Agrobacterium* is also chemotactic towards AS, a phenomenon dependent upon a functional *virA* gene (Ashby *et al.*, 1987).

2.2.2.3 *T-DNA transfer process.* Stachel *et al.* (1986b) used specialised Southern hybridisation techniques to characterise novel DNA structures homologous to the T-DNA in acetosyringone-induced agrobacteria. Their data, and those of others, indicated that an early event in the T-DNA transfer process is the nicking of the Ti plasmid at a specific site between the third and fourth base of the bottom strand of each 25 bp repeat. A single-stranded T-strand is then released from the Ti plasmid by replacement synthesis of DNA in a 5′–3′ direction, initiating from the nick in the right-hand 25 bp repeat sequence (Figure 2.5). This proposed mechanism accounted for the polarity and *cis*-acting nature of the border repeat sequences and also explained why the right-hand part of the border repeat sequences was essential for T-DNA transfer. Significantly, this mechanism is very similar to bacterial conjugation. The *virD* operon almost certainly encodes an endonuclease that produces the nicks in the border sequences. The functions of the other *vir* gene products are largely unknown. The VirE2 protein is a large single-stranded DNA-binding protein and may stabilise the T-strand during transport to the plant nucleus. Several gene products of the *virB* operon have been identified in the bacterial envelope, a location which suggests they may play a role in directing T-DNA transfer extracellularly (for recent reviews see Lichtenstein and Fuller, 1987, and Zambryski *et al.*, 1989).

Apart from Ti plasmid-encoded functions, several *Agrobacterium* chromosomal loci are known to affect virulence. These genes are concerned with the synthesis and secretion of glucans, cellulose fibrils and cell-surface proteins (reviewed in Zambryski *et al.*, 1989). These loci may have a more general role in bacterial–plant cell surface interactions, since all are constitutively expressed and several are conserved in other soil bacteria known to associate with plants.

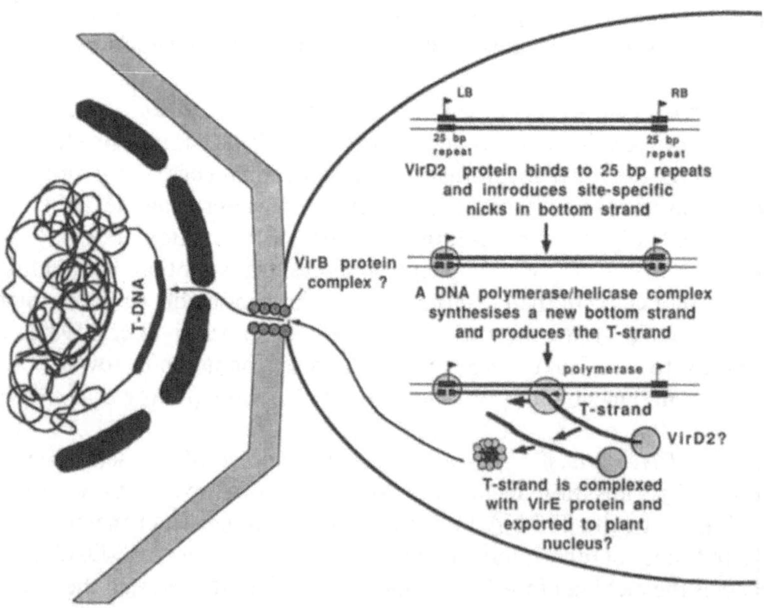

Figure 2.5. Molecular events in the transfer of the T-DNA to plant cells.

2.2.3 *Vectors based on the Ti plasmid*

2.2.3.1 *General features of Ti plasmids incorporated into plant transformation vectors.* The large size, and consequent absence of unique restriction enzyme sites, and the tumorigenic properties of Ti and Ri plasmids precluded their direct use as vectors. However, these plasmids have provided many useful

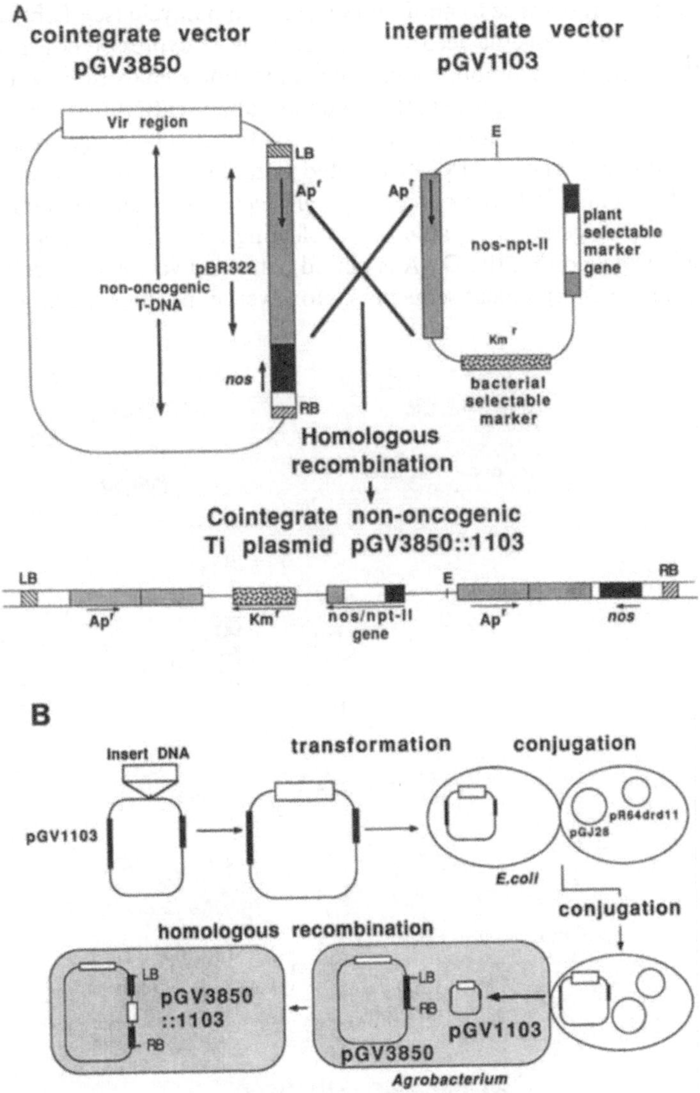

Figure 2.6. (A) Use of the cointegrate vector pGV3850 with the intermediate vector pGV1103. (B) Preparation of a cointegrative vector for the transfer of a desired DNA segment.

components which have been incorporated into modern plant transformation vectors.

Normal shoots cannot be regenerated from oncogenic cells, and thus the 'disarming' of Ti plasmids was an essential step in the development of useful plant transformation vectors. The oncogenic nature of the T-DNA was overcome simply by the removal of the tumour-inducing genes from the T-DNA; these were replaced by dominant selectable markers, mainly bacterial genes affording resistance to antibiotics such as kanamycin (see Table 2.2). In many early non-oncogenic vectors, these genes were expressed constitutively under the control of promoters and polyadenylation signals isolated from T-DNA genes such as octopine synthase and nopaline synthase. More recently, higher expression levels have been achieved by the use of stronger promoters, such as CaMV35S and CaMV19S, isolated from the cauliflower mosaic virus (e.g. Sanders *et al.*, 1987). In essence, any *Agrobacterium* strain containing a set of functional *vir* genes is capable of transferring any DNA sequence into a plant cell, provided that the DNA is placed between two correctly orientated T-DNA border 25 bp repeat sequences. However, in practice the large size of

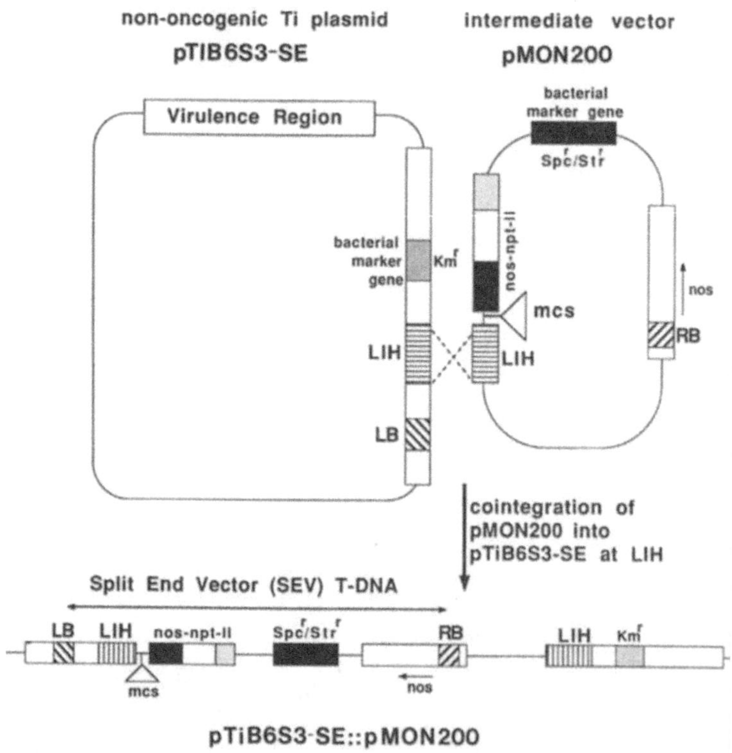

Figure 2.7. A specialised cointegrative strategy designed to minimise the transfer of unwanted vector sequences: the split end vector pMON200 and the non-oncogenic vector pTiB63-SE.

the Ti plasmid makes it impossible to insert DNA restriction fragments directly between the 25 bp repeats. Two strikingly different strategies have been devised to overcome this problem. Consequently, non-oncogenic plant transformation vectors can be divided into two types, *cis*- or *trans*-, depending on whether the T-DNA regions flanked by 25 bp repeat sequences are on the same or a different replicon from the *vir* functions (Figures 2.6 and 2.7). The former are often referred to as 'cointegrative' vectors; whilst the latter, with *trans*-acting *vir* genes, are known as 'binary' vectors.

2.2.3.2 *Cointegrative vector strategy.* Cointegrative, non-oncogenic Ti plasmids have the majority of the T-DNA removed and replaced by a section of DNA homologous to a small *E. coli* cloning vector (Figure 2.6A). These small plasmids (intermediate vectors) are used for all DNA manipulations in *E. coli* and are then conjugated into an *Agrobacterium* strain containing a cointegrative vector (Figure 2.6B); the intermediate vectors cannot replicate in *Agrobacterium*, and thus marker phenotypes can only be maintained if they cointegrate with the resident Ti plasmid at a site within the region of shared homology. One of the first cointegrative vectors, pGV3850 (Figure 2.6A), was developed from a nopaline-type Ti plasmid (C58) and has both T-DNA borders on the Ti plasmid (Zambryski *et al.*, 1983). The T-DNA has almost all been deleted and replaced by pBR322, a common small *E. coli* cloning vector. The intermediate vector (e.g. pGV1103; Hain *et al.*, 1985), based on pBR322, is conjugated into pGV3850 using the ColE1 transfer functions, where it cointegrates at a low frequency into the T-DNA at the region of pBR322 homology. After cointegration, the pGV3850::1103 T-DNA contains the whole of the intermediate vector, and the gene intended for transfer to plants is flanked by directly repeated plasmid sequences. This situation may limit the usefulness of these transformation vectors, since unwanted DNA sequences are introduced into transformed plants. In vectors such as pTiB6S3-SE (a 'split end vector'; Fraley *et al.*, 1985) the right border repeat is carried on the intermediate vector, whilst in pGV2260 (De Bleare *et al.*, 1985) the borders flank a cloning site on the intermediate vector; the net result of intermediate vector cointegration in both of these cases is the inclusion of only a small section of bacterial plasmid in the T-DNA (Figure 2.7).

The requirement for specific intermediate vector/non-oncogenic Ti plasmid combinations with particular 25 bp border repeat configurations and specific regions of mutual DNA homology somewhat restricts the flexibility of the *cis*-type vector strategy. Additionally, it is important to check the structure of the single-copy cointegrate plasmids in *Agrobacterium* to make certain that no unwanted recombination events have occurred; this requires Southern blotting analysis and is therefore a rather tedious affair when compared with the rapid DNA miniprep methods used to analyse plasmids in *E. coli*.

2.2.3.3 *Binary vector strategy.* The observation that the *vir* gene products function *in trans* suggested an alternative strategy to that of using cointegrative vectors; it is possible to provide *vir* functions on a 'helper' Ti plasmid having the whole of the T-DNA deleted (e.g. pAL4404) and to place the T-DNA on a separate wide-host-range cloning vector designed to replicate in both *E. coli* and *Agrobacterium*, and capable of conjugal transfer between the two bacterial species (Figure 2.6B). The cloning vehicle is often referred to as a binary vector and most contain replication and conjugation functions derived from the wide-host-range plasmid pRK252.

Many binary vectors have been developed since they were first conceived by Hoekema *et al.* (1983) and some of those designed for specialised purposes are described in Sections 2.4.3 and 2.4.4. Apart from an almost universal RK2 origin of replication and RK2 conjugation functions, binary vectors differ in size, source of 25 bp repeat sequences, plant selection marker, bacterial selection marker and cloning sites for the insertion of DNA for transfer to plants. pBin19 (Figure 2.8) became available in 1984 and is still a popular general-purpose binary vector, again based on the wide-host-range replicon pRK252 (Bevan, 1984). This vector contains a prokaryotic kanamycin resistance gene (APH-I) for selection in bacteria and T-DNA borders derived from pTiT37. Within the T-DNA, adjacent to the right border repeat, is a plant-selectable transformation marker consisting of the kanamycin resistance gene (neomycin phosphoryl transferase II; *npt II*) coding region

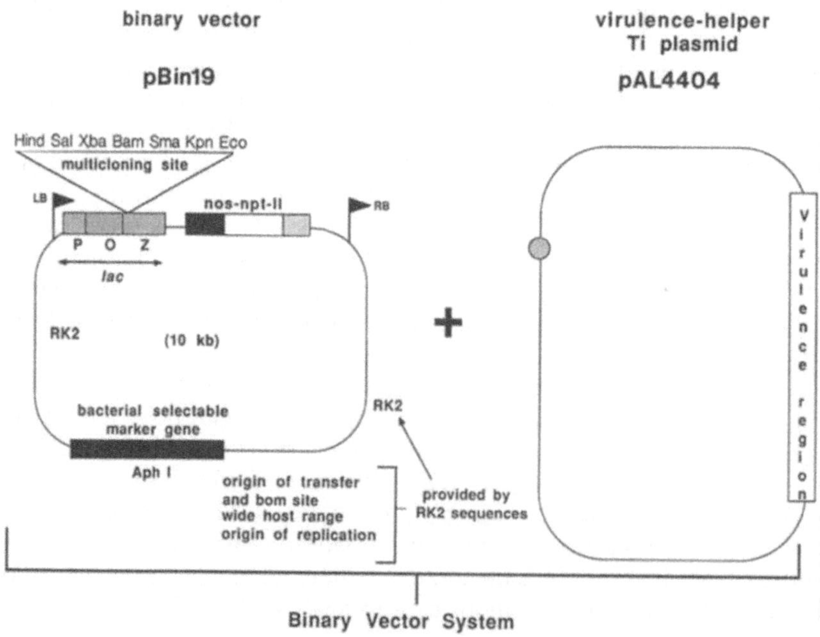

Figure 2.8. Typical binary vector strategy.

isolated from Tn5 placed under the control (transcriptional fusion) of the nopaline synthase (*nos*) promoter and polyadenylation signal. On the left-hand side of the T-DNA is a conventional multiple cloning site derived from pUC19, housed within the α-complementary region of β-galactosidase, which contains seven unique restriction enzyme sites for the insertion of passenger DNA; bacterial colonies containing recombinant pBin19 plasmids are recognised by loss of blue colour on IPTG/X-GAL plates. A typical non-oncogenic *vir* helper plasmid is pAL4404 (Hoekema *et al.*, 1983) which is derived from the octopine-type plasmid pTiAch5 (Figure 2.8).

Unlike the intermediate plasmids of the cointegrative system, binary vectors need have no homology with the resident Ti plasmid and provided selection is maintained, are capable of autonomous replication, usually in multiple copies, within *Agrobacterium*. This gives the binary a considerable advantage over the cointegrative system, since any binary vector can be used in conjunction with any *vir* helper strain; this may be an advantage, as many plant species are more susceptible to transformation by specific *Agrobacterium* strains (e.g. Byrne *et al.*, 1987). Since T-DNAs on separate plasmids are transferred independently (De Frammond *et al.*, 1986), it is even possible to use binary vectors with wild-type, oncogenic strains of *Agrobacterium tumefaciens* and *A. rhizogenes*, and to select transformed plants which have received only the binary vector T-DNA (Shahin *et al.*, 1986; Simpson *et al.*, 1986; Fillatti *et al.*, 1987; Hamil *et al.*, 1987). Non-oncogenic virulence helper plasmids have been developed from several of the common Ti plasmids, including nopaline (T37-SE, Fraley *et al.*, 1985), octopine (pAL4404; Hoekema *et al.*, 1983) and succinamopine strains (pEHA101; Hood *et al.*, 1986). As cointegration is not required, it is also much easier to select stable transconjugants following a bacterial mating, and it is also possible to isolate recombinant binary vectors by conventional plasmid 'miniprep' methods and to check their structure by restriction digest analysis on ethidium bromide-stained gels, before conjugation to *Agrobacterium*.

2.2.4 *Transformation techniques using* Agrobacterium *vectors*

2.2.4.1 *General species limitations for transformation by* Agrobacterium. The number of plant species to be transformed by *Agrobacterium* vectors has increased steadily over the past few years, and representatives of many taxonomically diverse genera have proved amenable to transformation (see Draper *et al.*, 1988, and Lindsey and Jones, 1989, for lists of species). Although the development of new marker genes has been important (see Section 2.2.4.5), the success of this approach for gene transfer has really resulted from improvements in tissue culture technology, particularly adventitious shoot regeneration in the crop plants concerned. Notable by their absence from this list are the majority of the major seed legumes. Perhaps more serious is that the insusceptibility of monocotyledons to transformation by *Agrobacterium* still remains a mystery. 'Agroinfection' studies (Grimsley *et al.*, 1987, 1988) indicate that T-DNA transfer to monocotyledonous cells occurs; the block to transformation by *Agrobacterium* may lie in the wound response of monocotyledonous cells and could reflect a lack of competence for T-DNA transport to the nucleus or T-DNA integration.

2.2.4.2 *Plant cell competence for transformation.* The important requirements for transformation by *Agrobacterium* are, firstly, that the plant explants must produce acetosyringone, or other active compounds, in order to induce the *vir* genes; where these compounds are absent *Agrobacterium* may be preinduced with synthetic acetosyringone. Secondly, the induced agrobacteria must have access to competent plant cells that are capable of regenerating adventitious shoots or somatic embryos at a reasonable frequency. Unfortunately, the specific factors affecting the competence of plant cells for transformation are largely unknown, but there is reasonable evidence to suggest that for gene transfer to occur cells must be replicating DNA or undergoing mitosis (Firoozabady and Galbraith, 1984; Meyer *et al.*, 1985; Okada *et al.*, 1986a). The majority of transformation experiments utilise either freshly explanted tissue sections, protoplasts in the process of reforming a cell wall and entering cell division, or callus/suspension-cultured cell clumps wounded by chopping/pipetting and stimulated into rapid cell division by the use of nurse cultures (see Section 2.2.4.4 and Draper *et al.*, 1988, for a fuller description of explants). Although it has yet to be proven, there is much circumstantial evidence that only wounded and dedifferentiating cells (those undergoing wound-induced callus formation) are competent for transformation. It is conceivable that the reactivation of the cell cycle and concomitant DNA repair and replication provide the nucleic acid biosynthetic enzyme machinery necessary for T-DNA second-strand synthesis and integration into the genomic DNA.

2.2.4.3 *Combining cell competence with totipotency.* Adventitious shoot production *in vitro* is common place for most important crop species; the major problem in the development of a transformation system is providing induced *Agrobacterium* with access to cells capable of dedifferentiation followed by regeneration. In many species, such as flax and sugar beet, adventitious shoot formation from some explants can be prolific. However, such explants are often unsuitable for transformation, since many of the shoots arise by direct primordia formation from subepidermal cells distant from the wound site, and thus not only do these cells not dedifferentiate, but they are also not readily accessible to *Agrobacterium* present in the external medium. Some plant tissues are hypersensitive to infection by *Agrobacterium* and simply die after a relatively short exposure. A still more frustrating problem was identified using *in situ* assays for marker genes such as β-glucuronidase (see Figure 2.9); it proved possible to recognise transformed tissue sectors on explants which

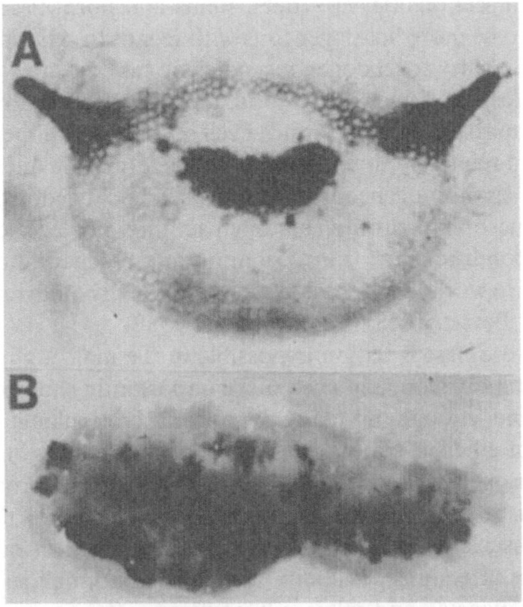

Figure 2.9. Histochemical localisation of *gus* gene activity in transformed plant tissues (A) Cross-section of a petiole from tobacco plant transformed with a CaMV35S-*gus* construct. (B) Section of tobacco leaf disc 10 days after transformation with CaMV35S-*gus*. Dark-staining, transformed cells are visible within callus tissue.

simply do not regenerate. Thus, in practice, it is often difficult to combine transformation competence with totipotency, even though in some species, such as tobacco, this combination can be achieved with relative ease.

There is some evidence that, under certain circumstances, *Agrobacterium* may be able to transform the relatively undifferentiated cells in developing embryos; in this situation one might expect chimeric plants to be produced, and in some cases it should prove possible to obtain non-chimeric transformed shoots by organogenesis or embryogenesis from transformed cell lineages in the organs concerned (e.g. McGranahan *et al.*, 1988). It has also been reported that germinating seeds can be inoculated with *Agrobacterium* and that transformed seedlings may be recovered at very low frequency (3.4×10^{-5}) in the F_1 generation (Feldmann and Marks, 1987). The origin of the transformed cells is not clear, and as yet the technique has not proved to be readily reproducible.

2.2.4.4 *Explants used for transformation by* Agrobacterium. Explants for transformation by *Agrobacterium* vary enormously in size, and range from individual cells (usually protoplasts), suspension-cultured cells and callus cell clumps (undifferentiated and proembryogenic), thin cell layers (e.g. epidermal strips), tissue slices (e.g. carrot phloem and potato tuber storage parenchyma) to whole organ sections (leaves, roots, stems and floral tissues). The larger explants tend to be more heterogeneous with regard to cell type and are more complex with regard to endogenous nutrient reserves and phytohormone gradients. In most cases such explants are more likely to contain at least some cells that are capable of wound-induced cell division and hopefully also shoot regeneration on relatively simple media. Callus formation and shoot regeneration in relatively homogeneous explants such as cotyledons, storage tissue sections and mesophyll protoplasts often require greater media specificity; thus defining conditions for transformation and regeneration may be more difficult, but once worked out these systems are usually quite reproducible (e.g. Sheerman and Bevan, 1988; Moloney *et al.*, 1989).

Although protoplast isolation is possible in the majority of plant species, reproducible cell division, and even more importantly shoot regeneration, is still not routine enough in many important crop plants to enable the development of efficient transformation systems. Another limitation of this particular approach is that shoots are regenerated via protoplast-derived callus, which is a relatively slow process and necessitates the exposure of transformed tissues to tissue culture-associated stresses for many cell generations, with the attendant possibility of genetic abnormalities. Many freshly initiated cell suspension and callus cultures from certain species exhibit stable regeneration via somatic embryogenesis for several months after their initiation. There is clear evidence that proembryogenic cells in such cultures are susceptible to transformation by *Agrobacterium* (Scott and Draper, 1987), but it is not yet clear if cells in somatic embryos are also competent for transformation.

2.2.4.5 *Recognising transformed shoots or calluses.* The majority of modern transformation vectors carry dominant marker genes designed to enable transformed cells to survive in media containing normally toxic levels of the associated selection agent, usually an antibiotic or herbicide. In model transformation systems, such as tobacco (Horsch *et al.*, 1985), it is indeed possible to place explants (in this case leaf sections) on regeneration medium and obtain transformed shoots directly in the presence of antibiotics such as kanamycin (Figure 2.10). However, this is not the case with most other species, and this has led to the development of alternative approaches for the generation of transformed plants; these generally involve either selection techniques, screening procedures, or a combination of both.

The selection of transformed cell lines developing from protoplast-derived cells (Marton *et al.*, 1979) or small suspension-cultured cell clumps is generally not a problem, provided the cell types concerned display a reasonable sensitivity to the selective agent (Hauptmann *et al.*, 1988). In some cases, plating at a fairly low density on selective media several days after exposure to *Agrobacterium* is the only necessary precaution; this helps to avoid 'cross-detoxification' effects and problems caused by polyphenol production from

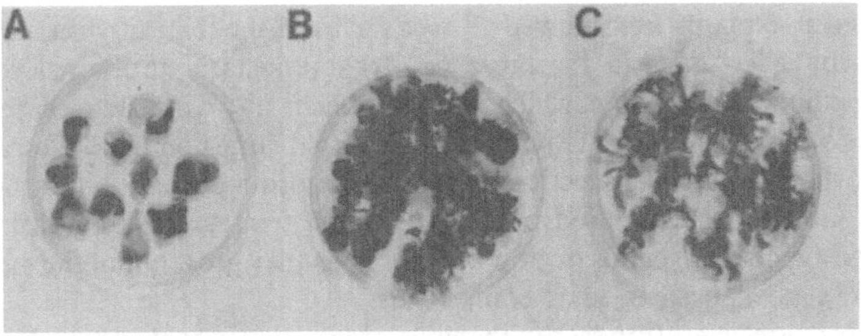

Figure 2.10. Leaf disc transformation of tobacco explants after 4 weeks on regeneration medium. (A) Control explants on kanamycin. (B) Control explants on media without kanamycin. (C) explants dipped in agrobacteria containing a modified T-DNA carrying a NOS-*npt* gene, and cultured in the presence of kanamycin.

the large proportion of dying, non-transformed cells. Kanamycin has proved particularly useful as a selective agent for transformed cells as it tends to bleach tissues first and inhibit further growth without causing rapid necrosis of plant tissues. Other antibiotics and herbicides may require more judicious use as, even at low concentrations, they can cause rapid cell death. Specialised plating techniques have been developed to aid the recovery of transformed cells, and with several protoplast or suspension-cultured cell systems transformation frequencies in the region of 60–70% (per plating unit) may be achieved (Scott and Draper, 1987). Following successful selection, the next obstacle is the regeneration of shoots from transformed calluses. With large heterogeneous explants, in order to ensure a good regeneration response and avoid genetic instability, the regeneration of transformed shoots should be as expeditious as possible. Further problems associated with the rescue of transformed shoots from such explants after incubation with *Agrobacterium* arise from the inherent difficulty of encouraging the growth of antibiotic/herbicide-resistant shoots, whilst preventing the excessive growth of non-transformed tissues. In such situations, the problems of polyphenol production by dying tissue cannot be avoided by low plating densities.

It is probably useful to describe a few possible reasons why *Nicotiana tabacum* has proved so amenable to transformation by *Agrobacterium* (e.g. Horsch *et al.*, 1985). In leaf tissue sections from tobacco, viable cells in the region of the peripheral wound form callus rapidly and shoot regeneration is evident after only a few cell divisions. What is important here is that the majority of adventitious shoots in the first wave of regeneration are derived from the wound-induced callus at the cut edge of the leaf explant. This represents the ideal situation for several reasons: the majority of the cells in unwounded leaf lamina simply bleach and do not form callus; the thinness of leaf explants means that the majority of the tissue is in contact with the medium and exposed to selective agents and thus very few non-transformed shoots 'escape' the selection procedure. Generally, superior regeneration is obtained if explants are allowed to recover from incubation with *Agrobacterium* for a few days in the presence of bacteriostatic antibiotics such as carbenicillin or cefotaxime, which interfere specifically with bacterial cell wall synthesis and are not toxic to plant cells. To avoid large numbers of 'escapes' it is important to impose selection as shoot primordia develop—adventitious shoots (and other organised organ structures) tend to be less sensitive to selective agents than callus or primordia and, as they grow out of the medium, they are less exposed to toxic compounds.

The rapid regeneration seen with many large complex explants, such as stem sections, often emanates from cells which have not undergone wound-induced cell divisions and thus, although regeneration may be profuse, the number of shoots derived from transformed cells will be very limited. Under these conditions, modification of the explant, media and inoculation conditions may be needed to inhibit shoot primordia formation directly from cells in the original explant tissues. This type of rapid regeneration is probably largely the result of pre-existing hormone gradients in the donor tissue. Fortunately, in many cases the sites of regeneration are predictable and therefore may either be removed from the explants or any shoots rising in such areas discarded. However, care should be exercised with the latter approach since, once regenerated, shoots rapidly synthesise auxins, which are then transported into the explant where they may alter the hormone balance to such an extent that further organogenesis from wound-induced callus is curtailed.

So far we have concentrated on the direct selection of transformed shoots. However, in many cases, alternative strategies have proved more fruitful for the recovery of transformed plants.

Adventitious root formation is very sensitive to antibiotics, and so it is useful to select putatively transformed shoots using threshold levels of selective agents (or not to impose selection at all) and then to recover transformants by their ability to root in the presence of antibiotics. Screening techniques offer an alternative in which no selection pressure at all is imposed on shoots developing from inoculated explants; samples of tissue from all regenerated shoots may then be tested for the expression of a marker gene (e.g. *npt-II, cat, gus, lux*) using rapid assay techniques (see Draper *et al*., 1988) or, alternatively, small samples of leaf tissue can be tested for the ability to callus in the presence of a selective agent. Screening in this way is particularly useful in species where profuse shoot regeneration is not routine; however a large amount of repetitive screening may be necessary in order to recover the occasional transformed shoot.

2.2.4.6 *Choice of vector system and marker gene.* The virtues of many marker gene and vector system combinations for the transformation of particular plant species are often extolled in the literature. Certain *vir* systems promote more efficient T-DNA transfer to some species than to others, and, similarly, the effectiveness of selective agents differs; there appears to be no rational hypotheses as to why these differences occur, and at present the development of efficient transformation procedures depends on the empirical testing of *vir* systems, marker genes, explant sources and preconditioning, regeneration conditions, selection and screening procedures.

Current lists of widely used selectable and screenable marker genes have been presented in Tables 2.2 and 2.3 respectively. Generally speaking, in many plant species, kanamycin resistance has proved most useful as a selection marker; some selection markers, particularly those affording herbicide resistance, have been more restricted in their distribution within the scientific community and, consequently, as they become more widely available they may prove effective for the transformation of a range of species. Of the screenable markers, β-glucuronidase certainly offers the greatest versatility, with fluorometric, colorimetric, immunolocalisation and *in situ* histochemical methods available for the detection of activity (Jefferson *et al.*, 1987; Draper *et al.*, 1988); as with most of the screenable marker genes it is important to realise that in some species certain tissue types may have background problems (enzymatic and chemical) which affect the GUS assay.

2.3 Physical DNA delivery methods

Agrobacterium-mediated gene transfer has already had a tremendous impact on the plant sciences, but this is somewhat tempered by the fact that the cereals (and therefore the most important food crops) are not amenable to transformation. Additionally, the tissue culture techniques required for the regeneration of transformed shoots from other groups of crop plants, such as the majority of the seed legumes, are still very rudimentary; this has led to an intensive search for alternative gene transfer techniques. Physical methods for the delivery of transforming DNA to eukaryotic cells have been available since the late 1970s. Most of these techniques were derived from studies examining the expression of virus genomes in tissue-cultured animal cells.

Besides the prerequisite of efficient DNA uptake, any direct gene transfer system must protect the vector from nuclease or mechanical degradation; such systems should also take into account any plant cell competence, be non-cytopathogenic and retain reasonable cell viability after DNA uptake. In general, physical methods for DNA transfer can be grouped according to the type of target cell. Some, such as the chemically stimulated endocytosis of plasmids or DNA-loaded liposomes and electroporation, are limited for use with protoplasts, whilst techniques such as microinjection and microprojectiles can theoretically be used with a much wider range of explants. It should be emphasised that, although it is possible to transform protoplasts from almost any species, the efficient regeneration of transformed plants from protoplast-derived tissue, especially that of cereals, remains problematical; recent progress with rice and maize (e.g. Rhodes *et al.*, 1988) is encouraging, but it

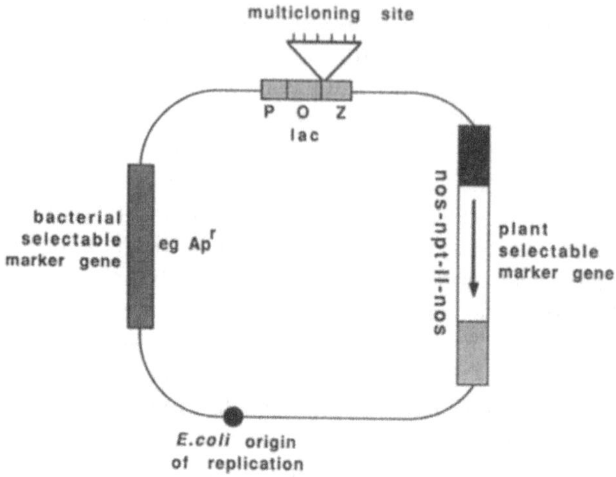

Figure 2.11. Generalised direct gene transfer vector.

remains to be seen if such techniques are applicable to a wide range of varieties and are free from fertility problems.

DNA delivery methods which are not limited only to protoplasts (microinjection, microprojectiles and 'macroinjection') aim to manipulate immature embryos, organ meristems, gametes and zygotes without adversely affecting their normal development. These techniques are currently under rigorous evaluation and, although some are promising, there is as yet insufficient evidence to suggest that this emerging technology is fully reproducible and efficient enough for general use.

Direct gene transfer techniques generally use small *E. coli* cloning plasmids as vectors. In theory, prokaryotic vectors are suitable for direct gene transfer (Figure 2.11) with only slight modification, that is the addition of an antibiotic resistance gene engineered for plant expression. The majority of these selection marker constructs have already been tested via *Agrobacterium* transformation and are known to be expressed efficiently in transformed cells; indeed many small intermediate vectors or binary vectors have been used successfully for direct gene transfer.

The delivery of vector DNA into synchronised protoplasts at S-phase and mitosis improved transformation efficiencies by several-fold (Meyer *et al.*, 1985; Okada *et al.*, 1986a); transformation of immobilised protoplasts by microinjection was also much improved if they were allowed to reform a cell wall for a day or so and thus reactivate the cell cycle prior to injection (Crossway *et al.*, 1986; Reich *et al.*, 1986). These data again strongly suggest that competence for transformation may be related to genome replication and repair in stressed cells re-entering cell division. Competent plant cells, like animal cells, will integrate any type of DNA introduced into the nucleus, and therefore specific DNA sequences designed to direct preferential integration of particular vector fragments rarely do so. Plants transformed by direct techniques often contain multiple copies of the vector sequences; these sequences are usually random subfragments of the vector DNA, or even whole linearised vector sequences, integrated into one or a few sites in the genome, either singly or as concatomers (e.g. Potrykus *et al.*, 1985; Riggs and Bates, 1986). The selection of transformants is only possible where transformed cells integrate at least one copy of the marker gene which has not been scrambled during DNA uptake and integration. There is some suggestion that more efficient transformation can be achieved using linear DNA (Riggs and Bates, 1986); vector DNA was found to be integrated in long concatomers. Under these circumstances the structure of the inserts was more predictable, and thus by linearising vectors at positions distant from the selection marker gene and passenger DNA it may be possible to predetermine, to a certain extent, the configuration of transferred DNA.

2.3.1 *Chemically stimulated plasmid uptake into protoplasts*

Many chemicals known to protect DNA and stimulate nucleic acid uptake into protoplasts unfortunately also cause major reductions in cell viability, and so it is not proposed to discuss their use in any detail. Such chemicals are often long-chain polycations, including polyamines (poly-L-ornithine and poly-L-lysine) or charged carbohydrate polymers such as dextran sulphate. These compounds bind tightly to negatively charged nucleic acid molecules and form more neutral complexes in aqueous solutions that are also relatively resistant to nuclease attack. As the plant membrane is generally also negatively charged, it may be supposed that such nucleic acid complexes will suffer less from charge repulsion by the membrane than naked DNA; there is also evidence that endocytosis is stimulated by these polymers and thus their net effect is to enhance DNA uptake into protoplasts. Liposome encapsulation of DNA (Deshayes *et al.*, 1985) certainly provides good protection from nuclease attack. However, the endocytosis of liposomes into protoplasts is achieved using the same chemical treatments used to stimulate the uptake of naked plasmid DNA; no apparent increase in transformation efficiency has been achieved by the use of liposomes and so they will not be discussed further.

Of the many chemicals tested for their ability to stimulate DNA uptake into protoplasts (Draper *et al.*, 1982, 1986; Freeman *et al.*, 1984), to date most success has been achieved using the long-chain hydrophobic polymer, polyethylene glycol (PEG 6000), often in combination with calcium phosphate-precipitated vector DNA (Draper *et al.*, 1982, 1986; Krens *et al.*, 1982; Freeman *et al.*, 1984; Paszkowski *et al.*, 1984; Hain *et al.*, 1985). PEG is also used to stimulate the uptake of liposomes (Deshayes *et al.*, 1985) and also to improve the efficiency of electroporation techniques for DNA delivery (see Section 2.3.2). Polyethylene glycol at high concentrations (15–25% PEG 6000) will precipitate ionic macromolecules, such as DNA, and stimulate their uptake by endocytosis without causing gross damage to protoplasts. Following DNA uptake, the protoplast populations are allowed to recover at high cell densities in normal protoplast growth medium. Once cell wall reformation has been completed and cell division initiated, the cells may then be plated out at low densities on selection medium.

Initial studies were limited to *Petunia* and *Nicotiana* protoplasts and used PEG in combination with pure Ti plasmid (Draper *et al.*, 1982), or calcium phosphate-precipitated Ti plasmid mixed with a carrier DNA (Krens *et al.*, 1982); transformation frequencies of $1-5 \times 10^6$ were reported. With the advent of efficiently expressed selectable marker genes, small 'direct gene transfer' vectors were developed (Paskowski *et al.*, 1984; Hain *et al.*, 1985) and the frequency of transformation was improved to 1 in 10^4, using kanamycin as a selection agent.

2.3.2 *Transformation of protoplasts by electroporation*

Electric impulses induce transient pores in the plasma membrane through which macromolecules, such as DNA, move into the cell. These pores appear to be around 30 nm in diameter and persist for several minutes after the pulse (Okada *et al.*, 1986b). The two main variables affecting the permeabilisation of membranes by electroporation are field strength (voltage gradient) and pulse duration (or decay time). The field strength required to induce electroporation is related to the diameter of the protoplast; with a 1 cm gap between electrodes, tobacco protoplasts of average diameter 40–44 μm require a field strength of 1.5 kV/cm^2 for efficient DNA internalisation (Shillito *et al.*, 1985).

The electroporation pulse is generated by discharging a capacitor across the

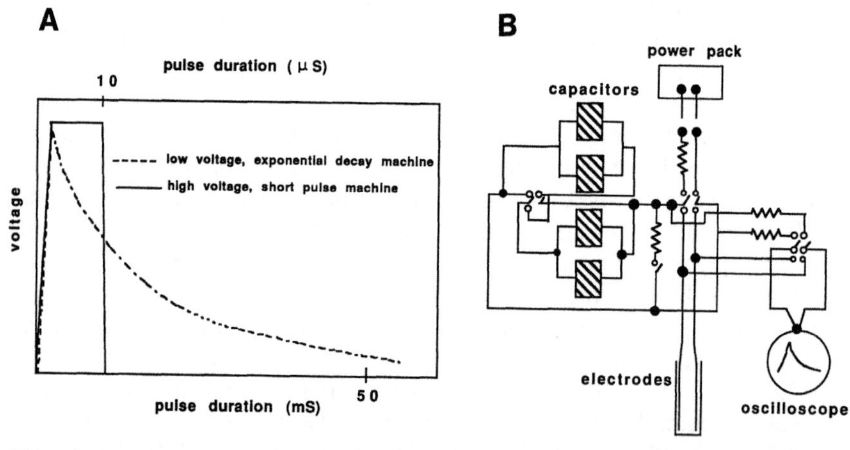

Figure 2.12. (A) Details of the output characteristics of the two principal types of electroporation devices in use. (B) Circuit diagram of a low-voltage, exponential decay machine.

electrodes in a specially designed electroporation chamber (Figure 2.12). Protoplasts have been successfully transformed using either a high-voltage rectangular wave pulse (Figure 2.12) of short duration generated by a commercial machine (e.g. 1.5 kV for 10 μs; Shillito *et al.*, 1985) or a lower voltage pulse of much longer duration (e.g. 350 V over 54 ms) generated with a home-made system (Figure 2.12; Fromm *et al.*, 1985). The protoplasts are suspended between the electrodes in an ionic solution containing the vector DNA, electroporated and then plated out as normal, and transformed colonies are selected as described for PEG-mediated gene transfer. The frequency of transformation of protoplasts by electroporation is improved in the presence of 13% PEG, if added after the DNA; other factors such as linearising the plasmid, use of carrier DNA, heat shocking (45°C, 5 min) prior to addition of vector and placing on ice after pulsing may all improve transformation frequency (Shillito *et al.*, 1985). Under optimum conditions, transformation frequencies of up to 2% (or greater) have been reported.

Stably transformed cell lines have been produced in a number of monocotyledonous species by electroporation, including maize (Fromm *et al.*, 1986), wheat (Lorz *et al.*, 1985) and rice (Uchimiya *et al.*, 1986). Recently it has proved possible to regenerate plants, which are in some cases fertile, from transformed cell lines of several cereals (Rhodes *et al.*, 1988; Zhang *et al.*, 1988; Shimamoto *et al.*, 1989). It remains to be seen if fertile plants free of somaclonal variation can be regenerated from a wider number of cereal varieties in the future.

Apart from stable transformation, electroporation is used widely in transient gene expression assays for many purposes (see Section 2.4.3), for example to study the properties of promoters or effects of modifications to DNA sequence that might modulate gene expression levels.

2.3.3 *Microinjection, 'macroinjection' and microprojectiles*

The difficulties experienced in regenerating genetically and phenotypically normal plants from tissue-cultured cells, particularly protoplasts, are unlikely to be resolved overnight. Tissue culture techniques that avoid cellular dedifferentiation but which encourage the continued development of immature structures (e.g. embryo rescue and meristem culture) provide alternative cellular targets for transformation. The main disadvantage of such an approach is that, unless it proves possible to transform a few cells in vegetative or preflowering meristems, any regenerated plants will almost certainly be chimeric. However, transformed plants of single-cell origin can theoretically be produced from chimeric plants when adventitious shoot/embryo production can be induced from the transformed cells, or if transformed cell lineages give rise to gametes. The direct genetic manipualtion of pollen prior to, or during, the fertilisation process is obviously an attractive proposition as tissue culture would be avoided entirely.

Some successes have been claimed recently using techniques which involve the microinjection of vectors into specific embryonic cells (Neuhaus *et al.*, 1987); the firing of DNA-loaded microprojectiles (Klein *et al.*, 1987; Christou *et al.*, 1988) into immature embryos (McCabe *et al.*, 1988a), suspension-cultured cells (Wang *et al.*, 1988) or pollen (McCabe *et al.*, 1988b); or the soaking of meristems (De la Pena *et al.*, 1987), dry and germinating pollen (Picard *et al.*, 1988) and severed pollinated stigmas (Luo and Wu, 1988) with solutions containing vector DNA. These methods are generally still at the exploratory stage in most laboratories, but it is undoubtedly true that, if successful, such procedures should allow the rapid transformation of any species, using techniques which largely avoid the use of tissue culture; this would represent a significant breakthrough and should make the genetic engineering of the majority of crop plants a realistic proposition in the future. This present section outlines the developing technology that may allow the transformation of intact plant tissue by direct gene transfer.

2.3.3.1 *Microinjection.* The microinjection technique is based on the use of glass micropipettes with a 0.5–10 μm diameter tip to effect the direct transfer of macromolecules into the cytoplasm or the nucleus of a recipient cell, which is either immobilised on a solid support, artificially bound to a substrate or held by a pipette under suction (Celis, 1984). Two great advantages of the microinjection technique for plant cell transformation relate to the fact that, firstly, very high frequencies of protoplast transformation have been reported (15–20%), and thus it may be useful where only a few cells/protoplasts/organs are available for use. Secondly, the technique does not necessarily require removal of the cell wall and so could be useful for plant species which at present do not regenerate from protoplast-derived colonies. The disadvantages of microinjection are that it is an extremely slow process, relatively expensive to set up and requires highly skilled and experienced personnel.

2.3.3.2. *Microinjection of protoplasts and single cells.* Recently the feasibility of plant cell transformation by direct microinjection of vector DNA into protoplasts has been shown by Crossway *et al.* (1985, 1986) and Reich *et al.* (1986). The success of these experiments was mainly due to the development of microinjection procedures that enabled the localisation of the nucleus and orientation of the protoplasts in such a way that DNA could be injected directly into the nucleus without drastically reducing viability. For injection, plant cells (usually 1- to 2-day-old protoplasts which have started to form a cell wall) are either immobilised on coverslips with poly-L-lysine or embedded in a thin layer of agarose. An alternative procedure has been developed by Crossway *et al.* (1986) which involves the immobilisation of protoplasts using a specially designed 'holding pipette' (Figure 2.13). A protoplast can be sucked up or released by developing positive or negative pressure in the holding pipette connected to a syringe. This method allows manipulation of the

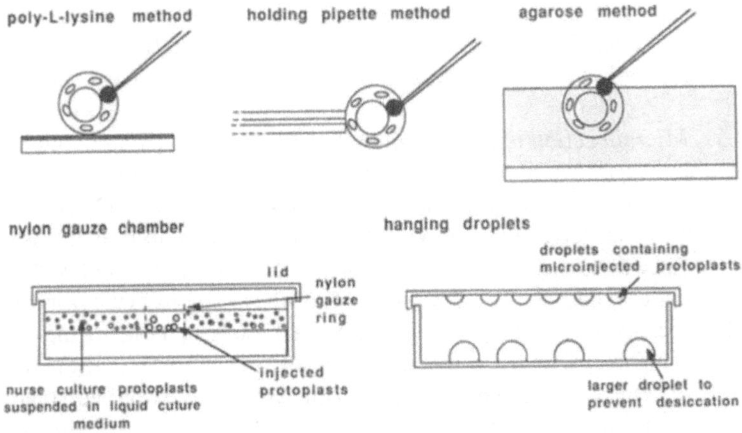

Figure 2.13. Methods for the microinjection and culture of plant protoplasts.

protoplasts for optimal orientation of the nucleus relative to the injection pipette.

An inverted microscope equipped with differential interference ('Normarski') optics is normally used to facilitate visualisation of the nucleus. Alternatively, nuclei may be stained with DNA-specific vital fluorescent dyes (e.g. Hoechst 33258) and visualised under an inverted microscope equipped with epifluorescence optics (Reich *et al.*, 1986). The micropipette is first filled with the vector DNA solution and then directed into the immobilised target protoplast or cell with the aid of a 'micromanipulator'. When in position within the nucleus, an appropriate volume of solution is injected by gentle air, or hydraulic pressure, exerted by a syringe connected to the micropipette at one end and to a pressure regulator system at the other end. If cells are immobilised on coverslips, or in agarose, then nuclei are usually inoculated after the cells have had time to recover, usually late on the first day after attachment or embedding. For convenience, cells are commonly immobilised on a special microscope slide with a grid pattern. This allows individual cells to be located at any time before or after injection, by filming on video. Hence, their development can be followed and appropriate action can be taken to ensure their proper growth in culture. If 'holding pipettes' are used then protoplasts have to be isolated from liquid protoplast cultures at a stage competent for transformation (2–3 days old), injected and then cultured under conditions that allow their recognition.

Specialised 'nurse culture' or microdrop culture techniques are required to recover microinjected cells (Figure 2.13). In most instances, the inoculated protoplasts are cultured singly on 'nurse cultures' or in microdrops (see Figure 2.13). The plating efficiency after injection is usually reduced by anything from 10 to 70% but, providing efficient microculture techniques are available, this is more than compensated for by the very high transformation frequencies (15–25% on average!). Southern blot analyses have shown that transformants resulting from microinjected cells have patterns of transferred DNA organisation similar to those found in other direct gene transfer systems.

2.3.3.3 *Microinjection of multicellular structures.* In order to avoid some of the problems associated with protoplasts, alternative targets for microinjection have been sought, such as the cells of immature embryos (Neuhaus *et al.*, 1987). Unfortunately, this approach produces only chimeric transformed plants, and consequently, to date, success has only been reported with pollen-derived somatic embryos of *Brassica napus* where secondary embryogenesis enables the recovery of wholly transformed plants. Transformation was optimal (10–30%) with 12 cell-stage embryos, but transformation was also possible at later developmental stages prior to the heart-shaped stage. Such high transformation frequencies effectively obviate the requirement for selection markers as transformants can be screened for without too much effort.

The development of zygotic embryo isolation and microinjection techniques for a wider range of species is under way in several laboratories, but as yet there is no published evidence that non-chimeric transformed plants can be generated in any other species. However, if successful, such techniques will clearly be of tremendous importance, since zygotic embryo rescue techniques are available for the majority of important crop plants, including seed legumes and cereals.

Alternative targets under investigation by groups around the world include cells such as immature pollen, germinating pollen and multicellular structures such as embryogenic suspension-cultured cells, isolated ovules and cells in floral and vegetative meristems. Microinjection into the nuclei of germinating pollen is relatively straightforward, and has been attempted by many groups, but to date transformed plants have not been produced by fertilisation with injected gametophytes. The cells of immature zygotic embryos can just about be visualised using inverted microscopes, but the target cells in large multicellular structures, such as meristems and embryogenic suspension-cultured cells tissues, are rather difficult to see; perhaps different optical systems might prove useful in the future.

2.3.3.4 *Microprojectiles and macroinjection.* It is possible to deliver nucleic acids into plant cells on the surface of high-velocity tungsten or gold microprojectiles (Klein *et al.*, 1987; McCabe *et al.*, 1988a; Wang *et al.*, 1988). The microprojectiles are normally 1–3 μm in diameter and are accelerated into living plant cells by either an explosive charge (Klein *et al.*, 1987) or using shock waves initiated by a high-voltage electric discharge (McCabe *et al.*, 1988). Schematic diagrams of the two microprojectile acceleration devices are given in Figure 2.14. The technology basically involves loading tiny tungsten or gold spheres with vector DNA and then spreading the particles on the

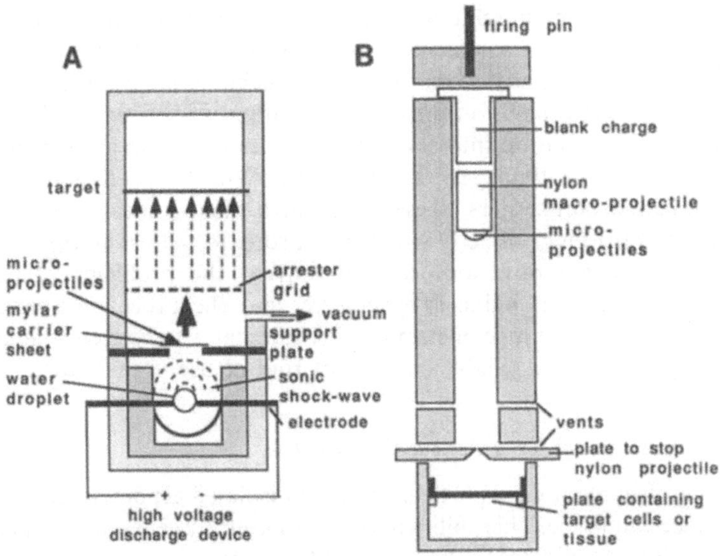

Figure 2.14. Microprojectile acceleration devices. (A) Electrostatic. (B) Ballistic.

surface of a mobile plate or plastic 'bullet'. Then, under a partial vacuum, the 'macroprojectile' is fired against a retaining plate or mesh, either by a shock wave caused by vaporising a water droplet, or by a cordite explosion; the 'macroprojectile' decelerates instantly, whilst the momentum and small size of the dense microprojectiles causes them to be thrown from the surface of the 'macroprojectile' and to penetrate the target plant tissue. To date there have been several reports of transient gene expression (usually GUS activity) following the microprojectile bombardment of several types of plant tissue (callus, embryos, epidermis). McCabe *et al.*, (1988) recovered a single transformed plant in the first seed generation from only one of many chimeric transformed soybean plants. Transformed tobacco plants have been recovered recently by several laboratories. A recent patent application filed by Agracetus (McCabe *et al.*, 1988a) describes a method for the delivery of vector DNA into dehisced maize pollen, using the electrostatic microprojectile device, and the recovery of several transformed plants following the pollination of maize plants with treated pollen. At the time of writing there have been no further reports of the successful transformation of pollen from seed legumes or cereals using microprojectiles. However, this type of transformation technology promises to be so widely applicable that it is likely that there will be considerable effort to improve such procedures in the very near future.

In 1987, data presented by De la Pena *et al.*, suggested that it was possible to transform rye archesporial cells by simply allowing them to soak up a solution of vector DNA. However, despite many attempts to repeat the procedure, now termed 'macroinjection', no further successful reports have been published.

Nevertheless, the manipulation of the pollination/fertilisation process for gene transfer remains a very attractive proposition. Recently, two groups have provided some evidence that this might be possible, either by enclosing wheat stigmas in a vector solution prior to pollination (Picard *et al.*, 1988) or by injecting vector solution into severed rice styles following growth of pollen tubes almost into the eggs sac (Luo and Wu, 1988). If confirmed by independent experimenters, these procedures offer a means of obtaining transformed seedlings directly without resorting to the use of tissue culture techniques. Such innovations will circumvent the problems of possible somaclonal variation induced by culture, and the procedures should be applicable to a wide range of crop species. It remains to be seen whether this technology will prove reliable enough for routine use in the future.

2.3.4 *Virus vectors for gene transfer to plants*

Many purified plant virus particles, and their isolated genomes, are infectious on intact plant tissue. This ability to transfer naturally genes into plants made them obvious candidates for use as plant transformation vectors in the early 1980s. However, viral genomes do not integrate into plant DNA, and so viral vectors may only be useful for transient gene expression studies in infected plant tissues. Plant viruses vary enormously in terms of their genome structure, modes of infection, replication, host range and basic lifestyle; broadly speaking viruses fall into a number of groups depending on whether their genome is composed of double-stranded DNA (e.g. caulimoviruses), single-stranded DNA (e.g. gemini viruses) or RNA (e.g. tobacco mosaic virus). All three categories of virus have been shown to be capable of delivering genes into intact plant tissues where they are expressed.

In caulimoviruses, such as CaMV, the strategy has been to replace a small section of the genome, not required for virus propagation, with a passenger gene small enough not to interfere with the packaging of the genome into the virion particle (e.g. Brisson *et al.*, 1984). However, the amount of passenger DNA amounts to only a few hundred base pairs, which imposes a great constraint on the use of CaMV as a vector. Isolated gemini virus genomes are infective as cloned double-stranded linear forms. DNA in the same form may also be delivered into plant cells as part of a Ti plasmid T-DNA via *Agrobacterium*-mediated gene transfer, a technique commonly referred to as 'agroinfection' (Grimsley *et al.*, 1987). Some gemini virus vectors have been used to express cloned genes in plant tissue, but there are several reports of genetic instability caused by recombination (e.g. MacDowell *et al.*, 1986). Recently a vector has been designed by Hayes *et al.*, (1988) which, following delivery by agroinfection, is capable of amplifying to very high levels. This type of vector may be useful for the production of specific proteins in large populations of plants.

Single-stranded RNA viruses, such as TMV, have been used as vectors to express foreign genes in plants (e.g. Takamatsu *et al.*, 1987). cDNA clones of the TMV genome are manipulated to insert the passenger genes at a site which does not affect virus replication (usually replacing the coat protein); genomic RNA is then synthesised from the cDNA clones *in vitro* and virion particles reconstituted *in vitro* by packaging in viral coat protein. However, most of these engineered virions lack sequences coding for coat proteins and therefore cannot spread systemically, which reduces their usefulness for many purposes. Many other RNA viruses have multipartite genomes, making infection with engineered and reconstituted virions difficult. It is also known that errors frequently occur during RNA-dependent RNA synthesis, bringing into question the usefulness of RNA viruses for high-level gene expression in infected tissues where multiple rounds of replication are required.

In conclusion, whilst there is little doubt that gene transfer to plants via virus vectors will have some very useful applications in the future, this may not include the production of stably transformed plants.

2.4 Uses of gene transfer technology

2.4.1 *Properties of transformed plants*

The majority of genetically transformed plants studied to date have been generated via organogenesis in tissue culture. In many instances, such plants have been used to examine a particular aspect of gene expression or molecular biology, and there have been few comprehensive studies on the genetic stability of such plants. However, this is set to change now that the commercial sector has begun field trials to evaluate the performance of genetically engineered plants. In most instances the aim of a genetic engineering experiment is to transfer a single gene to an existing plant variety without altering its major characteristics. Apart from uncontrollable genetic changes associated with tissue culture, individual transformed plants vary with regard to the transferred gene(s) copy number, organisation and expression.

The number of transferred DNA inserts is variable with both *Agrobacterium*-mediated and direct gene transfer techniques, although the former is generally more predictable, usually varying between one and three copies. Multiple gene copies are often present as large concatemeric inserts; in terms of genetic linkage these will generally act as a single locus, even though several copies of the transferred DNA are present. Generally, direct gene transfer methods afford less control over the scrambling of genes during transfer, and this may lead to problems in identifying the active loci, except by genetic analysis. It is even possible that problems with stability could occur as a result of recombination between homologous sections of the transferred DNA. However, to date most transferred genes have exhibited normal Mendelian inheritance. There is a wealth of evidence which suggests that individual transformed plants with random, single-copy, intact T-DNA inserts of the same gene exhibit wide variation in expression levels (e.g. Jones *et al.*, 1985). This phenomenon is usually referred to as a genome 'position effect' and may be related to the 'context' of the DNA surrounding the T-DNA insert; insertion into repetitive versus unique DNA, degree of DNA methylation, chromosome containing the insert, chromosomal position, transcriptional activity of neighbouring genes and locality of transcriptional enhancers and silencers are just some of the factors that may affect the level of gene expression from any particular T-DNA.

2.4.2 *Plant variety improvement: addition of useful traits to crop species*

The factors which affect gene expression in transformed plants outlined in Section 2.4.1 are largely uncontrollable, with the exception that transferred DNA structure is usually more predictable with *Agrobacterium*-mediated transformation. Thus, a general rule for variety improvement is that many independent transformed plants should be screened for appropriate expression of the transferred gene. As many physiological and morphological abnormalities exhibited in plants derived from tissue culture are 'epigenetic' in origin, it is important to concentrate any studies on gene expression in seedling populations derived from the original transformant plants. Such genetically engineered plants will only be commercially useful if the new gene functions in the desired way and if the plant is free of any genetic modifications to the important characteristics of the original variety.

Although transformed plants have been obtained in many species, reasonably efficient transformation procedures exist for only a few (e.g. alfalfa, *Arabidopsis*, carrot, poplar, potato, tobacco and tomato). The host range for transformation is still largely related to the availability of tissue culture techniques for any particular species and there is still room for improvement in many other crops besides the obvious bottlenecks of the seed legumes and cereals.

The number of 'useful' traits that might be engineered into crop plants is growing slowly, but in the near future an enormous amount of effort will be required to develop novel strategies for identifying genes controlling many important crop characteristics, for example disease resistance, flowering characteristics, fertility, incompatibility and morphology. Some of the biochemical, molecular, physiological and genetic studies aimed at tackling the basic biology of several of these complex traits are outlined in this and the companion volume.

The containment of genetically engineered plants, particularly the release of transgenic pollen into the environment, is a subject of intense debate at present with guidelines and legislation under preparation in most countries. However, some companies and government institutes have already been granted permission to grow transgenic plants outside under 'contained' and very closely supervised conditions, and field trials are already under way for herbicide-, virus- and insect-resistant plants in several countries.

2.4.3 *Analysis of gene regulation and basic biochemical and molecular studies*

Currently, a major application for transgenic plants relates to their utility for the detailed analysis of the regulatory elements that control gene expression; specialised promoterless 'reporter gene' vectors have been developed to assist these types of experiments (e.g. An, 1986; Jefferson *et al.*, 1987). These experiments commonly involve the construction of a transcriptional gene fusion between a DNA sequence with putative regulatory properties (e.g. tissue-specific or environmentally regulated) and a reporter gene coding for β-glucuronidase or chloramphenicol acetyltransferase, which are both enzymes for which a simple and sensitive assay exists. Such studies have identified regions upstream (5′) of several genes which confer developmentally and environmentally regulated gene expression (e.g. Herrera-Estrella *et al.*, 1984; Sengupta-Gopalan *et al.*, 1985); the further dissection of such DNA is defining specific nucleotide sequence 'motifs' associated with the binding of various *trans*-acting factors that regulate transcription levels (see, for example, Green *et al.*, 1988, and Chapter 1). Although most promoters studied retain cell or environmental specificity when moved into heterologous species, there is often a quantitative reduction in mRNA levels; this fact might be directly related to the absence of the correct *trans*-acting factors in the heterologous situation. A few studies examining the control of transcript polyadenylation (Barta *et al.*, 1986; Keith and Chua, 1986), splicing (Sengupta-Gopalan *et al.*, 1985; Barta *et al.*, 1986; Rosahl *et al.*, 1987) and translation initation (Taylor *et al.*, 1987) have been undertaken. Unlike the regulation of transcription levels, the correct maturation of eukaryotic transcripts does not always occur in transgenic plants; whether or not a particular transcript will be spliced or polyadenylated as expected is less predictable. In addition to the methods for the permanent alteration of gene expression, transient gene expression systems are useful in probing promoter function, testing for the correct engineering of a gene construct and so on. These systems normally utilise efficient DNA-uptake methods based on polyethylene glycol in conjunction with electroporation to deliver vectors into large populations of protoplasts (Werr *et al.*, 1986; Howard *et al.*, 1987). The use of protoplasts makes it rather difficult to study developmentally or tissue-specific regulatory functions, since specialised cells frequently dedifferentiate when converted to protoplasts. However, in future it may be possible to use microprojectile gene transfer techniques to study the transient expression of genes in more complex, intact plant tissues.

Besides studying the regulation of expression of individual genes in a heterologous genomic background, transgenic plants also enable the investigator to study the effects of manipulating particular enzymes, or other proteins, on basic plant biochemistry and metabolism. Alterations that may be achieved include nullifying the expression of particular enzymes using antisense RNA techniques (Van der Krol *et al.*, 1988; Smith *et al.*, 1988) or homologous recombination (Paszkowski *et al.*, 1988). Alteration of the developmental timing, subcellular location or abundance of a particular protein, or the directed mutagenesis of a polypeptide to modify the amino acid sequence before transferring to plants, is already possible. Specific expression cassettes have been developed to drive either constitutive (e.g. Sanders *et al.*, 1987), organ-specific (e.g. Jensen *et al.*, 1985; Chen *et al.*, 1986) or environmentally regulated gene expression (e.g. Nagy *et al.*, 1985; Thornberg *et al.*, 1988).

Transgenic plants have applications in the study of protein transport in plants. In particular, hybrid genes with altered *N*-terminal transit signal and peptides have been constructed to examine protein targeting into organelles (Kuntz *et al.*, 1986; Boutry *et al.*, 1987, and Chapter 6) and routing of seed storage proteins through the endomembrane system into protein bodies (Iturriaga *et al.*, 1989, and Chapter 7).

A more specialised use of *Agrobacterium* transformation is the delivery of virus genomes into intact plants by a process known as 'agroinfection' (Grimsley *et al.*, 1987). In these experiments the T-DNA contains a dimer of the viral genome, which, when delivered to the plant cell, produces an active viral genome, either by homologous recombination and excision from the T-DNA or by synthesis of molecules which act as replicative intermediates. Although agroinfection depends on a functional *vir* gene system, T-DNA integration into the plant nuclear genome is not required; indeed agroinfection studies using a dimer of maize streak virus, provided the first evidence that T-DNA delivery into the cytoplasm of cereals was possible (Grimsley *et al.*, 1987). From a commercial point of view, agroinfection techniques can be used to introduce rapidly viral genomes to a large number of host plants to test for their susceptibility to infection by particular strains. This technique is particularly useful with viruses which are normally transmitted by an insect vector, or which are normally impossible to inoculate by any techniques other than grafting.

2.4.4 *Genetic mapping and gene cloning*

Although some plant genes have been isolated using heterologous probes from bacteria, fungi or mammals, the majority of plant genes cloned to date were identified as cDNA clones using 'molecular biological' approaches based on the use of some knowledge of the gene product (use of expression vectors and antibody; peptide sequencing and DNA oligomers; immunoprecipitation of polysomes; enzymatic complementation in bacteria), or the fact that transcripts associated with a particular trait were present in certain cell types at specific stages of development, or after applying a particular stimulus (differential screening strategies, techniques based on *in vitro* translation analysis). The majority of these approaches are satisfactory for isolating clones which are relatively abundant in cDNA libraries, however they are mostly inadequate for the isolation of genes whose transcripts fall into the very low abundance category. Additionally, such technology cannot be used effectively to isolate genes contributing to genetic traits for which there are no known gene products; classically the link between genotype and phenotype in these circumstances is achieved by mutagenesis studies.

Mutagenesis using radiation or chemical mutagens does not offer an easy route for the isolation of mutated alleles. In contrast, mutagenesis by the insertion of a known piece of DNA into a functional gene allows the easy recovery of affected genes (Figure 1.15). Insertional mutagenesis strategies depend on the availability of a controllable, random DNA insertion element, the genetic identification of target loci and a means of identifying cells/plants with particular mutations. Transgenic plant technology offers at least two basic techniques for the isolation of genes via insertion mutagenesis. The random insertion of modified T-DNA sequences might be used to inactivate particular genes. This approach requires either the transformation of large numbers of single cells and selection of a particular mutant trait in clonally-derived transformed colonies, or the regeneration of large numbers of independently transformed plants and their screening/selection for particular mutant phenotypes. To date these approaches have been hampered by the

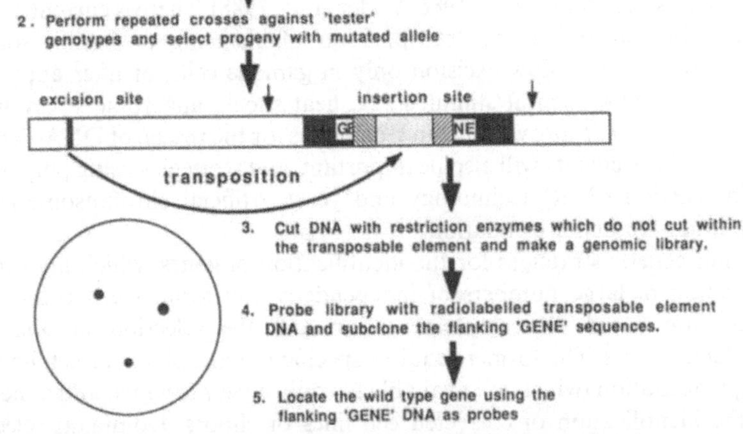

Figure 2.15. Strategy for cloning genes following transposon mutagenesis.

paucity of selection procedures that can be imposed *in vivo* and by limitations on the sheer numbers of transformed plants that can be efficiently regenerated in most species. However, recently Yanofsky *et al.*, (1989) have shown in *Arabidopsis* that it is possible to use random T-DNA insertions to mutate, and subsequently clone, a homeotic gene (*Agamous*) affecting flower development. A cosmid clone containing the wild-type gene has been used to complement an *Agamous* mutant, thus proving that the cloned gene was indeed responsible for regulating the development of flower organ primordia.

Random T-DNA insertion has been proposed as a means of isolating DNA sequences with a particular promoter activity (Andre *et al.*, 1986; Teeri *et al.*, 1988). These experiments use specialised vectors which contain promoterless selectable marker genes close to the right T-DNA border; the insertion of the T-DNA in the right orientation close to a promoter may create a gene fusion, or drive gene expression by 'read-through', thus allowing the expression of the dominant marker gene and the growth of the transformed colony on selection plates.

An alternative strategy for insertion mutagenesis is to utilise a transposable element from another species and attempt to create a mutation by transposition into a functional allele (Figure 2.15). This approach has the same genetic requirements as T-DNA insertion but has several advantages related to the fact that mutants are screened in seedlings: (1) a proportion of the transposable elements should move in each generation and thus repeated transformations are not required; (2) the mutant selection is performed on seed populations and thus it is possible first to locate transposon inserts that are closely linked to a genetic trait before attempting to mutagenise the locus concerned; (3) in some cases achieving tight linkage of a transposable element with a trait, rather than mutagenesis, might be sufficient to allow the isolation of a gene by chromosome walking techniques. To date it has been shown that several transposable elements continue to be mobile in heterologous transgenic plants (e.g. Baker *et al.*, 1986; Yoder *et al.*, 1988). There is currently much interest in controlling the transposition of elements to inhibit somatic transpositions and allow excision only in gametic cells, or after applying a particular environmental stimulus (e.g. heat shock, anaerobic stress, heavy metal solutions). Improvements in techniques for the rescue of DNA-flanking transposable elements will also be important, and strategies using polymerase chain reaction (PCR) technology and yeast artificial chromosome (YAC) techniques are under evaluation.

Other genetic strategies for the identification of genes, which involve the generation of large numbers of independent transformed cell colonies or plants, are based on complementation and the selection of dominant transferred traits. The former requires specific mutant plants or cell lines for complementation (which are available for only a very few loci) and a method for the identification of corrected cell lines or shoots. Dominant selection strategies depend on the target gene being screened/selected for as a dominant phenotype (e.g. disease resistance allele) in a particular genetic background. Bearing in mind these constraints, plant transformation vectors have been engineered to contain cosmid sequences which allow the cloning of large random fragments of genomic DNA for transfer to plants (Klee *et al.*, 1987a). Apart from the genetic problems outlined above, these strategies depend heavily on the routine generation of tens of thousands (or more) of independent transformants, and are thus potentially very labour-intensive. However, the recent development of comprehensive restriction fragment length polymorphism (RFLP) genetic maps (e.g. Goodman *et al.*, 1989), coupled with the use of YAC vector technology for handling very large DNA fragments (>400 kb), means that it is now theoretically possible to identify two RFLP markers flanking a genetic trait of interest and then to screen a YAC library of the species concerned in order to find a clone containing both RFLP markers; the section of DNA containing the gene of interest might then be identified by complementation-like techniques, using a bank of YAC subclones (contained in specialised *Agrobacterium* binary cosmid vectors) to transform plant lines lacking the particular genetic trait (Meyerowitz, 1987).

'Reverse genetic' strategies for the isolation of genes and investigation of genetic function are theoretically now possible using transgenic plants. Reverse genetic experiments usually start with a cDNA, often corresponding to a transcript with an interesting pattern of expression, and then attempt to ascribe a phenotypic function to the gene. The function is generally investigated by using the cDNA sequence to create a mutation of the wild-type allele in a transgenic plant. The phenotypic alteration may, in theory at least, be brought about by one of three methods. A target gene can be grossly down-regulated by the expression of an antisense mRNA which binds to the wild-type transcript, either interfering with splicing, causing it to be degraded, or inhibiting translation (Van der Krol *et al.*, 1988). This has recently been achieved for a few well-characterised genes such as rubisco, chalcone synthase and polygalacturonase (e.g. Rodermel *et al.*, 1988; Smith *et al.*, 1988), but there is no reason why a similar approach could not be used to deactivate transcripts of unknown functions and then to look for changes in the phenotype. A second theoretical approach to elucidate the genetic function of a cloned gene is to exchange the wild-type allele for an inactive mutant allele by homologous recombination (Paszkowski *et al.*, 1988). A final potential strategy is to use several site-directed mutagenised forms of a full-length cDNA to overproduce mutant proteins in a transgenic plant (Herskowitz, 1987). In this case it is hoped that at least one of the abundant mutant proteins will interfere with the normal function of the wild-type protein, for example by competing for receptor sites, competing for substrates or interfering with DNA–protein or protein–protein interactions. These latter two approaches are currently under investigation in several laboratories.

Thus, from the above discussion, it can be seen that transgenic plants should play a important role in the future, not only in gene expression work, but also in helping to provide methods for linking genotype to phenotype in order to identify new genes for both fundamental and commercially orientated studies.

References

An, G. (1986) Development of plant promoter expression vectors and their use for analysis of differential activity of nopaline synthase promoter in transformed tobacco cells. *Plant Physiol.* **81**: 86.

Andre, D., Colau, D., Schell, J., Van Montagu, M. and Hernalsteens, J. (1986) Gene tagging in plants by a T-DNA insertion mutagen that generates APH (3)II–plant gene fusions. *Mol. Gen. Genet.* **204**: 512.

Ashby, A.M., Watson, M.D. and Shaw, C.H. (1987) A Ti plasmid determined function is responsible for chemotaxis of *Agrobacterium tumefaciens* towards the plant wound product acetosyringone. *FEMS Microbiol. Lett.* **41**: 189.

Baker, B., Schell, J., Lorz, H. and Federoff, N. (1986) Transposition of the maize controlling element "activator" in tobacco. *Proc. Natl Acad. Sci. USA* **83**: 4844.

Barta, A., Sommergruber, K., Thompson, D., Hartmuth, K., Matzke, M.A. and Matzke, A.J.M. (1986) The expression of nopaline synthase–human growth hormone chimeric gene in transformed tobacco and sunflower callus. *Plant. Mol. Biol.* **6**: 347.

Bevan, M.W. (1984) Binary vectors for plant transformation. *Nucl. Acids Res.* **12**: 8711.

Bevan, M.W., Flavell, R.B., and Chilton, M.-D. (1983) A chimeric antibiotic resistance gene as a selectable marker for plant cell transformation. *Nature* **304**: 184.

Boutry, M., Nagy, F., Poulson, C., Aoyagi, K. and Chua, N-H. (1987) Targeting of bacterial chloramphenicol acetyl transferase to mitochondria in transgenic plants. *Nature* **328**: 340.

Brisson, N., Paszkowski, J., Penswick, J., Gronenborn, B., Potrykus, I. and Hohn, T. (1984) Expression of a bacterial gene in plant cells using a viral vector. *Nature* **310**: 511.

Byrne, M.C., McDonnell, R.E., Wright, M. and Carnes, M.G. (1987) Strain and cultivar specificity in the *Agrobacterium*–soybean interaction. *Plant Cell, Tissue and Organ Culture* **8**: 3.

Celis, J.E. (1984) Microinjection of somatic cells with micropipettes: Comparison with other transfer techniques. *Biochem. J.* **233**: 281.

Chen, Z.-L., Pan, N.-S. and Beachy, R.N. (1988) A DNA sequence that confers seed-specific enhancement to a constitutive promoter. *EMBO J.* **7**: 297.

Comai, L., Facciotti, D., Hiatt, W.R., Thompson, G., Rose, R.E. and Stalker, D.M. (1985) Expression in plants of a mutant *aroA* gene from *Salmonella typhimurium* confers resistance to glyphosate. *Nature* **317**: 741.

Cristou, P., McCabe, D.E. and Swain, W.F. (1988) Stable transformation of soybean callus by DNA-coated gold particles. *Plant Physiol.* **87**: 671.

Crossway, A., Oakes, J.V., Irvine, J.M., Ward, B., Knauf, V.C. and Shewmaker, C.K. (1985) Integration of foreign DNA following microinjection of tobacco mesophyll protoplasts. *Mol. Gen. Genet.* **202**: 179.

Crossway, A., Hauptli, H., Houck, C.M., Irvine, J.M., Oakes, J.V. and Perani, L.A. (1986) Micromanipulation techniques in plant biotechnology. *Biotechniques* **4**: 320.

Debleare, R., Bytebier, B., De Greve, H., Schell, J., Van Montagu, M. and Leemans, J. (1985) Efficient octopine Ti plasmid-derived vectors for *Agrobacterium*-mediated gene transfer to plants. *Nucl. Acids Res.* **13**: 4777.

De Block, M., Botterman, J., Vandewiele, M., Dockx, J., Thoen, C., Gossele, V., Rao, D., Movva, N., Thompson, C., van Montagu, M, and Leemans, J. (1987) Engineering herbicide resistance in plants by expression of a detoxifying enzyme. *EMBO J.* **6**: 2513.

De Frammond, A.J., Back, E.W., Chilton, W.S., Kayes, L. and Chilton M.-D. (1986) Two unlinked T-DNAs can transform the same tobacco plant cell and segregate in the F_1 generation. *Mol. Gen Genet.* **202**: 125.

De la Pena, A., Lorz, H. and Schell, J. (1987) Transgenic rye plants obtained by injecting DNA into young floral tillers. *Nature* **325**: 274.

Deshayes, A., Herrera-Estrella, L. and Caboche, M. (1985) Liposome-mediated transformation of tobacco mesophyll protoplasts by an *E. coli* plasmid. *EMBO J.* **4**: 2731.

Draper, J., Davey, M.R. Freeman, J.P., Cocking, E.C. and Cox, B.J. (1982) Ti plasmid homologous sequences present in tissue from *Agrobacterium* plasmid-transformed *Petunia* protoplasts. *Plant and Cell Physiol.* **23**: 451.

Draper, J., Davey, M.R., and Freeman, J.P. (1986) Delivery systems for the transformation of plant protoplasts. In *Plant Cell Culture Technology*, ed. Yeoman, M.M., Blackwell Scientific Publications, Oxford, 271.

Draper, J., Scott, R., Armitage, P. and Walden, R. (1988) *Plant Genetic Transformation and Gene Expression: a Laboratory Manual.* Blackwell Scientific Publications, Oxford.

Evans D.A. and Sharp, W.R. (1986) Applications of somaclonal variation. *Biotechnology* **4**: 528.

Feldmann, K.A. and Marks, M.D. (1987) *Agrobacterium*-mediated transformation of germinating seeds of *Arabidopsis thaliana*: A non-tissue culture approach. *Mol. Gen. Genet.* **208**: 1.

Fillatti, J.J., Sellmer, J., McCown, B., Haissig, B. and Comai, L. (1987) *Agrobacterium* mediated transformation and regeneration of *Populus. Mol. Gen. Genet.* **206**: 192.

Firoozabady, E. and Galbraith, D.W. (1984) Presence of a plant cell wall is not required for transformation of *Nicotiana* by *Agrobacterium tumefaciens. Plant Cell and Tissue Culture* **3**: 175.

Fraley, R.T., Rogers, S.G., Horsch, R.B., Eichholtz, D.A., Flick, C.L., Hoffman, N.L. and Saunders, P.R. (1985) The SEV system: A new disarmed Ti plasmid vector system for plant transformation. *Biotechnology* **3**: 629.

Freeman, J.P., Draper, J., Davey, M.R., Cocking, E.C., Gartland, K.M.A., Harding, K. and Pental, D. (1984) A comparison of methods for plasmid delivery into plant protoplasts. *Plant Cell Physiol.* **25**: 1353.

Fromm, M.E., Taylor, L.P., and Walbot, V. (1985) Expression of genes transferred into monocot and dicot plant cells by electroporation. *Proc. Natl Acad. Sci. USA* **82**: 5824.

Fromm, M.E., Taylor, L.P. and Walbot, V. (1986) Stable transformation of maize after gene transfer by electroporation. *Nature* **319**: 791.

Gelvin, S.B. and Schilperoort, R.A. (1988) *Plant Molecular Biology Manual.* Kluwer Academic.

Gheysen, G., Dhaese, P., Van Montagu, M. and Schell, J. (1985) DNA flux across genetic barriers: the crown gall phenomenon. In *Genetic Flux in Plants. Advances in Plant Gene Research*, vol. 2, eds. Hohn, B. and Davis, E.S. Springer Verlag, Berlin, 11.

Green, P.J., Yong, M.-H., Cuozzo, M., Kano-Murakami, Y., Silverstein, P. and Chua, N.-H. (1988) Binging site requirement for pea nuclear protein factor GT-1 correlate with sequences required for light-dependent transcriptional activation of the *rbcS-3A* gene. *EMBO J.* **7**: 4035.

Grimsley, N.H., Hohn, T., Davies, J.W., Hohn, B. (1987) *Agrobacterium*-mediated delivery of infectious maize streak virus into maize plants. *Nature* **325**: 177.

Grimsley, N.H., Romos, C., Hein, T. and Hohn, B. (1988) Meristematic tissues of maize plants are most susceptible to agroinfection with maize streak virus. *Biotechnology* **6**: 185.

Hain, R., Stabel, P., Czernilofsky, A.P., Steinbis, H.H., Herrera-Estrella, L. and Schell, J. (1985) Uptake, integration, expression and genetic transmission of a selectable chimaeric gene by plant protoplasts. *Mol. Gen. Genet.* **199**: 161.

Hamil, J.D., Prescott, A. and Martin, C. (1987) Assessment of the efficiency of cotransformation of the T-DNA of disarmed binary vectors derived from *Agrobacterium tumefaciens* and the T-DNA of *A. rhizogenes. Plant Mol. Biol.* **9**: 573.

Hammer, R.E., Pursel, V.G., Rexnroad, C.E., Wall, R.J., Bolt, D.J., Ebert, K.M., Palmiter, R.D. and Brinster, R.L. (1985) Production of transgenic rabbits, sheep and pigs by microinjection. *Nature* **315**: 680.

Hauptmann, R.M., Vasil, V., Ozias-Akins, P., Tabaeizadeh, Z., Rogers, S.G., Fraley, R.T., Horsch, R.B. and Vasil, I.K. (1988) Evaluation of selectable markers for obtaining stable transformants in the Gramineae. *Plant Physiol.* **86**: 602.

Hayes, R.J., Petty, I.T.D., Coutts, R.H.A. and Buck, K.W. (1988) Gene amplification and expression in plants by a replicating geminivirus vector. *Nature* **334**: 179.

Helmer, G., Casadaban, M., Bevan, M., Kayes, L. and Chilton, M.-D. (1984) A new chimeric gene as a marker for plant transformation: The expression of *Escherichia coli* β-galacotosidase in sunflower and tobacco cells. *Biotechnology* **2**: 520.

Herskowitz, I. (1987) Functional inactivation of genes by dominant negative mutations. *Nature* **329**: 219.

Herrera-Estrella, L., De Block, M., Messens, E., Hernalsteens, J-P., Van Montagu, M. and Schell, J. (1983*a*) Chimeric genes as dominant selectable markers in plant cells. *EMBO J.* **2**: 987.

Herrera-Estrella, L., Depicker, A., Van Montagu, M. and Schell, J. (1983*b*) Expression of chimeric genes transferred into plant cells using a Ti-plasmid-derived vector. *Nature* **303**: 209.

Hille, J., Verheggen, F., Roelvink, P., Fransen, H., Van Kammen, A. and Zabel, P. (1986)

Bleomycin resistance: a new dominant selectable marker for plant cell transformation. *Plant Mol. Biol.* **7**: 171.

Hoekema, A., Hirsch, P.R., Hooykaas, P.J.J. and Schilperoort, R.A. (1983) A binary plant vector strategy based on separation of *vir* and T-region of the *A. tumefaciens* Ti-plasmid. *Nature* **303**: 179.

Hood, E.E., Helmer, G.L., Fraley, R.T., Chilton, M-D. (1986) The hypervirulence of *Agrobacterium tumefaciens* A281 is encoded in a region of pTiBo542 outside of T-DNA. *J. Bacteriol* **168**: 1291.

Horsch, R.B., Fry, J.E., Hoffmann, N.L., Eichholtz, D., Rogers, S.G. and Fraley, R.T. (1985) A simple and general method for transferring genes into plants. *Science* **227**: 1229.

Howard, E.A., Walker, J.C., Dennis, E.S. and Peacock, W.J. (1987) Regulated expression of an alcohol dehydrogenase 1 chimeric gene introduced into maize protoplasts. *Planta* **170**: 535.

Iturriaga, G., Jefferson, R.A. and Bevan, M. (1989) Endoplasmic reticulum targeting and glycosylation of hybrid proteins in transgenic tobacco. *Plant Cell* **1**: 381.

Jefferson, R.A., Kavanagh, T.A. and Bevan, M.W. (1987) GUS fusions: β-glucuronidase as a sensitive and versatile gene fusion marker in higher plants. *EMBO J.* **6**: 3901.

Jensen, J.S., Marcker, K.A., Otten, L. and Schell, J. (1986) Nodule-specific expression of a chimaeric soybean leghaemoglobin gene in transgenic *Lotus corniculatus. Nature* **321**: 669.

Jones, J.D.G., Dunsmuir, P. and Bedbrook, J. (1985) High level expression of introduced chimeric genes in regenerated transformed plants. *EMBO J.* **4**: 2419.

Keith, B. and Chua, N-H. (1986) Monocot and dicot pre-mRNAs are processed with different efficiencies in transgenic tobacco. *EMBO J.* **5**: 2419.

Klee, H, Horsch, R. and Rogers, S. (1987*a*) *Agrobacterium*-mediated plant transformation and its further applications to plant biology. *Ann. Rev. Plant Physiol.* **38**: 467.

Klee, H.J., Hayford, M.B. and Rogers, S. (1987*b*) Gene rescue in plants: A model system for "shotgun" cloning by retransformation. *Mol. Gen. Genet.* **210**: 282.

Klein, T.M., Wolf, E.D., Wu, R. and Sanford, J.C. (1987) High-velocity microprojectiles for delivering nucleic acids into living cells. *Nature* **327**: 70.

Koncz, C., Olsson, O., Landridge, W.H.R., Schell, J., and Szalay, A.A. (1987) Expression and assembly of functional bacterial luciferase in plants. *Proc. Natl Acad. Sci. USA* **84**: 131.

Krens, F.A., Molendijk, L., Wullems, G.J. and Schilperoort, R.A. (1982) *In vitro* transformation of plant protoplasts with Ti plasmid DNA. *Nature* **296**: 72.

Kuntz, M., Simons, A., Schell, J. and Schreier, P. (1986).Targeting of a protein to chloroplasts in transgenic tissue by fusion with a mutated transit peptide. *Mol. Gen. Genet.* **205**: 454.

Leroux, B., Yanofsky, M.F., Winans, S.C., Ward, J.E., Ziegler, S.F. and Nester, E.W. (1987) Characterisation of the *virA* locus of *Agrobacterium tumefaciens*: a transcriptional regulator and host range determinant. *EMBO J.* **6**: 849.

Lichtenstein, C.P. and Fuller, S.L. (1987) Vectors for the genetic engineering of plants. In *Genetic Engineering*, vol. 6. Academic Press, London.

Lindsey, K. and Jones, M.G.K. (1989) *Plant Biotechnology in Agriculture.* Open University Press, Milton Keynes.

Lorz, H., Baker, B. and Schell, J. (1985) Gene transfer to cereal cells mediated by protoplast transformation. *Mol. Gen. Genet.* **199**: 179.

Luo, Z. and Wu, R. (1988) A simple method for the transformation of rice via the pollen-tube pathway. *Plant Mol. Biol. Reporter* **6**: 165.

McCabe, D.E., Swain, W.F., Martinell, B.J. and Christou, P. (1988*a*) Stable transformation of soybean (*Glycine max*) by particle acceleration. *Biotechnology* **6**: 923.

McCabe, D.E., Swain, W.F. and Martinell, B.J. (1988*b*) Pollen-mediated plant transformation, European Patent Application No. 88156530.

McDonnell, R.E., Clark, R.D., Smith, W.A. and Hinchee, M.A. (1987) A simplified method for the detection of neomycin phosphotransferase II activity in transformed plant tissue. *Plant Mol. Biol. Reporter* **5**: 380.

MacDowell, S.W., Coutts, R.H.A., and Buck, K.W. (1986) Molecular characterisation of subgenomic single-stranded DNA forms from plants infected with tomato golden mosaic virus. *Nucl. Acids Res.* **14**: 7967.

McGranahan, G.H., Leslie, C.A., Uratusu, S.L., Martin, L.A. and Dandekar, A.M. (1988) *Agrobacterium*-mediated transformation of walnut somatic embryos and regeneration of transgenic plants. *Bio-Technology* **6**: 800.

Marton, L., Wullems, G.J., Molendijk, L. and Schilperoort, R.A. (1979) *In vitro* transformation of cultured cells from *Nicotiana tabacum* by *Agrobacterium tumefaciens*. *Nature* **227**: 129.

Meyer, P., Walgenbach, E., Bussmann, K., Hombrecher, G. and Saedler, H. (1985) Synchronized tobacco protoplasts are efficiently transformed by DNA. *Mol. Gen. Genet.* **201**: 513.

Meyerowitz, E.M. (1987) *Arabidopsis thaliana*. *Ann. Rev. Genet.* **21**: 93.

Moloney, M.M., Walker, J.W. and Sharma, K.K. (1989) High efficiency transformation of *Brassica napus* using *Agrobacterium* vectors. *Plant Cell Reports* **8**: 238.

Nagy, F., Morelli, G., Fraley, R.T., Rogers, S.G. and Chua, N.-H. (1985) Photoregulated expression of a pea *rbcS* gene in leaves of transgenic plants. *EMBO J.* **4**: 3063.

Nam, H.-G., Giraudat, J., Boer, B., Moonan, F., Loos, W.D.B., Hauge, B.M. and Goodman, H.M. (1989) Restriction fragment polymorphism linkage map of *Arabidopsis thaliana*. *Plant Cell* **1**: 699.

Neuhaus, G., Spangenberg, G., Mittelsten-Scheid, O. and Schweiger, H-G. (1987) Transformation of microspore-derived embryos of *Brassica napus* by microinjection. *Theor. Appl. Genet.* **75**: 30.

Okada, K., Takebe, I. and Nagata, T. (1986*a*) Expression and integration of genes introduced into highly synchronised plant protoplasts. *Mol. Gen. Genet.* **205**: 398.

Okada, K., Nagata, T. and Takebe, I. (1986b) Introduction of functional RNA into plant protoplasts by electroporation. *Plant Cell Physiol.* **27**: 619.

Ow, D.W., Wood, K.V., Deluca, M., De Wet, J.R., Helsinki, D.R. and Howell, S.H. (1986) Transient and stable expression of the firefly luciferase gene in plant cells and transgenic plants. *Science* **234**: 856.

Paszkowski, J., Shillito, R.D., Saul, M., Mandak, V., Hohn, T., Hohn, B. and Potrykus, I. (1984) Direct gene transfer to plants. *EMBO J.* **3**: 2717.

Paszkowski, J., Baur, M., Bogucki, A. and Potrykus, I. (1988) Gene targeting in plants. *EMBO J.* **7**: 4021.

Petit, A., Berkaloff, A. and Tempe, J. (1986) Multiple transformation of plant cells by *Agrobacterium* may be responsible for the complex organisation of T-DNA in crown gall and hairy root. *Mol. Gen. Genet.* **202**: 388.

Picard, E., Jacquemin, F., Granier, M., Bobin and Forgeois, P. (1988) Genetic transformation of wheat (*Triticum aestivum*) by plasmid DNA uptake during pollen tube germination. In *Proceedings of the Seventh International Wheat Genetics Symposium, Cambridge, UK* **1**: 779.

Potrykus, I., Paskowski, J., Saul, M., Petruska, J. and Shillito, R.D. (1985) Molecular and general genetics of a hybrid foreign gene introduced into tobacco by direct gene transfer. *Mol. Gen. Genet.* **199**: 169.

Reich, T.J., Holbrook, L.A., Miki, B.L.A. and Iyer, V.N. (1984) Development of intranuclear microinjection into plant protoplasts. *J. Cell Biochem.* **8**: 60.

Reich, T.J., Iver, V.N. and Miki, B.L. (1986) Efficient transformation of alfalfa protoplasts by the intranuclear microinjection of Ti plasmids. *Biotechnology* **4**: 1001.

Reiss, B., Sprengel, R., Will, H. and Schaler, H. (1984) A new sensitive method for quantitative and qualitative assay of neomycin phosphotransferase in crude cell extracts. *Gene* **30**: 211.

Rhodes, C.A., Pierce, D.A., Mettler, I.J., Mascerenhas, D. and Detmer, J.J. (1988) Genetically transformed maize plants from protoplasts. *Science* **240**: 204.

Riggs, C.D. and Bates, G.W. (1986) Stable transformation of tobacco by electroporation: Evidence for plasmid concatation. *Proc. Natl Acad. Sci. USA* **83**: 5602.

Rodermel, S.R., Abbott, M.S. and Bogorad, L. (1988) Nuclear–organelle interactions: Nuclear antisense gene inhibits ribulose bisphosphate carboxylase enzyme levels in transformed tobacco plants. *Cell.* **55**: 673.

Rosahl, S., Schell, J. and Willmitzer, L. (1987) Expression of a tuber-specific storage protein in transgenic plants: demonstration of esterase activity. *EMBO J.* **6**: 1155.

Rubin, R.A. (1986) Genetic studies on the role of octopine T-DNA border regions in crown gall tumor formation. *Mol. Gen. Genet.* **202**: 312.

Sanders, P.R., Winter, J.A., Barnason, A.R., Rogers, S.G. and Fraley, R.T. (1987) Comparison of cauliflower mosaic virus 35S and nopaline synthase promoters in transgenic plants. *Nucl. Acids Res.* **15**: 1543.

Scott, R.J. and Draper, J. (1987) Transformation of carrot tissues derived from proembryogenic suspension cells: A useful model system for gene expression studies in plants. *Plant Mol. Biol.* **8**: 265.

Sengupta-Gopalan, C., Reichert, N.A., Barker, R.F., Hall, T.C. and Kemp, J.D. (1985)

Developmentally regulated expression of the bean β-phaseolin gene in tobacco seed. *Proc. Natl Acad. Sci. USA* **82**: 3320.

Shah, D.M., Horsch, R.B., Klee, H.J., Kishorer, G.M., Winter, J.A., Tumer, N.E., Hironaka, C.M., Sanders, P.R., Gasser, C.S., Aykent, S., Siegel, N.R., Rogers, S.G. and Fraley, R.T. (1986) Engineering herbicide tolerance in transgenic plants. *Science* **223**: 478.

Shahin, E.A., Sukhapinda, K., Simpson, R.B. and Spivey, R. (1986) Transformation of cultivated tomato by a binary vector in *Agrobacterium rhizogenes*; transgenic plants with normal phenotypes harbour binary vector T-DNA but no Ri plasmid T-DNA. *Theor. Appl. Genet.* **72**: 770.

Sheerman, S. and Bevan, M.W. (1988) A rapid transformation method for *Solanum tuberosum* using binary *Agrobacterium* vectors. *Plant Cell Reports* **7**: 13.

Shillito, R.D., Saul, M.W., Paszkowski, J., Muller, M. and Potrykus, I. (1985) High efficiency direct gene transfer to plants. *Biotechnology* **3**: 1099.

Shimamoto, K., Terada, R., Izawa, T. and Fujimoto, H. (1989) Fertile transgenic rice plants regenerated from transformed protoplasts. *Nature* **338**: 274.

Simoens, C., Alliote, T.H., Mendel, R., Muller, A., Schiemann, J., Van Lijsebettens, M., Schell, J., Van Montagu, M. and Inze, D. (1986) A binary vector for transferring genomic libraries to plants. *Nucl. Acids Res.* **14**: 8073.

Simpson, R.B., Spielmann, A., Margossian, L. and McKnight, T.D. (1986) A disarmed binary vector from *Agrobacterium tumefaciens* functions in *Agrobacterium rhizogenes*: Frequent co-transformation of two distinct T-DNAs. *Plant Mol. Biol.* **6**: 403.

Skoog, F. and Miller, C.O. (1957) Chemical regulation of growth and organ formation in plant tissues cultured *in vitro*. *Symp. Soc. Exp. Biol.* **11**: 118.

Smith, C.J.S., Watson, C.F., Ray, J., Bird, C.R., Morris, P.C., Schuch, W. and Grierson, D. (1988) Antisense RNA inhibition of polygalacturonase gene expression in transgenic tomatoes. *Nature* **334**: 724.

Stachel, S.E., Nester, E.W. and Zambryski, P.C. (1986*a*) A plant cell factor induces *Agrobacterium tumefaciens vir* gene expression. *Proc. Natl Acad. Sci. USA* **83**: 379.

Stachel, S.E., Timmerman, B. and Zambryski, P. (1986*b*) Generation of single stranded T-DNA molecules during the initial stages of T-DNA transfer from *Agrobacterium tumefaciens* to plant cells. *Nature* **322**: 706.

Stachel, S.E. and Zambryski, P.G. (1986) *Agrobacterium tumefaciens* and susceptible plant cell: A novel adaption of extracellular recognition and DNA conjugation. *Cell* **47**: 155.

Takamatsu, N., Ishikawa, M., Meshi, T. and Okada, Y. (1987) Expression of bacterial chloramphenicol acetyletransferase gene in tobacco plants mediated by TMV-RNA. *EMBO J.* **15**: 307.

Taylor, J.L., Jones, J.D.G, Sandler, S., Mueller, G.M., Bedbrook, J. and Dunsmuir, P. (1987) Optimising the expression of chimeric genes in plant cells. *Mol. Gen. Genet.* **210**, 572.

Teeri, T.H., Herrera-Estrella, L., Depicker, A., Van Montagu, M. and Palva, E.T. (1986) Identification of plant promoters *in situ* by T-DNA-mediated transcriptional fusions to the *npt-II* gene. *EMBO J.* **5**: 1755.

Thornberg, R.W., An, G., Cleveland, T.E., Johnson, R. and Ryan, C.A. (1987) Wound-inducible expression of a potato protease inhibitor II-chloramphenicol acetyl transferase gene fusion in transgenic tobacco plants. *Proc. Natl Acad. Sci. USA* **84**: 744.

Uchimiya, H., Fushimi, T., Hashimoto, H., Harada, H., Syono, K. and Sugawara, Y. (1986) Expression of a foreign gene in callus derived from DNA-treated protoplasts of rice (*Oryza sativa* L.) *Mol. Gen. Genet.* **204**: 204.

Van den Elzen, P.J.M., Townsend, J., Lee, K.Y. and Bedbrook, J.R. (1985) A chimaeric hygromycin resistance gene as a selectable marker in plant cells. *Plant Mol. Biol.* **5**: 299.

Van der Krol, A.R., Mol, J.N.M. and Stuitje, A.R. (1988) Antisense genes in plants: an overview. *Gene* **72**: 45.

Wang, K., Herrera-Estrella, L., Van Montagu, M. and Zambryski P. (1984) Right 25 bp terminus sequence of the nopaline T-DNA is essential for and determines direction of DNA transfer from *Agrobacterium* to the plant genome. *Cell* **38**: 455.

Wang, Y.-C., Klein, T.M., Fromm, M., Cao, J., Sanford, J.C. and Wu, R. (1988) Transient expression of foreign gene in rice, wheat and soybean cell following particle bombardment. *Plant Mol. Biol.* **11**: 433.

Werr, W. and Lorz, H. (1986) Transient gene expression in a Gramineae cell line. A rapid procedure for studying plant promoters. *Mol. Gen. Genet.* **202**: 471.

Yanofsky, M., Ma, H., Bowman, J., Drews, G., Vijayraghavan, U., Feldmann, K. and Meyerowitz, E. (1989) Characterisation of *Agamous*: A homeotic gene from *Arabidopsis thaliana*. *Proceedings of a Conference on 'The General and Molecular Biology of Arabidopsis'*, Indiana University, Bloomington, Indiana, USA, 12–15 October 1989.

Yoder, J.I., Palys, J., Alpert, K. and Lassner, M. (1988) *Ac* transposition in transgenic tomato plants. *Mol. Gen. Genet.* **213**: 291.

Zambryski, P., Joos, H., Genetello, C., Leemans, J., Van Montagu, M. and Schell, J. (1983) Ti-plasmid vector for the introduction of DNA into plant cells without alteration of their normal regeneration capacity. *EMBO J.* **2**: 2143.

Zambryski, P., Tempe, J. and Schell, J. (1989) Transfer and function of T-DNA genes from *Agrobacterium* Ti and Ri plasmids in plants. *Cell* **56**: 193.

Zhang, H.M., Yang, H., Rech, E.L., Golds, A.S., Davis, A.S., Mulligan, B.J., Cocking, E.C. and Davey, M.R. (1988) Transgenic rice plants produced by electroporation-mediated plasmid uptake into protoplasts. *Plant Cell Reports* **7**: 379.

3 Developing herbicide resistance in crops by gene transfer technology

D.M. STALKER

3.1 Introduction

Herbicides have become an indispensable tool of modern agriculture in that they allow economically superior weed control and are more labour- and energy-efficient than manual or mechanical cultivation methods. Although many agronomically useful compounds pollute the environment, others are reasonably safe because they are rapidly degraded in the biosphere and/or are non-toxic to animals. The information available elucidating the mode of action of several compounds has rapidly increased in the past few years. This fact, coupled with the progress achieved in the genetic manipulation of plants and the ability to transfer foreign genetic material from a variety of sources to plants (see Chapter 2), has opened up an exciting the productive area of research. These observations are further enhanced in that many possible herbicide resistance gene determinants are single dominant traits, making them amenable to gene transfer techniques. The ensuing result will hopefully be the expanded use of environmentally safer herbicides which favour a broader weed control spectrum as concern increases over contamination of the environment, toxicity to animals and persistence of residues in soil and water. Another exciting aspect of this research is the potential for the genetic manipulation of new herbicide resistance determinants that complement new classes of broad-spectrum biorational herbicidal compounds.

Many herbicides exert their mode of action by inactivating 'target proteins' (usually enzymes) essential for vital functions such as the photosynthetic or other plant biosynthetic pathways. Since crop plants usually share these processes with competing weeds, herbicides are often non-selective. Other compounds can be used selectively on tolerant crops, mainly as a consequence of a differential uptake or metabolism of the herbicide or by a precise localisation of the herbicide application. An alternative is to confer to crops resistance against broad-spectrum compounds. This would permit use of broad-spectrum herbicides and the selection from this range of compounds of those which are environmentally acceptable and exhibit no toxicity.

At present, two approaches have been developed for the engineering of herbicide resistance in plants: (1) the modification of a plant enzyme or other sensitive biochemical target of herbicidal action to render it insensitive to the herbicide or by inducing the overproduction of the unmodified target protein permitting normal metabolism to occur despite the presence of the herbicide; (2) the introduction of an enzyme or enzyme system to degrade and/or detoxify the herbicide in the plant prior to its action. Plants modified in these ways may be obtained either by selection for resistance against a specific herbicide or by applying gene transfer techniques utilising genes encoding herbicide resistance determinants. Genetic engineering of plants may be brought about by the use of gene transfer vectors derived from the soil bacteria *Agrobacterium tumefaciens* and *A. rhizogenes* or by direct gene transfer of DNA into plant cells or protoplasts and subsequent regeneration to whole plants (Chapter 2).

This chapter describes the present status and achievements of these different strategies that have been applied to introduce resistance against several important herbicides in crops and future perspectives of crops with herbicide-resistant traits.

3.2 Modification of the target of herbicide action

This approach depends on the identification in biochemical/molecular terms of the site of herbicide action in the plant cell. Considerable effort has been

Table 3.1. Herbicides affecting biochemical targets.

Compound/ herbicide	Chemical formula	Inhibited pathway	Target protein
Glyphosate (Roundup)	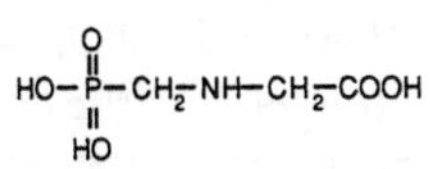	Aromatic amino acids	5′-Enolpyruvyl shikimate 3′-phosphate synthase
Chlorsulphuron (Glean)	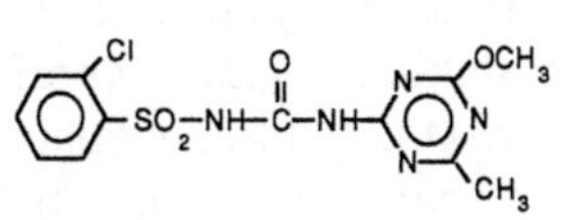	Branched-chain amino acids	Acetolactate synthase
AC 243, 997 (Arsenal)	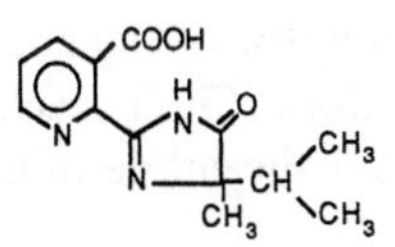	Branched-chain amino acids	Acetolactate synthase
Phosphinothricin (Basta)	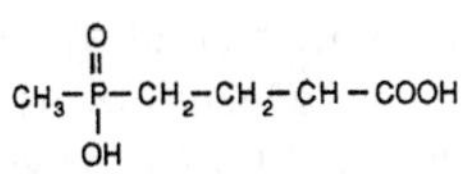	Nitrogen assimilation	Glutamine synthase
Atrazine (Lasso)	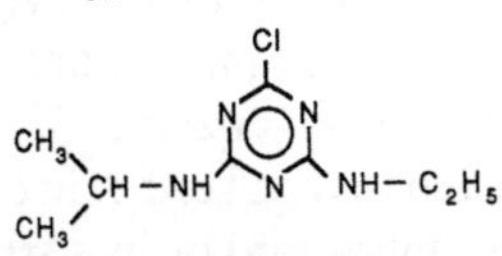	Photosynthesis	32 kDa protein

devoted to understanding the mode of action of herbicides through physiological, biochemical and genetic studies. Some herbicides have been shown to disrupt amino acid biosynthetic pathways (LaRossa and Falco, 1984) and others to interfere with photosynthesis (Arntzen, *et al.*, 1982). Genes encoding herbicide-sensitive or -insensitive target proteins have been isolated from both plants and micro-organisms. The following examples indicate how these resistance determinants have been utilised to genetically engineer herbicide tolerance in model crop systems (Table 3.1).

3.2.1 *Glyphosate*

The herbicide glyphosate (*N*-phosphonomethylglycine) has a simple molecular structure as it is an analogue of the amino acid glycine. Plant physiologists and biochemists have studied the effect of glyphosate on cellular metabolism, and different modes of action have been proposed (Hoagland and Duke, 1982). Present research indicates strong evidence for a single biochemical target site, although other targets cannot be ruled out.

Elucidation of the target site began when Jaworski (1972) observed that in *Lemna* aromatic amino acid levels were reduced after glyphosate treatment. Furthermore, growth inhibition by glyphosate was suppressed by the addition of aromatic amino acids to the medium. Based on these results, it was hypothesised that a metabolic step in the pathway leading to the production of aromatic amino acids may be inhibited. A number of researchers attempted to relieve glyphosate poisoning of different plant species with aromatic amino acids (Haderlie *et al.*, 1977). These results were encouraging but not definitive, as they ranged from complete suppression of glyphosate inhibition to no effect (Hoagland and Duke, 1982). Between 1980 and 1983, Amrhein and his colleagues published a series of articles characterising the effect of glyphosate on aromatic amino acid biosynthesis (Amrhein *et al.*, 1980, 1983). It was noted from these studies that inhibition of the pathway appeared to be at a step prior to chorismate synthesis and that shikimic acid, a chorismate precursor, accumulated in the vacuole of plant cells treated with glyphosate. These results were supported by the observation that in cell-free extracts of *Klebsiella aerogenes* glyphosate acted as a specific inhibitor of the enzyme enolpyruvyl shikimate 3-phosphate synthase (EPSP synthase) (Steinruecken and Amrhein, 1984). This represented strong evidence that the mode of glyphosate action was by inhibiting EPSP synthase.

Salmonella typhimurium provided an experimental system based on the availability of defined mutations in the genetic locus (*aro*A) coding for EPSP synthase. If EPSP synthase is the biochemical target for glyphosate, then it should be possible to isolate glyphosate-resistant mutants mapping at the *aro*A locus and encoding a glyphosate-resistant enzyme. Based on the models described above, mutations were expected to result either in overproduction of the target enzyme (via an up-promoter mutation) or in a structurally altered

EPSP synthase enzyme. Both types of mutations were obtained from ethylmethanesulphate-mutagenised *S. typhimurium* cultures (Comai *et al.*, 1983) and a mutant and the wild-type *aro*A genes were cloned and subjected to nucleotide sequence analysis. Comparison of the wild-type mutant gene sequences revealed a cytosine-to-thymine transition mutation that resulted in a Pro-to-Ser amino acid substitution at position 101 of the 349 amino acid polypeptide (Stalker *et al.*, 1985). It is also likely that other substitutions at different positions of the polypeptide confer resistance. It was observed that intragenic suppressor mutations, obtained by selecting for recovery of EPSP synthase activity from different point mutants deficient in activity, frequently confer glyphosate resistance (L. Comai, personal communication). Since a very high frequency of glyphosate-resistant mutants mapping at the *aro*A locus was obtained, it is unlikely that other biochemical targets are affected by glyphosate (at least in *S. typhimurium*). This conclusion was also supported by the observation that overexpression of the *aro*A locus in *E. coli*, by cloning the *E. coli aro*A gene on a multicopy plasmid, resulted in bacterial glyphosate resistance (Rogers *et al.*, 1983). These experiments established EPSP synthase as the primary, and perhaps sole, biochemical target in bacteria. The glyphosate-resistant and wild-type *S. typhimurium* EPSP synthases were purified and the kinetics of inhibition by glyphosate determined. When assayed at pH 5.0, the mutant *S. typhimurium* EPSP synthase was 100-fold less sensitive to glyphosate than the wild-type enzyme. At pH 7.0, the mutant enzyme was threefold less sensitive than the wild type and exhibited a K_i for glyphosate of 5.1 mM while the wild-type enzyme revealed a K_i of 15 mM (Thompson and McDonald, unpublished observations). Although the differences in inhibition for the wild-type and mutant *S. typhimurium* enzymes were much closer at pH 7.0 than pH 5.0, both the wild-type and mutant were much less efficient in binding glyphosate than EPSP synthases isolated from a number of plants species (Boocock and Coggins, 1983; Mousdale and Coggins, 1984; Rubin *et al.*, 1984). Glyphosate K_i values for the plant enzymes in these studies ranged from 0.08 to 1 mM.

The *Salmonella* glyphosate-resistant *aro*A gene was introduced in tobacco (*Nicotiana tabacum*) (Comai *et al.*, 1985) and tomato (*Lycopersicon esculentum*) plants (Fillatti *et al.*, 1987) using chimeric gene constructs in Ti plasmid vectors and *Agrobacterium*-mediated T-DNA transformation. Expression of the chimeric *aro*A gene under the control of the octopine synthase (*ocs*) promoter or manopine synthase (*mas*) promoter in transgenic plants resulted in the presence of a bacterial EPSP synthase in the plant cell cytoplasm. Transgenic plants were up to three times more tolerant to glyphosate than the control plants depending on the level of bacterial *aro*A expression obtained in independent transgenic events. This tolerance was partial and was dependent on the level of expression of the bacterial *aro*A gene, since plants having higher levels of bacterial EPSP synthase were more tolerant. The level of tolerance exhibited by these transformed plants is not sufficient for agronomic use, as plants (especially in the case of tobacco) were significantly stunted using a rate of 1.0 kg/ha of the isopropylamine salt of glyphosate. This is probably due to the fact that glyphosate is translocated after spraying and concentrated in the growing regions (shoot tips and root tips) of the plant. In the experiments described above, expression of the chimeric bacterial *aro*A gene in plants resulted in the localisation of the bacterial EPSP synthase in the cell cytoplasm. However, the majority of the plant EPSP synthase is found in the chloroplast (Feierabend and Brassel, 1977; Bickel and Schultz, 1978), which represents a major site of aromatic amino acid biosynthesis. A cytoplasmic form of the plant enzyme may also be present, but it constitutes a minority of the total assayable EPSP synthase activity (Mousdale and Coggins, 1985) in the plant cell.

In a parallel approach, glyphosate-tolerant cell cultures of *Cordalis sempervirens* and a *Petunia hybrida* cell line were obtained after selection on glyphosate (Amrhein *et al.*, 1983; Steinrucken *et al.*, 1986). The petunia cell line overproduced EPSP as a result of a 20-fold amplification of the gene encoding the plant EPSP synthase. A cDNA clone encoding EPSP synthase was isolated and the corresponding nuclear gene was shown to encode a cytoplasmic precursor with an amino terminal peptide sequence that permits post-translational transfer of the enzyme from the cytoplasm to the chloroplasts (Della-Cioppa *et al.*, 1986). Since amino acid biosynthesis is usually conducted by enzymes encoded by nuclear genes that function in plastids, it is required to target the gene product to the chloroplasts. The transfer of foreign proteins encoded by nuclear genes to chloroplasts has been demonstrated when the protein is linked to chloroplast-specific transit peptides (Schreier *et al.*, 1985; Van der Broek *et al.*, 1985; Chapter 6).

A chimeric EPSP gene encoding wild-type precursor EPSP synthase under the control of the cauliflower mosaic virus (CaMV) 35S promoter was introduced into petunia plants (Shah *et al.*, 1986a). The CaMV 35S promoter has been shown to direct high-level expression of foreign genes in transformed plants (Odell *et al.*, 1985). Transgenic plants were significantly tolerant when sprayed with glyphosate and showed increased expression of endogenous EPSP synthase that was localised in the chloroplasts. Similarly, a gene fusion between the transit peptide sequence of the petunia cDNA clone and an *E. coli* *aro*A gene, glyphosate-resistant EPSP synthase, yielded transgenic tobacco plants that showed higher tolerance to glyphosate than plants overexpressing the wild-type native plant EPSP synthase enzyme. (Della-Cioppa *et al.*, 1987; Fraley *et al.*, 1987). Recently, the *Salmonella aro*A gene encoding a glyphosate-resistant enzyme has been fused to a portion of the gene encoding the tobacco ribulose bisphosphate (RUBP) carboxylase small subunit (SSU). This gene fusion consists of the SSU transit peptide and part of the mature SSU-coding sequence coupled to the *aro*A-coding region and allows the glyphosate-resistance bacterial EPSP synthase to be efficiently transported into chloroplasts *in vitro* and *in vivo* (Comai *et al.*, 1988). Transgenic plants expressing this fusion protein in plant plastids resulted in improved tolerance compared with plants expressing the bacterial enzyme in the plant cell cytosol (Figure 3.1). In all cases described above, tolerance was still not adequate to give commercial levels of glyphosate resistance, as plants were still stunted when sprayed with glyphosate (L. Comai, personal communication). Field trials of transgenic tomato plants with a chloroplast-directed glyphosate-resistant bacterial EPSP synthase indicated that these plants are slightly stunted (compared with unsprayed controls) when sprayed with glyphosate at 1.0 kg/ha. These plants recover and give normal yields of fruit, however (see Figure 3.2).

Since glyphosate is concentrated in the plant growing regions, it is likely that increased and localized expression of reintroduced EPSP synthase in these regions is required, and/or a more kinetically efficient glyphosate-resistant enzyme is needed. Even these improvements may not be sufficient for product-level glyphosate resistance if targets other than EPSP synthase are slightly affected under the high glyphosate concentrations that exist in meristematic regions.

3.2.2 *Sulphonylureas and imidazolinones*

The sulphonylurea herbicides irreversibly inhibit the enzyme acetolactate synthase (ALS) from either bacteria, yeast or plants (Chaleff and Mauvais, 1984; Ray, 1984; Falco and Dumas, 1985). ALS is involved in the biosynthesis of the branched-chain amino acids leucine, isoleucine and valine. A chemically unrelated class of herbicides, the imidazolinones, also inhibit ALS (Shaner, *et al.*, 1984; Muhitch *et al.*, 1987). Dominant mutants that confer resistance to the herbicide sulphometuron methyl in *S. typhimurium* (LaRossa and Shloss, 1984), *Saccharomyces cerevisiae* (Falco and Dumas, 1984), *E. coli* (Yadav *et al.*, 1986) and *Chlamydomonas rheinhardii* (Hartnett *et al.*, 1987) have been obtained, and these mutations map to the structural gene encoding ALS. Nucleotide sequences of these mutant genes revealed single nucleotide differences, and the corresponding enzymes carry single amino acid substitutions and vary in ALS activity and in herbicide sensitivity (Yadav *et al.*, 1986).

Chlorsulphuron- or sulphometuron-resistant plants have been obtained by selection in cultures of haploid tobacco (*Nicotiana tabacum*) protoplasts (Chaleff and Ray, 1984) and in mutagenised *Arabidopsis thaliana* seedlings (Haughn and Somerville, 1986). In contrast to the situation with glyphosate, overproduction of the target enzyme has not been reported in plants resistant to sulphonylurea- or imidazoline-type herbicides. Diploid tobacco plants were regenerated and the resistant phenotype was shown to be expressed at the whole-plant level. Through genetic crosses it was established that single nuclear mutations in two unlinked loci were co-segregating with a herbicide-insensitive ALS activity (Chaleff and Mauvais, 1984; Chaleff and Bascomb, 1987). It has also been demonstrated that a single dominant nuclear mutation at one locus in *Arabidopsis* conferred sulphonylurea resistance (Haughn and

Somerville, 1986). However, not all resistant phenotypes against sulphonylurea-type herbicides are correlated with a modified ALS activity. Field screening has yielded chlorsulphuron-resistant alfalfa seedlings (*Medicago sativa*) (Stannard and Fay, 1987), and mutagenised soybean seeds (*Glycine max*) have been selected for chlorsulphuron tolerance (Sebastian *et al.*, 1987). In both cases, none of the mutants were resistant at the level of the ALS enzyme activity. In the soybean mutants, tolerance was conditioned by a single recessive gene. Selection applied to haploid anther cells and protoplasts derived from these microspores of *Brassica napus* yielded chlorsulphuron-resistant mutants. It is entirely possible that these resistant plants have developed alternative resistance mechanisms such as impermeability or detoxification. The genes encoding the wild-type and mutant ALS enzymes have been isolated from *N. tabacum* and from *A. thaliana* (Falco *et al.*, 1987; Mazur *et al.*, 1987). The tobacco line HRA, which is mutated at the *surB* locus, and the second tobacco line, C3, which carries a mutation at the *surA* locus, have been used as sources of mutant ALS genes. Four genes, representing all the ALS loci from the two mutant tobacco lines, have been isolated (Falco *et al.*, 1987). A single ALS gene is present in *A. thaliana* (Mazur *et al.*, 1987). A mutant ALS gene, *crs*1, has also been obtained from *Arabidopsis* (Haughn *et al.*, 1988). These plant ALS genes are nuclear-encoded and contain a chloroplast transit peptide at the amino-terminal portion of the polypeptide. In resistant tobacco and *Arabidopsis* ALS-encoding genes, different sites in the

---

Figure 3.1. Glyphosate-resistant tomato plants (*Lycopersicon esculentum* cv. UC82B). Plants grown in the flats shown in the foreground were sprayed at the two- to three- leaf stage with a commercial formulation (Roundup) at a rate of 1.33 lb/acre (1.5 kg/ha) of the isopropylamine salt of glyphosate. The photograph was taken 17 days after herbicide application. The flat on the left contained control non-transformed tomato plants. The flat in the centre contained transformed tomato plants where expression of the glyphosate-resistant bacterial EPSP synthase was directed to the plant cell cytosol. The flat on the right contained transformed tomato plants where expression of the glyphosate-resistant bacterial EPSP synthase was directed to plant cell plastids. Flats containing plants in the background are unsprayed, control plants from each of the above-described constructs. Plants were transformed via *Agrobacterium* binary vectors containing the following chimeric genes. (1) The *S. typhimurium* glyphosate-resistant *aro*A under the control of the CaMV 35S promoter and a 3′ region specified by the *tml* locus. This chimeric gene directs expression of the bacterial EPSP synthase to the plant cell cytosol. (2) The *S. typhimurium* glyphosate-resistant *aro*A gene fused to a portion of the tobacco rubisco small subunit gene (*rbcS*) encoding the transit peptide and an amino terminal portion of the small subunit polypeptide. This fusion is expressed under the control of the CaMV 35S promoter and a 3′ region specified by the *tml* locus. This chimeric gene directs the bacterial EPSP synthase to the chloroplast. Individual tomato transformants were selected and plants expressing the above-described *aro*A chimeric genes that expressed similar levels of the bacterial EPSP synthase were identified. Seeds (T_2) from these individual transformants were placed in flats and grown in the greenhouse.

Figure 3.2. Field trials of glyphosate-resistant transgenic tomato (*Lycopersicon esculentum* cv. UC82B) plants. Seed from homozygous-derived second-generation plants was planted under normal field conditions. Plants were sprayed at the two- to three-leaf stage with a commercial formulation of glyphosate (Roundup) at a rate equal to 1.0 lb/acre (1.1 kg/ha) of the isopropylamine salt. The photograph was taken 61 days after application. The plants in the left rows are transformed, unsprayed plants, while the plants in the right row are transformed and sprayed with glyphosate.

Figure 3.1

Figure 3.2

Figure 3.3

Figure 3.4

protein chain where mutations occurred have been observed. For example, a proline-to-serine substitution by a single nucleotide change in the *Arabidopsis* ALS gene (position 197) occurs at the same position and is identical to an amino acid substitution resulting in sulphonylurea resistance for the yeast ALS gene (position 192) (Yadav *et al.*, 1986; Haughn *et al.*, 1988). Transgenic tobacco (*N. tabacum*) and tomato (*Lycopersicon esculentum*) plants expressing a sulphonylurea-resistant ALS mutant gene exhibited high-level tolerance to sulphonylurea herbicides (Falco *et al.*, 1987). Reintroduction of mutant genes isolated from either locus of tobacco also conferred resistance in transformed cells and plants regenerated from protoplasts transformed with the *surB*-Hra gene are resistant to field application rates of sulphonylurea herbicides (Falco *et al.*, 1987). Similarly, tobacco plants transformed with the *Arabidopsis crs*1 gene were resistant to chlorsulphuron (Haughn *et al.*, 1988).

Tolerance to the imidazolinone-type compounds has been demonstrated in maize (*Zea mays*) cell lines that were selected in tissue cell culture. The resistant phenotype was expressed in plants regenerated from these resistant lines and was also due to an altered ALS enzyme (Shaner and Anderson, 1985). Plants homozygous for the resistant gene were tolerant to 300 times higher concentrations of several imidazolinone herbicides (Anderson and Georgeson, 1986). This is the first example of selection and regeneration of herbicide-tolerant lines in monocotyledonous crops such as maize. A major drawback in obtaining herbicide-resistant corn plants has been the inability to regenerate plants from monocotyledon tissue culture (Rhodes *et al.*, 1988).

Cross-resistance of mutants to sulphonylurea and imidazolinones might be related to the origin of the herbicide binding site of ALS (Schloss *et al.*, 1988). This phenomenon has been observed in the imidazolinone-resistant maize, but selective resistance of the products might be possible (Haughn and Somerville, 1987).

3.2.3 L-*Phosphinothricin*

L-Phosphinothricin (PPT) is an analogue of glutamate which was initially characterised as part of a tripeptide antibiotic produced by *Streptomyces*

Figure 3.3. Bromoxynil-resistant tomato plants (*Lycopersicon esculentum* cv. UC82B). Plants were transformed via *Agrobacterium* binary vectors containing the gene (*bxn*) encoding the bromoxynil-specific nitrilase under the control of the manopine synthase (*mas*) promoter. Seed from a positive primary transformant expressing the enzyme was grown in pots in the growth chamber and sprayed with a commercial formulation of bromoxynil (Buctril) at concentrations of 2.0 and 4.0 lb/acre (2.25 and 4.5 kg/ha). Plants on the left are sprayed, non-transformed, control tomato plants.

Figure 3.4. Field trials of bromoxynil-resistant transgenic tomato (*Lycopersicon esculentum* cv. UC82B) plants. Seed from homozygous-derived second-generation plants was planted and grown under normal field conditions and sprayed with a commercial formulation of bromoxynil (Buctril) at the two- to three-leaf stage. The spray rate was 1.5 lb/acre (1.65 kg/ha). The rows on the left show transformed unsprayed tomato plants, while the rows on the right show transformed tomato plants sprayed with Buctril. The photograph was taken 61 days after application.

viridochromogenes (Bayer *et al.*, 1972). It is an irreversible inhibitor of glutamine synthetase (GS) in bacteria and in plants (Colanduoni and Villefranca, 1986; Manderscheid and Wild, 1986). Two forms of GS with a cytosolic or chloroplastic compartmentalisation have been identified in plants. GS is involved in the primary assimilation of ammonia produced by nitrate reduction of nitrogen fixation in the roots as well as the reassimilation of ammonia released by photorespiration in plants. In higher plants, PPT causes rapid accumulation of ammonia and inhibits photosynthesis, which leads to death of the plant cell (Tachibana *et al.*, 1986). A PPT-tolerant alfalfa (*Medicago sativa*) suspension culture line was selected on increasing PPT concentrations (Donn *et al.*, 1984). Gene amplification of GS resulted in a fivefold enhanced GS expression, yielding a cell line which is about 20 times more resistant than the parental line. Further screening showed that the copy number of the amplified gene was correlated to the level of resistance. A GS cDNA clone was isolated from the tolerant line (Tischer *et al.*, 1986), and expression of this GS cDNA under the control of the CaMV 35S promoter conferred low-level tolerance to PPT when this chimeric gene was introduced into transgenic tobacco plants (Eckes and Wengemayer, 1987). It has also been shown that this plant GS gene complements a mutant GS gene (*glnA*) in *E. coli* (Das Sarma *et al.*, 1986). The system can thus be utilised to select for other resistant plant GS mutants in *E. coli*. Specific mutations which result in PPT resistance are a glycine to serine, cysteine or arginine substitution at position 245, or arginine to lysine at position 332 (Goodman, unpublished results).

3.2.4 *Atrazine*

The *S*-triazine-type herbicides like atrazine and the chemically unrelated ureas like diuron inhibit photosynthesis. They inhibit electron transport on the reducing site of photosystem II (Arntzen *et al.*, 1982). The mechanism of action involves the binding of the herbicide to chloroplast thylakoid membranes and blocking of electron transport at the second electron acceptor of photosystem II. This protein, the Q_B protein, is encoded by the *psbA* gene located in the plant chloroplast genome (Zurwaski *et al.*, 1982). It is expressed as a precursor polypeptide which is processed to a 32 kDa mature form in the chloroplast thylakoid membranes (Mattoo and Edelman, 1987). In agricultural areas where atrazine has been extensively used, atrazine-resistant biotypes of many weed species have appeared. Resistance was found to be maternally inherited and was correlated with mutations in the *psbA* gene (Souza-Machado, 1982). The *psbA* DNA sequence of a resistant weed species, *Amaranthus hybridus*, revealed three nucleotide differences. Two mutations were silent, whereas the third mutation resulted in a serine-to-glycine amino acid substitution at residue 264 (Hirschberg and McIntosh, 1983). Mutant *psbA* genes isolated from other atrazine-resistant weeds biotypes also showed a highly conserved sequence among different species and revealed single amino acid substitutions at the same position in the 32 kDa protein chain (Goloubinoff and Edelman, 1984; Barber, 1987). These mutations resulted in at least 10-fold reduced flow of electrons through the pathway. Of the more than 50 weed species that evolved resistance to atrazine, only *Abutilon* has been shown to possess a detoxification pathway as a mode of resistance (Anderson and Gronwald, 1987). A mutation conferring triazine resistance in the blue-green algae *Chlamydomonas rheinhardii* was caused by a serine-to-alanine amino acid substitution at position 264 of the *psbA*-coding sequence (Erickson *et al.*, 1985), while a diuron-resistant mutant has an isoleucine-to-valine change at amino acid position 219. These algal mutants also exhibited varying cross-resistances towards different photosystem II-inhibiting herbicides (Galloway and Mets, 1984). Similarly, naturally occurring weed biotypes with varying atrazine resistance levels showed variable cross-resistances (Arntzen *et al.*, 1982) to a variety of photosystem II-inhibiting compounds. Although genes encoding atrazine-resistant 32 kDa proteins have been available, the engineering via recombinant gene transfer of atrazine-resistant plants has not been totally successful, primarily because of the lack of a reproducible system for integration of foreign DNA into plant chloroplasts (Cornelissen *et al.*, 1987). Recently, the *A. hybridus psbA* gene encoding the atrazine-resistant 32 kDa protein has been converted to a nuclear gene by fusion to the RUBP carboxylase small subunit promoter and small subunit transit peptide-encoding DNA sequences. This chimeric gene has been reintroduced and expressed in transgenic tobacco plants and limited atrazine tolerance observed (Cheung *et al.*, 1988). Eventually, these plants are bleached by atrazine due to the relatively small number of atrazine-resistant Q_B protein reaction centres in relation to the sensitive 32 kDa polypeptides in the population.

Utilising a classical breeding approach, the maternally inherited triazine resistance from two weeds has been used for introducing the trait into related crop varieties. The triazine-tolerant trait was transferred from a resistant weed biotype of *Brassica campestris* to cultivated rapeseed cultivars from *B. napus* and *B. campestris* (Beversdorff *et al.*, 1980). Resistance from *Setaria viridis* was crossed into the crop *S. italica* (Darmency and Pernes, 1985). However, many of the atrazine-tolerant commercial lines fail to perform agronomically at levels comparable to non-tolerant lines (Souza-Machado, 1982). Although it is not clear if a slower flow of electrons decreases photosynthesis, there is generally a 20% loss of yield and a considerably lower competitive fitness (Gressel and Ben-Sinai, 1985; Ricroch *et al.*, 1985; Vaughn, 1986).

3.2.5 *Conclusion*

Although these examples illustrate that it is feasible that modified targets, or an increased expression of a target, might confer herbicide tolerance, some considerations must be noted. To reduce the complexity in generating such engineered plants the action of the herbicide is preferably confined to a unique target. Since an altered affinity of the target for the herbicide is required without a concomitant decrease in normal functional activity, the generation of a mutant form is likely to be a rare event. This is presumably only achieved by a very specific genetic alteration. In some cases mutant target enzymes are found (e.g. ALS) or amplified genes are generated (e.g. GS and EPSP synthase) (Amrhein *et al.*, 1983; Donn *et al.*, 1984; Smith *et al.*, 1986; Steinrucken *et al.*, 1986). Single-site mutations are clearly the easiest to obtain, in both plants and micro-organisms. The frequency of resistance mutations versus gene amplification events might be totally dependent on the characteristics of the enzyme system. For example, in multimeric enzyme complexes like GS, which contains eight subunits, a selection system most likely favours the isolation of mutants with altered expression levels, rather than structural mutants. Resistant mutants can be isolated after selection from a population of whole plants, or from cell culture; the latter is simpler, however, as selection on whole plants is difficult, since the population size becomes a limiting factor. Cell culture systems permit the screening of a large number of cells. However, only a limited number of crop species can be regenerated from cell cultures. Thus, herbicide resistance is often achieved in 'non-cultivar' material and must be transferred to commercial lines. This requires numerous back-crosses to introduce the resistance in suitable breeding material. Even with two generations per year, it takes at least five years before a herbicide-resistant plant variety can be developed.

One advantage of a genetic engineering approach is that the gene can be used to transform different plant species. In order to apply plant transformation, the gene encoding the target has to be available, and transformation procedures for the crop in question are required. Since target components are frequently compartmentalised within the cell, or because dual pathways in separate plant cell components may occur, a correct expression of the herbicide resistance determinant is difficult. Moreover, the wild-type gene cannot easily be deleted from the plant genome, and consequently a mixture of wild-type and mutant gene products in transgenic plants may yield an inconsistent resistance phenotype.

3.3 Detoxification or degradation of the herbicide

The second approach for developing herbicide resistance gene determinants is to identify and isolate genes encoding enzymes involved in herbicide detoxification. In many instances, a detoxification approach may have considerable advantages over 'altered target' resistance mechanisms. This strategy is preferred when the biochemical site of action of a given herbicide is not known or when the biochemical target is difficult to engineer. Many biochemical pathways are localised within one of the several plant cell compartments, making the correct placement of the herbicide resistance gene exceedingly difficult. Furthermore, many structural cellular components may also be difficult to engineer because they function as polymers (tubulins and actins are specific examples). Therefore, genes encoding proteins involved in herbicide detoxification may prove highly useful because these enzymes, when expressed in a herbicide-sensitive plant, would inactivate the herbicide prior to inhibition of the biochemical target. Genes encoding potential herbicide-detoxifying enzymes should, however, conform to certain criteria. The detoxifying enzyme should be encoded by a single gene, not require complex co-factors for activity, exhibit suitable kinetic parameters, and result in a metabolite(s) that exhibits little or no toxicity to the plant cell. The detoxification enzyme should also exhibit strict specificity for the herbicide, so that any related plant compounds required for plant development will not be recognised. Finally, it should be noted that the ability to isolate genes and transfer foreign genetic material to plants via gene transfer technology renders any organism a source of germplasm for agronomically important traits. Herbicide detoxification or degradation pathways may occur in plant species tolerant to specific herbicides or may be isolated from micro-organisms.

3.3.1 *Plant detoxifying enzymes*

The observation that many crop species are naturally tolerant to specific herbicides allowed the use of these herbicides for selective control of weeds by exploiting the difference in phytotoxicity without significantly reducing crop yields. Herbicide selectivity in many crops is determined by several factors, such as rates of absorption in the plant, translocation, subcellular localization, variation in target site sensitivity and metabolic detoxification to non-phytotoxic derivatives. The selective properties of herbicides often result from a complex interaction of these factors. Differential metabolism most likely constitutes the most important factor, and genes encoding the enzymes responsible might be explored as herbicide resistance determinants. A variety of enzymes such as mixed-function oxidases, amidases, decarboxylases, and thiol, sugar and amino acid conjugative enzyme systems were identified in herbicide-tolerant plant species. Some mechanisms involve the hydroxylation of aryl groups, oxidation of alkyl substitutes or sulphur moieties, hydrolysis of carboxylic acid ester groups, reductive deamination and glycosidation (Owen, 1987). A brief review of the mechanisms by which plant species transform foreign compounds will be given here. A well-known example is the detoxification of atrazine and alachlor in tolerant maize lines (*Zea mays*) which involves the conjugation with the tripeptide glutathione by the enzyme glutathione *S*-transferase (Shimabukuro *et al.*, 1971). The conjugate is non-toxic to the plant cell and is eventually further metabolised. The enzyme is present at high levels, and seems to contain at least three polypeptide components, which are encoded by a small multigene family (Moore *et al.*, 1986; Shah *et al.*, 1986b; Grove *et al.*, 1988). In metribuzin-tolerant tomatoes, detoxification involves the conjugation with UDP glucose by an *N*-glucosyltransferase (Frear *et al.*, 1983). Tolerant tomato cultivars contain more transferase activity than sensitive lines, and tolerance is controlled by one gene locus (Souza-Machado and Ditto, 1982). Mixed-function oxidases are involved in the detoxification of 2,4-dichlorophenoxyacetic acid (2,4-D) in pea (*Pisum sativum*) and dicamba in tolerant barley (*Hordeum vulgare*) (Comai and Stalker, 1986). Other herbicides, such as the bipyridinium herbicide paraquat, are known to generate active oxygen species which lipoxidise plant membranes, causing death of the plant cell. Enhanced levels of the enzymes that detoxify oxygen radicals might be correlated with resistance to such herbicides. At least three enzymes, a superoxide dismutase, a catalase and a peroxidase, are most likely to be involved in resistance against paraquat in *Conyza* species (Shaaltiel and Gressel, 1986, 1987). In some cases corresponding genes for these plant enzymes have been identified, but since most of the above examples deal with complex reactions and/or multienzyme systems it is not straightforward to isolate these genes for use as herbicide resistance determinants. Transfer of a plant detoxifying pathway by breeding or gene transfer has not yet been demonstrated.

3.3.2 *Bacterial detoxifying enzymes*

An alternative approach for obtaining herbicide-resistant genes is to isolate them from herbicide-degrading bacteria. Many herbicides undergo rapid biotransformation in herbicide-contaminated environments, suggesting involvement of soil microflora in their metabolism. Genetic and molecular manipulations are easier in bacteria than in plants and, in most cases, elucidation of metabolic pathways and biochemical identification and purification of degradative/detoxification enzymes are simpler. It has already been demonstrated that proteins encoded by bacterial genes function in a plant cell when these genes are transferred to plants. These considerations make the tailoring of bacterial detoxification genes as herbicide resistance determinants extremely attractive. A number of microbial systems have been identified for obtaining bacterial genes encoding detoxification/degradation enzymes. Bacterial genera such as *Pseudomonas, Arthrobacter, Alcaligenes* and *Acinetobacter* have been shown to metabolise the herbicide 2,4-D (Don and Pemberton, 1981). *Pseudomonas* and *Flavobacterium* species which use glyphosate as a sole phosphate source by cleaving the phosphate moiety from the compound have been identified (Balthazar and Hallas, 1986; Kishore and Jacob, 1987). The metabolic pathway of glyphosate has been characterised, but genes encoding glyphosate-degrading enzymes have not been cloned. Microbes such as *Bacillus sphaericus* and *Anacystis nidulans* have also been shown to degrade urea and phenylcarbamate herbicides (Table 3.2) (Wallnofer and Bader, 1970; Wright and Maule, 1982).

A bacterial degradative/detoxification system has been identified for the

Table 3.2. Bacterial reactions detoxifying herbicides.

Compound herbicide	Enzyme	Metabolite	Organism
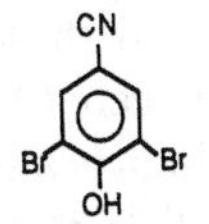 Bromoxynil (Buctril)	Nitrilase	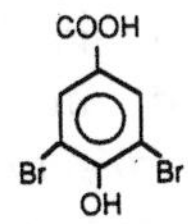	*Klebsiella ozaenae*
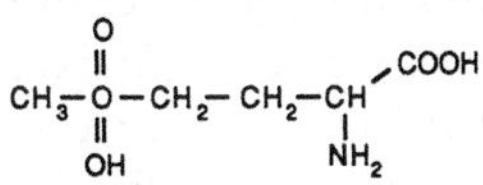 Phosphinothricin (Basta)	Acetyl-transferase	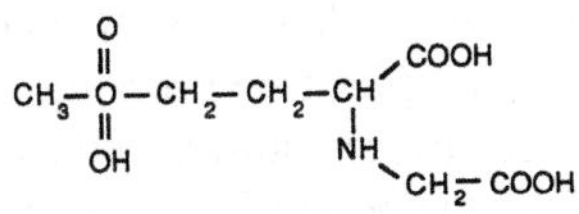	*Streptomyces hydroscopicus*
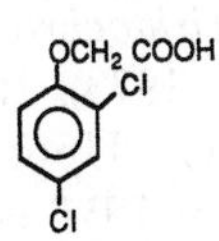 2,4-D	Mono-oxygenase	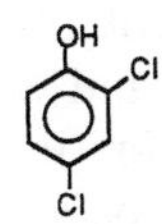	*Alcaligenes eutrophus*

herbicide bromoxynil (3,5-dibromo 4-hydroxybenzonitrile). Bromoxynil is a broad-leaf weed killer registered for use on a variety of tolerant crop species. Metabolism studies in tolerant wheat plants (Buckland *et al.*, 1973) revealed that a primary metabolic pathway for bromoxynil is the conversion of the herbicide to 3,5-dibromo 4-hydroxybenzamide with the following conversion to 3,5-dibromo 4-hydroxybenzoic acid. This metabolite is degraded further by dehalogenating and lactonising enzyme activities. It is known that the two metabolites are at least 100-fold less toxic to plant cells than bromoxynil, therefore conversion of bromoxynil to either of these metabolites should inactivate the compound and result in a bromoxynil-tolerant phenotype. Bromoxynil exhibits a rapid conversion in soil, utilising a similar pathway found in tolerant plants (Collins, 1973; Ingram and Pullin, 1974). In soil, the conversion bypasses the amide derivative of bromoxynil and proceeds directly to 3,5-dibromo 4-hydroxybenzoic acid in a single enzyme-catalysed reaction (Hsu and Camper, 1975). The enzyme that acts on the cyano moiety of the molecule is termed nitrilase. Nitrilases are ubiquitous in nature and act on a variety of cyano-substituted aromatic compounds (Bandyopadhyay *et al.*, 1986; Harper, 1977), but no nitrilase had previously been identified exhibiting a high specificity for bromoxynil as substrate. An organism (*Klebsiella ozaenae*) has been isolated from bromoxynil-treated soil that utilizes bromoxynil as the sole nitrogen source and expresses a nitrilase with a strict specificity for bromoxynil as substrate in relation to other benzonitrile compounds

(McBride *et al.*, 1986). The nitrilase was plasmid-encoded in *K. ozaenae* (Stalker and McBride, 1987), and a gene *bxn*, that encodes a 37 000 molecular weight polypeptide was cloned and expressed in *E. coli* (Stalker *et al.*, 1988a). The *bxn* gene was spliced to plant promoters such as the tobacco RUBP carboxylase small subunit (SSU) and the CaMV 35S promoters, and these chimeric genes were introduced into regenerated transgenic tobacco (*N. tabacum*) and tomato plants (*L. esculentum*). As shown in Figure 3.3, high levels of bromoxynil (BUCTRIL) resistance were displayed (up to 10-fold field rate) by plants expressing the bromoxynil-specific nitrilase in photosynthetic plant tissues (Stalker *et al.*, 1988b). Transgenic tomato and transgenic cotton (*Gyossipium hirsutum*) plants expressing the nitrilase were field tested (Stalker, unpublished observations), and no effect on growth or yield (as measured by fruit and boll yields respectively) was observed (see Figure 3.4). Detoxification enzymes can also be found in biosynthesis pathways of naturally produced components with herbicide activity. L-Phosphinothricyl-L-alanyl-L-alanine (bialaphos) is a tripeptide antibiotic produced by *Streptomyces viridochromogenes* (Bayer *et al.*, 1972) and *S. hygroscopicus* (Kondo *et al.*, 1973). It consists of L-phosphinothricin (PPT) and two L-alanine residues. Upon removal of these residues by peptidases, the herbicidal component PPT is released. In the bialaphos biosynthesis pathways, a resistance gene (*bar*) was isolated and characterised (Murakami *et al.*, 1986). It encodes an acetyltransferase (PAT) that acetylates the free amino group of PPT or intermediates in the pathway and protects the micro-organism from autotoxicity (Thompson *et al.*, 1987). Purified PAT enzyme showed high substrate specificity for PPT. The *bar* gene under the control of the CaMV 35S promoter was expressed in regenerated transgenic tobacco (*N. tabacum*), potato (*Solanum tuberosum*) and tomato (*L. esculentum*) plants (De Block *et al.*, 1987). Greenhouse-grown plants showed complete resistance towards doses of PPT and bialaphos up to 10 times the doses recommended by the manufacturer. The resistance was reproducible in field-tested tobacco and potato plants. Two tobacco lines with a 100-fold difference in *bar* gene expression were both resistant to high herbicide levels and there was no significant effect on yield, as measured by leaf length (Leemans *et al.*, 1987). The fate of the acetylated metabolite in transgenic plants requires further study.

Recently, a gene (*tfd*A) encoding a 2,4-dichlorophenoxyacetate monooxygenase (DPAM) that degrades the herbicide 2,4-D was identified and cloned from *Alcaligenes eutrophus* strain JMP 134 (Streber *et al.*, 1987). The *tfd*A gene encodes a single polypeptide, and the enzymatic activity removes the phenoxyacetate moiety of 2,4-D to form the metabolite 2,4-dichlorophenol, which exhibits limited plant toxicity at 10 mg/l concentrations. The *tfd*A gene encoding the mono-oxygenase has been expressed in transgenic tobacco (*N. tabacum*) under the control of the CaMV 35S and RUBP small subunit promoters, and three- to fivefold resistance to 2,4-D has been observed (Streber and Willmitzer, 1989). It will be interesting to see whether this resistance mechanism translates into complete resistance in the field or whether the residual toxicity of the metabolite 2,4-dichlorophenol becomes a problem at higher spray rates.

3.3.3 *Conclusion*

Genes obtained from microbial sources and transferred and expressed in plants have conferred complete field resistance to a number of herbicides. With the above-described examples, it is also clear that detoxification is a simpler and more effective method for developing herbicide resistance by gene transfer technology then engineering 'altered biochemical target' type mechanisms.

3.4 Perspectives

In the past few years, rapid progess in conferring herbicide resistance via gene transfer to crops has been reported. Field trials over the next two or three years with engineered crops will impart more information on the practical value of the scientific results summarised in this chapter. At present, gene transfer systems for higher plants based on *Agrobacterium* T-DNA transformation or on free DNA transfer have enabled researchers to obtain transgenic plants for about 20 plant species (Fraley *et al.*, 1986; Chapter 2). It can be expected that in the future all major crops will be accessible to genetic improvement by this technology. As the methods for gene transfer and for incorporation of agronomically useful traits in commercial germplasm become routine, the engineering of selective herbicide tolerance will become an accepted and essential strategy and will influence weed control practices. Moreover, the growing knowledge of herbicide mode of action and selectivity and the identification and characterisation of herbicide-detoxifying enzymes will stimulate genetic manipulation of crops and increase the use of such herbicides. It will also affect the classical procedure followed to develop new herbicides. An integrated research programme will involve the engineering of herbicides, the engineering of targets and the characterisation of detoxifying enzymes.

Herbicide resistance genes can also be used as dominant selectable markers for both basic research and plant breeding. For basic research, the resistance genes will serve as analogues of the antibiotic resistance genes presently in use. For plant breeding, the resistance genes can be physically linked to other genes conferring agronomically useful traits which are not easy to detect phenotypically during transformation and regeneration procedures. They are introduced into plants via transformation using selection for resistance to the herbicide. Subsequently, in a plant breeding programme, the trait can be transferred to various cultivars through standard genetic crosses by following the easily assayable herbicide resistance phenotype associated with the linked genetic marker.

Finally, before herbicide-resistant plants will be commercialised, several aspects must be considered:

(1) The performance, the cost and environmental acceptability of the herbicide. When herbicides with well-elucidated weed control properties and toxicology are dealt with, and if it can be shown that residues of the herbicide match with those found in crops where the herbicide is already used, registration requirements can be simple.
(2) The nature of the resistance mechanism and the potential undesired agronomic characteristics associated with the resistance trait itself. Most herbicide resistance traits being developed are for herbicides with single enzyme sites of action. This might lead to an increase of the use of herbicides with single target sites. The development of weed resistance to a specific herbicide by mutation or natural selection can be expected to be highest for this class of herbicides. An additional problem is that crops potentially have the ability to cross genetically with closely related weeds species. The risk for inter- and intraspecific transfer of herbicide tolerance should thus be evaluated.
(3) The accessibility to commerical germplasm and the time period required to introduce a resistance gene into a significant percentage of the germplasm for a given crop.
(4) The quality and quantity characteristics of the engineered crop compared with other competitive products. Commercialisation will be justified if the manipulated crop solves an uncontrolled weed problem or brings additional benefits to the crop.
(5) Given the time periods required from feasible demonstration in the laboratory to the marketable crop, an important factor will be patent life and the post-patent price for a particular herbicide; indeed, the best cost-effective weed control is obtained with herbicides that are off-patent.
(6) The registration procedure and cost of a herbicide for use on an engineered resistant crop will depend largely on the final destination of the crop: food or non-food. Most naturally resistant crops degrade the selective herbicides and no residues appear in the consumed portions. Regulatory issues claim that there should be no pesticide residues in foods, even if they are non-toxic. In engineered plants, the herbicide may remain unchanged. For example, there is no information available on the fate of glyphosate in resistant plants (Gressel, 1988).
(7) Crops can pose serious problems as volunteer weeds, especially in crop rotation programmes. For example, maize appearing in a soybean field has to be considered as a weed. Herbicide-tolerant cultivars might show increased problems as volunteer weeds.

In conclusion, herbicide-resistant cultivars will not cause a major green revolution, but niches will be created where particularly troublesome weeds cannot be controlled by existing compounds, or in minor crops with high added value for which no adequate herbicides are developed. It would also be imprudent to invest in herbicide-tolerant cultivars for compounds whose future is in doubt on the basis of their toxicology (e.g. triazines and bipyridilliums). On the other hand, the withdrawal of a range of herbicides could leave important gaps in classical chemical weed control which might be covered by herbicide-resistant cultivars. Gene transfer technology will accelerate the trend of herbicide market evolution to a mature industry commercialising fewer, but more effective, less expensive and environmentally more acceptable products.

Acknowledgements

The author would like thank Deidre Walker for typing, Beth Gabor and Jack Kiser for tables and photographs and Jan Leemans for perspectives. The author is also indebted to his longtime colleague, Luca Comai, for his inception of and commitment to the glyphosate project.

References

Amrhein, N., Deus, B., Gehrke, P. and Steinruecken, H.C. (1980) The site of the inhibition of the shikimate pathway by glyphosate. II. Interference of glyphosate with chorismate formation *in vivo* and *in vitro*. *Plant Physiol.* **66**: 830.

Amrhein, N., Johanning, D., Scab, J. and Schulz, A. (1983) Biochemical basis for glyphosate tolerance in a bacterium and a plant tissue culture. *FEBS Lett.* **157**: 191.

Andersen, R.M. and Gronwald, J (1987) Non-cytoplasmic inheritance of atrazine-tolerance in velvet leaf (*Abutilon theorphrasti*). *Weed Sci.* **35**: 496.

Anderson, P.C. and Georgeson, M. (1986) Selection and characterization of imidazolinone tolerant mutants of maize. In *Biochemical Basis of Herbicide Action*, 27th Harden Conference Program and Abstracts. Wye College, Ashford, UK.

Arntzen, C.J., Pfister, K. and Steinback, K. (1982) The mechanism of chloroplast triazine resistance: alterations in the site of herbicide action. In *Herbicide Resistance in Plants* (eds. Le Baron, H.M. and Gressel, J.) Wiley, New York, 185.

Baltazhar, T.M. and Hallas, L.E. (1986) Glyphosate degrading microorganisms from industrial activated sludge. *Appl. Environ. Microbiol.* **91**: 432.

Bandyopadhyay, A.K., Nagasawa, T., Asano, Y., Fujishiro, K., Tani, Y. and Yamada, H. (1986) Purification and characterization of benzonitrilases from *Arthrobacter* sp. strain J-1. *Appl. Environ. Microbiol.* **51**: 302.

Barber, J. (1987) Photosynthetic reaction centers: a common link. *Trends Biochem. Sci.* **12**: 321.

Bayer, E., Gugel, K.H., Haegele, K., Hagenmaier, H., Jessipow, S., Koenig, W.A. and Zaehner, Z. (1972) Stoffwechselphrodukte von Microoganismen 98. Mitteilung: Phosphinothricin und Phosphinothricyl-Alanyl-Alanin. *Helvet. Chim. Acta* **55**: 224.

Beversdorff, W.D., Weiss-Lerman, J., Erickson, L.R. and Souza-Machado, V. (1980) Transfer of cytoplasmatically-inherited triazine resistance from bird's rape to cultivated *Brassica campestris* and *Brassica napus*. *Can. J. Genet. Ctyol.* **22**: 167.

Bickel, H. and Schultz, G. (1978) Shimikate pathway regulation in suspensions of isolated spinach chloroplasts. *Phytochemistry* **18**: 498.

Bickel, H., Palme, L. and Schultz, G. (1978) Incorporation of shikimate and other precursors into aromatic amino acids and prenylquinones of isolated spinach chloroplasts. *Phytochemistry* **17**: 119.

Boocock, M.R. and Coggins, J.R. (1983) Kinetics of 5-enolpyruvylshikimate 3-phosphate synthase inhibition by glyphosate. *FEBS Lett.* **154**: 127.

Buckland, J.L., Collins, R.F. and Pullin, E.M. (1973) Metabolism of bromoxynil octanoate in growing wheat. *Pestic. Sci.* **4**: 149.

Chaleff, R.C. and Mauvais, C.J. (1984) Acetolactate synthase is the site of action of the herbicides chlorsulfuron and sulfometuron methyl in higher plants. *Science* **224**: 1442.

Chaleff, R.S. and Bascomb, N.F. (1987) Genetic and biochemical evidence for multiple forms of acetolactate synthase in *Nicotiana tabacum. Molec. Gen. Genet.* **210**: 33.

Chaleff, R.S. and Ray, T.B. (1984) Herbicide-resistant mutants from tobacco cell cultures. *Science* **223**: 1148.

Cheung, A.Y., Bogorad, L., Van Montagu, M. and Schell, J. (1988) Relocating a gene for herbicide tolerance: a chloroplast gene is converted into a nuclear gene. *Proc. Natl Acad. Sci. USA* **85**: 391.

Colanduoni, J.A. and Villafranca, J.J. (1986) Inhibition of *Escherichia coli* glutamine synthetase by phosphinothricin. *Biorganic Chem.* **14**: 163.

Cole, D.J., Edwards, R., Owen, W.J. (1987) The role of metabolism in herbicide selectivity. In *Progress in Pest. Biochem. Toxicol*, vol. 6, eds. Huston, D.H. and Roberts, T.R. Chichester, Wiley, 57.

Collins, R.F. (1973) Perfusion studies with bromoxynil octanoate in soil, *Pestic. Sci.* **4**: 181.

Comai, L. and Stalker, D.M. (1986) Mechanism of action of herbicides and their molecular manipulation. *Oxford Surveys of Plant Molecular and Cell Biology* **3**: 167.

Comai, L., Sen, L.C. and Stalker, D.M. (1983) An altered *aro*A gene product confers resistance to the herbicide glyphosate. *Science* **221**: 370.

Comai, L., Facciotti, D., Hiatt, W.R., Thompson, G., Rose, R.E. and Stalker, D.M. (1985) Expression in plants of a mutant *aro*A gene from *Salmonella typhimurium* confers tolerance to glyphosate. *Nature* **317**: 741.

Comai, L., Larson-Kelly, N., Kiser, J., Mau., D.J-D., Pokalsky, A. Shewmaker, C.K., McBride, K., Jones, A. and Stalker D.M. (1988) Chloroplast transport of a ribulose bisphosphate carboxylase small subunit-5-enolpyruvyl 3-phosphoshikimate synthase chimeric protein requires part of the mature small subunit in addition to the transit peptide. *J. Biol. Chem.* **263**: 15104.

Cornelissen, M.J., De Block, M., Van Montagu, M., Leemans, J., Schreier, P.H. and Schell, J. (1987) Plastid transformation: a progress report. In *Plant Gene Research: Plant DNA Infectious Agents*, eds. Hohn, T.H. and Schell, J. Springer Verlag, Vienna, 311.

Darmency, H.M. and Pernes, J (1985) Use of wild *Setaria viridis* to improve triazine resistance in cultivated *S. italica* by hybridization. *Weed Res.* **25**: 175.

Dassarma, S., Tischer, E. and Goodman, H.M. (1986) Plant glutamine synthetase complements a *gln*A mutation in *Escherichia coli. Science* **232**: 1242.

De Block, M., Botterman, J., Vandewiele, M., Dockx, J., Thoen, C., Gossele, V., Movva, N., Thompson, C., Van Montagu, M. and Leemans, J. (1987) Engineering herbicide resistance in plants by expression of a detoxifying enzyme. *EMBO J.* **6**: 2513.

Della-Cioppa, G. and Kishore, G.M. (1988) Import of a precursor protein into chloroplasts is inhibited by the herbicide glyphosate. *EMBO J.* **7**: 1299.

Della-Cioppa, G., Bauer, S.C., Klein, B.K., Shah, D.M., Fraley, R.T. and Kishore, G.M. (1986) Translocation of the precursor of 5-enolpyruvylshikimate-3-phosphate synthase into chloroplasts of higher plants *in vitro. Proc. Natl Acad. Sci. USA* **83**: 6873.

Della-Cioppa, G., Bauer, S.C., Taylor M.L., Rochester, D.E., Klein, B.K., Shah, D.M., Fraley, R.T. and Kishmore, G.M. (1987) Targeting a herbicide-resistant enzyme from *Escherichia coli* to chloroplasts of higher plants. *BiotecholOlgy* **5**: 579.

Don, R.H. and Pemberton, J.M. (1981) Properties of six pesticide degradation plasmids isolated from *Alcaligenes paradoxus* and *Alcaligenes eutrophus. J. Bacteriol.* **145**: 681.

Donn, G., Tischer, E. Smith, J.A. and Goodman, H.M. (1984) Herbicide-resistant alfalfa cells: an example of gene amplicifation in plants. *J. Molec. Appl. Genet.* **2**: 621.

Eckes, P. and Wengemayer (1987) In *Regulation of Plant Expression*, 29th Harden Conference Program and Abstracts. Wye College, Ashford, UK.

Erickson, J.M., Rahire, M. and Rochaix, J.D. (1985) Herbicide resistance and cross-resistance: changes at three distinct sites in the herbicide-binding protein. *Science* **228**: 204.

Falco, S.C. and Dumas, K.S. (1985) Genetic analysis of mutants of *Saccharomyces cerevisiae* resistant to the herbicide sulfometuron methyl. *Genetics* **109**: 21.

Falco, S.C., Knowlton, S., Larossa, R.A., Smith, J.K. and Mazur, B.J. (1987) Herbicides that inhibit amino acid biosynthesis in the sulphonylureas–a case study. In *1987 British Crop Protection Conference–Weeds.* BCPC Publications, Surrey, 149.

Feierabend, J., and Brassel, D. (1977) Subcellular localization of shikimate dehydrogenase in higher plants, *Z. Pflanzenphysiol.* **82**: 334.

Fillatti, J.J., Kiser, J., Rose, R and Comai, L. (1987) Efficient transfer of a glyphosate tolerance gene into tomato using a binary *Agrobacterium tumefaciens* vector. *Biotechnology* **5**: 726.

Fraley, R.T., Rogers, S.G. and Horsch R.B. (1986) Genetic transformation in higher plants. *CRC Crit. Rev. Plant Sci.* **4**: 1.

Fraley, R., Kishore, G., Gasser, C., Padgette, S., Horsch, R., Rogers, S., Della-Cioppa, G. and Shah, D. (1987) Genetically-engineered herbicide tolerance–technical and commercial considerations. In *1987 British Crop Protection Conference–Weeds.* BCPC Publications, Surrey, 463.

Frear, D.S., Manager, E.R., Swanson, H.R. and Tanaka, F.S. (1983) Metribuzin metabolism in tomato: isolation and identification of N-glucoside conjugates. *Pestic. Biochem. Physiol.* **19**: 270.

Galloway, R.E. and Mets, L.J. (1984) Atrazine, bromacil and diuron resistance in *Chlamydomonas. Plant Physiol.* **74**: 4699.

Goloubinoff, P. and Edelman, M. (1984) Chloroplast-coded atrazine resistance in *Solanum nigrum: psbA* loci from susceptible and resistant biotypes are isogenic expect for a single codon change. *Nucl. Acids Res.* **12**: 9489.

Gressel, J. (1988) Conferring herbicide resistance in susceptible crops. In *Herbicide and Plant Metabolism*, ed. Dodge, A.D. Cambridge, University Press, in press.

Gressel, J. and Ben-Sinai, G. (1985) Low intra-specific competitive fitness in a triazine resistant, nearly isogenic line of *Brassica napus. Plant Sci.* **38**: 29.

Grove, G., Zarlenga, R.P., Timmerman, K.P., Li, N., Tam, M.E. and Tu, C.-P.D. (1988) Characterization and heterospecific expression of cDNA clones of genes in the maize GSH S-transferase multigene family. *Nucl. Acids Res.* **16**: 425.

Haderlie, L.C., Widholm, J.M. and Slife, F.W. (1977) Effects of glyphosate on carrot and tobacco cells. *Plant Physiol.* **60**: 40.

Harper, D.B. (1977) Fungal degradation of aromatic nitriles: Enzymology of C-N cleavage by *Fusarium solani. Biochem. J.* **167**: 685.

Hartnett, M.D., Newcomb, J.R. and Hodson, R.C. (1987) Mutation in *Chlamydomonas rheinhardii* conferring resistance to the herbicide sulfometuron methyl. *Plant Physiol.* **85**: 898.

Haughn, G.W. and Somerville, C. (1986) Sulfonylurea-resistant mutants of *Arabidopsis thaliana. Molec. Gen. Genet.* **204**: 430.

Haughn, G.W., Smith, J., Mazur, B. and Somerville, C. (1988) Transformation with a mutant *Arabidopsis* acetolactate synthase gene renders tobacco resistant to sulfonylurea herbicides. *Molec. Gen. Genet.* **211**: 266.

Hickok, L.G. (1987) Application of *in vitro* selection systems: whole plant selection using the haploid phase of *Ceraptoteris.* In *Biotechnology in Agricultural Chemistry*, eds. LeBaron, H.M., Mumma, R.O., Honeycutt, R.C. and Duesing, J.H. American Chemical Society, Washington DC, 53.

Hirschberg, J. and Mcintosh, L. (1983) Molecular basis of herbicide resistance in *Amaranthus hybridus. Science* **222**: 1346.

Hoagland, R.E. and Duke, S.O. (1982) Biochemical effects of glyphosate. In *Biochemical Responses Induced by Herbicides*, eds. Moreland, J.E., St John, J.B. and Hess, F.D. American Chemical Society, Washington.

Hsu, J.C. and Camper, N.D. (1975) Isolation of ioxynil degraders from soil-enrichment cultures. *Can J. Microbiol.* **22**: 537.

Ingram, G.H. and Pullin, E.M. (1974) Persistence of bromoxynil in three soil types. *Pestic. Sci.* **5**: 287.

Jaworski, E.G. (1972) Mode of action of *N*-phosphonomethylglycine: inhibition of aromatic amino acid biosynthesis. *J. Agric. Food Chem.* **20**: 1195.

Kishore, G.M. and Jacob, G.S. (1987) Degradation of glyphosate by Pseudomonas sp. PG2982 via a sarcosine intermediate. *J. Biol. Chem.* **262**: 12164.

Kondo, Y., Shomura T., Ogawa, Y., Tsuruoka, T., Watanabe, H., Totukawa, K., Suzuki, T., Moriyama, C., Yoshida, J., Inouye, S. and Niida, T. (1973) Studies on a new antibiotic SF-1293. 1. Isolation and physico-chemical and biological characterization of SF-1293 substances. *Sci. Rep. Meiji Seika* **13**: 34.

Larossa, R.A. and Schloss, J.V. (1984) The sulfonylurea herbicide sulfometuron methyl is an extremely potent and selective inhibitor of acetolactate synthase in *Salmonella typhimurium. J. Biol. Chem.* **259**: 8753.

Lebaron, H.M. and Gressel, J. (1982) *Herbicide Resistance in Plants.* John Wiley and Sons, New York.

Leemans, J., De Block, M., D'Halluin, K., Botterman, J. and De Greef, W. (1987) The use of glufosinate as a selective herbicide on genetically engineered resistant tobacco plants. In *1987 British Crop Protection Conference—Weeds* BCPC Publication, Surrey, 867.

McBride, K.E., Kenny, J.W. and Stalker, D.M. (1986) Metabolism of the herbicide bromoxynil by *Klebsiella pneumoniae* subsp. *ozaenae*. *Appl. Environ. Microbiol.* **52**: 325.

Manderscheid, R. and Wild, A. (1986) Studies of the mechanism of inhibition by phosphinothricin of glutamine synthetase isolated from *Triticum aestivum* L. *J. Plant Physiol.* **123**: 132.

Mattoo, A.K. and Edelman, M. (1987) Intramembrane translocation and posttranslational palmitoylation of the chloroplast 32-kDa herbicide-binding protein. *Proc. Natl Acad. Sci. USA* **84**: 1497.

Mazur, B.J., Chui, C.F., and Smith, J.K. (1987) Isolation and characterization of plant genes coding for acetolactate synthase, the target enzyme for two classes of herbicides. *Plant Physiol.* **85**: 1110.

Moore, R.E. Davies, M.S., O'Connell, K.M., Harding, E.I. Weigand, R.C. and Tiemeier, D.C. (1986) Cloning and expression of a cDNA encoding a maize glutathione-*S*-transferase in *E. coli. Nucl. Acids Res.* **14**: 7227.

Mousdale, D.M. and Coggins, J.R. (1984) Purification and properties of 5-enolpyruvylshikimate 3-phosphate synthase from seedlings of *Pisum sativum* L. *Planta* **160**: 78.

Mousdale, D.M. and Coggins, J.R. (1985) Subcellular localization of the common shikimate pathway enzymes in *Pisum sativum* L. *Planta* **163**: 241.

Muhitch, M.J., Shaner, D.L. and Stidham, M.A. (1987) Imidazolinones and acetohydroxyacid synthase from higher plants. *Plant Physiol.* **83**: 451.

Murakami, T., Anzai, H., Imai, S., Satoh, A., Nagaoka, K. and Thompson, C.J. (1986) The bialaphos biosynthetic genes of *Streptomyces hygroscopicus:* molecular cloning and characterization of the gene cluster. *Molec. Gen. Genet.* **205**: 42.

Odell, J.F., Nagy, F. and Chua, N.-H. (1985) Identification of DNA sequences required for activity of the cauliflower mosaic virus 35S promoter. *Nature* **313**: 810.

Owen, W.J. (1987) Herbicide detoxification and selectivity. In *1987 British Crop Protection Conference—Weeds*. BCPC Publications, Surrey, 309.

Ray, T.B. (1984) Site action of chlorsulfuron: inhibition of valine and isoleucine biosynthesis in plants. *Plant Physiol.* **75**: 827.

Reid, W.J. (1987) Biotechnology and breeding team up in agriculture. *Biotechnology* **5**: 899.

Rhodes, C.A., Lowe, K.S. and Ruby, K.L. (1988) Plant regeneration from protoplasts isolated from embryogenic maize cell cultures. *Biotechnoloty* **6**: 56.

Ricroch, A., Mousseau, M., Darmency, H. and Pernes, J. (1987) Comparison of triazine-resistant and -susceptible cultivated *Setaria italica* growth and photosynthetic capacity. *Plant Physiol. Biochem.* **25**: 29.

Rogers, S.G., Brand, L.A., Holder, S.B., Sharps, E.S. and Brackin, M.J. (1983) Amplification of the *aroA* gene from *E. coli* results in tolerance to the herbicide glyphosate. *Appl. Environ. Microbiol.* **46**: 37.

Rubin, J.L., Gaines, C.G. and Jensen, R.A. (1984) Glyphosate inhibition of 5-enolpyruvylshikimate 3-phosphate synthase from suspension-cultured cells of *Nicotiana sylvestris. Plant Physiol.* **75**: 839.

Rueppel, M.L. Brightwell, B.B., Schaefer, J., Marvel, J.T. (1977) Metabolism and degradation of glyphosate in soil and water. *J. Agric. Food Chem.* **25**: 517.

Schloss, J.V., Ciskanik, L.M. and Van Dyk, D.E. (1988) Origin of the herbicide binding site of acetolactate synthase. *Nature* **331**: 360.

Schreier, P.H., Seftor, E.A., Schell, J., and Bohnert, J.H. (1985) The use of nuclear encoded sequences to direct the light regulated synthesis and transport of a foreign protein into plant chloroplast. *EMBO J.* **4**: 25.

Sebastian, S.A. and Chaleff, R.S. (1987) Soybean mutants with increased tolerance for sulfonyurea herbicides. *Crop Sci.* **27**: 948.

Shaaltiel, Y. and Gressel, J. (1986) Multienzyme oxygen radical detoxifying system correlated with paraquat resistance in *Conyza bonariensis. Pest Biochem. Physiol.* **26**: 22.

Shah, D.M., Hironaka, C.M. Wiegand, R.C., Harding, E.I. Krivi, G.G. and Tiemeir, D.C. (1986) Structural analysis of a maize gene coding for glutathione *S*-transferase involved in herbicide detoxification. *Plant Molec. Biol.* **6**: 203.

Shah, D.M., Horsch, R.B., Klee, H.J., Kishore, G.M., Winter, J.A., Tumer, N.E., Hironaka, C.M.,

Sanders, P.R., Gasser, C.S., Aykent, S., Siegel, N.R., Rogers, S.G. and Fraley, R.T. (1986) Engineering herbicide tolerance in transgenic plants. *Science* **233**: 478.

Shaner, D.L., Anderson, P.C. and Stidham, M.A. (1984) Imidazolinones, potent inhibitors of acetohydroxyacid synthase. *Plant Physiol.* **76**: 545.

Shaner, D.L. and Anderson, P.C. (1985) In *Biotechnology in Plant Science. Relevance to Agriculture in the Eighties* eds. Zaitlin, M., Day, P. and Hollaender, A. Academic Press, New York, 287.

Shimabukuro, R.H. Frear, D.S., Swanson, H.R. and Walsh, W.C. (1971) Glutathione conjugation: an enzymatic basis for atrazine resistance in corn. *Plant Physiol.* **47**: 10.

Smith, C.M., Pratt, D., and Thompson, G.A. (1986) Increased 5-enolpyruvyl-shikimic acid 3-phosphate synthase activity in a glyphosate tolerant variant of tomato cells. *Plant Cell Rep.* **5**: 298.

Souza-Machado, V. (1982) Inheritance and breeding potential of traizine tolerance and resistance in plants. In *Herbicide Resistance in Plants*, eds. Le Baron, H.M. and Gressel, J. Wiley, New York, 257.

Stalker, D.M., Hiatt, W.R., and Comai (1985) A single amino acid substitution in the enzyme 5-enolpyruvylshikimate-3-phosphate synthase confers resistance to the herbicide glyphosate. *J. Biol. Chem.* **260**: 4724.

Stalker, D.M. and McBride, K.E. (1987) Cloning and expression in *Escherichia coli* of a *Klebsiella ozaenae* plasmid-borne gene encoding a nitrilase specific for the herbicide bromoxynil. *J. Bacteriol.* **169**: 955.

Stalker, D.M., Malyj, L.D. and McBride, K. (1988a) Purification and properties of a nitrilase specific for the herbicide bromoxynil and corresponding nucleotide sequence analysis for the *bxn* gene. *J. Biol. Chem.* **263**: 6310.

Stalker, D.M., McBride, K.E. and Malyj, L.D. (1988b) Herbicide resistance in transgenic plants expressing a bacterial detoxification gene. *Science* **242**: 419.

Stannard, M.D. and Fay, P.K. (1987) *Selection of Alfalfa Seedlings for Tolerance to Chlorosulfuron.* Weed Science Society of America Abstracts, St Louis, 61.

Steinruecken, H.C. and Amrhein, N. (1984) EPSP synthase of *Klebsiella pneumoniae.* 2. Inhibition by glyphosate, *Eur. J. Biochem.* **143**: 351.

Steinruecken, H.C., Schulz, A., Amrhein, N., Porter, C.A. and Fraley, R.T. (1986) Overproduction of 5-enopyruvylshikimate-3-phosphate synthase in glyphosate-tolerant petunia hybrid cell line. *Arch. Biochem. Biophys.* **244**: 169.

Streber, W.R. and Willmitzer, L. (1989) Transgenic tobacco plants expressing a bacterial detoxifying enzyme are resistant to 2,4-D. *Biotechnology* **7**: 811.

Streber, W.S., Timmis, K.N. and Zenk, M.H. (1987) Analysis, cloning, and high-level expressing of 2,4-dichlorophenoxyacetate monoxygenase gene *tfd*A of *Alcaligenes eutrophus* JMP134. *J. Bacteriol.* **169**: 2950.

Swanson, E.B., Coumans, M.P., Brown, G.L., Patel, J.D. and Beversdorff, W.D. (1988) The characterization of herbicide tolerant plants in *Brassica napus* after *in vitro* selection of microspores and protoplasts. *Plant Cell Rep.* **7**: 83.

Tachibana, K., Watanabe, T., Sekizuwa, Y. and Takematsu, T. (1986) Accumulation of ammonia in plants treated with bialaphos. *J. Pestic. Sci.* **11**: 33.

Thompson, C., Movva, N., Tizard, R., Crameri, R., Davies, J., Lauwereys, M. and Botterman, J. (1987) Characterization of the herbicide-resistance gene *bar* from *Streptomyces hygroscopicus. EMBO J.* **6**: 2519.

Tischer, E., Dassarma, S. and Goodman, H.M. (1986) Nucleotide sequence of an alfalfa glutamine synthetase gene. *Molec. Gen. Genet.* **203**: 221.

Van der Broek, G., Timko, M.P., Kausch, A.P., Cashmore, A.R., Van Montagu, M. and Herrera-Estrella, L. (1985) Targeting of a foreign protein to chloroplasts by fusion to the transit peptide from the small subunit of ribulose 1,5-biphosphate carboxylase, *Nature* **313**: 358.

Vaughn, K.C. (1986) Characterization of triazine-resistant and susceptible isolines of canola (*Brassica napus*). *Plant Physiol.* **82**: 859.

Wallnofer, P.R. and Bader, J. (1970) Degradation of urea herbicides by cell-free extracts of *Bacillus sphaericus. Appl. Microbiol.* **19**: 714.

Wright, S.J.L., and Maule, A. (1982) Transformation of the herbicides propanil and chlorpropham by micro-algae. *Pestic. Sci.* **13**: 253.

Yadav, N., McDevitt, R.E., Bernard, S. and Falco, S.C. (1986) Single amino acid substitutions in enzyme acetolactate synthase confer resistance to the herbicide sulfometuron methyl. *Proc. Natl Acad. Sci. USA* **83**: 4418.

Zurwarski, G., Bohnert, H.J., Whitfield, P.R. and Bottomley, W. (1982) Nucleotide sequence of the gene for the M_r 32,000 thylakoid membrane protein from *Spinacia olearacea* and *Nicotiana deneyi* predicts a totally conserved primarily translation product of M_r 38,950. *Proc. Natl Acad. Sci. USA* **79**: 7699.

4 Genetic engineering of plants for insect resistance

J.A. GATEHOUSE, V.A. HILDER and A.M.R. GATEHOUSE

4.1 Introduction

Plants and insects have been co-evolving for at least 300 million years; during this time plant species have been under continuous selection pressure from herbivorous insect predators, and, in the absence of the ability to avoid predators by movement, have been forced to rely on physical and chemical defensive mechanisms. Physical mechanisms include the presence of hairs on the surfaces of plant tissues, spines, exuded gums and toughened surface layers, such as hardened seed coats. Chemical defensive mechanisms include all the cases where a plant metabolite present in the plant tissues is toxic to predators, or interferes with the normal growth and development of predators, or deters predators by its taste or smell (Norris and Kogan, 1980).

Many plant species have built up a formidable arsenal of defensive strategies against predators. However, in the wild a balance exists between plant hosts and their insect predators. Such a balance is achieved partly by having relatively low population densities of both hosts and predators, and by many plant species being able to defend themselves against most predator species. A typical situation is where a plant species contains defences effective against all but one, or a few, more or less specialised insect predators, which have evolved to overcome the plant's defensive strategies. The more effective the plant's defences against general predation, the greater the selection pressure for a highly specialised predator to evolve; indeed, the specialised predator may be able to utilise the plant's chemical defences for its own purposes, and may become completely dependent on its chosen host. Such specialised predation allows survival of a high diversity of both plant and insect species. Those plant species that are heavily defended against predation often contain highly toxic chemicals. Another strategy for the plant is to have less effective defences, to tolerate higher levels of general predation in order to avoid producing a specialised predator and to have a shorter life cycle; alternatively, certain parts of the plant (e.g. seeds) can be protected at the expense of other parts (e.g. leaves).

The balance between plant and predator achieved in natural ecosystems is very often not achieved under agricultural conditions. The threshold for 'economic damage' to a field crop, generally through a reduction in yield, may

Figure 4.1. Examples of insect pests of major agronomic importance. (A) *Pectinophora gossypiella* (cotton bollworm; Lepidoptera); larvae on damaged cotton boll. (B) *Diabrotica undecimpunctata* (corn root worm; Coleoptera); adult. (C) *Locusta migratoria* (migratory locust; Orthoptera); adult. (D) *Nilaparvata lugens* (brown rice plant hopper; Hemiptera); adults and nymphs on rice plant.

be lower than that which would seriously affect a natural ecosystem. More importantly, in the absence of effective intervention there tends to be a catastrophic build-up of insect pest populations to levels that may devastate the crops. Some insect pests of major agronomic importance are illustrated in Figure 4.1.

A combination of factors contributes to the ecological basis of major outbreaks of pest insects. Amongst these the following may be identified:

4.1.1 *Monoculture*

It has long been recognised that extensive cultivation of certain crops, to the exclusion of other plants, may favour drastic increases in the populations of insects which feed on those crops (Marchal, 1908). The number of herbivores exploiting a host plant species has been demonstrated to increase with the area occupied by that host (Strong, 1979). Although the build-up of pests to a specific host plant is generally reduced in diverse plant communities (Cromartie, 1981), it can actually be exacerbated by such agricultural practices as combining crops which act as hosts for a particular pest, e.g. cotton and maize, both of which are attacked by *Heliothis zea* (Pimental *et al.*, 1977).

A related situation occurs during storage of food crops. The accumulation of vast quantities of seeds means that any insect which can utilise them as food is almost certain to undergo a population explosion. Losses of crops in storage to insects can be very significant—for example, in West Africa up to 100% of the stored cowpea crop may be damaged within five months due to attack by the bruchid beetle *Callosobruchus maculatus* (Singh, 1978).

4.1.2 *Absence of inherent resistance*

The selection of plants as agricultural crops by primitive man was made largely on the basis of their edibility. This selection automatically excluded many species that were effectively defended against general predation by highly toxic compounds, since many defences that are effective against insect predation are also often effective against would-be mammalian predators. Many food crops are thus inherently deficient in defence mechanisms. An additional problem is introduced when crop plants are exposed to insect pests to which they have not had the opportunity to evolve any defence. Many crops are grown in locations where they are not native—for example, the introduction of the potato from Bolivia into the south-western United States exposed it to the Colorado beetle (*Leptinotarsa decemlineata*), towards which it had no inherent resistance. The Colorado beetle has subsequently established itself as the most serious pest of potato worldwide. In some instances pests have been introduced into new areas, often as a result of increasing world movement of commodities—for example, the severe losses of European grape vines in the nineteenth century by *Phylloxera* introduced from America.

In some crops, inherent resistance has been lost in new, 'improved' varieties of the crop as a result of a lack of emphasis on pest resistance in plant breeding programmes. In the recent past the prime goal of the plant breeder has been to increase yield, usually measured under ideal growing conditions, and this has frequently resulted in cultivars which have lost most of their natural resistance to pests. In a few cases the compounds involved in resistance to pests have been deliberately bred out, in order to improve the utilisation or ease of processing of the crop; for example, a cotton breeding programme was undertaken with the aim of producing glandless lines devoid of the complex phenolic, gossypol, which is toxic to mammals and prevented the utilisation of cotton seed meal as an animal feed. Unfortunately, the gossypol-lacking lines were vulnerable to attack by various insects, including the cotton budworm, *Heliothis virescens*, towards which gossypol is toxic, forming part of the wild-type plant's defences against these predators (Berardi and Goldblatt, 1980).

Table 4.1. Insecticide expenditure, by territory and pest order.

	Cotton					Maize			Rice			
	USSR	USA	China	India	Other	USA	Europe + USSR	Other	Japan	China	India	Other
Insecticide expenditure*	275	305	180	175	595	235	120	145	450	148	128	319
Application expenditure†	53	200	24	23	86	35	40	25	165	20	17	45
Total world cost			1916				600			1292		
Major pest order			Lepidoptera				Coleoptera			Homoptera		

Source: *Nat West 1988 expenditure, values given are $\$ \times 10^{-6}$.
†AGC Ltd (estimates), 1987 expenditure.

4.1.3 *Agrochemicals*

Since the earliest recorded uses of topically applied pesticides (nicotine, for control of aphids, was used as early as the eighteenth century) western agriculture has become increasingly dependent on chemical insecticides. The extent of this dependence in financial terms may be judged from expenditure on insecticides for the three major crops grown worldwide, shown in Table 4.1.

The widespread use of agrochemicals may actually aggravate problems with certain pest insects. The most obvious risk is that predator species which contribute significantly to the control of pest populations are at least as sensitive as the pests themselves to certain insecticides. The effect of pesticides may be to replace one pest species with another, as has happened in some Asian countries, where the brown rice hopper has been elevated from a minor pest of rice to a major one, after indiscriminate use of DDT had removed the previous major pest species (Heinrichs and Mockida, 1983). Certain other agrochemicals have side-effects on the physiology of crop plants, rendering them more susceptible to pests; for example, herbicide-treated maize plants have been shown to be infested with twice as many aphids, and nearly twice as many corn borers, as untreated plants (Oka and Pimental, 1976).

Although many features of modern agricultural practice actually contribute towards worsening the problem of insect attack on crop plants, even if sensible and ecologically sound practices were followed insect pests would still be a problem. To contain this problem it seems likely that both exogenously applied pesticides and improved inherent resistance mechanisms will be required. This review addresses itself to the latter area. Effective defensive mechanisms for agricultural crops against insect pests may be found either in wild types of the species, on the assumption that these mechanisms have been lost from, or bred out of, the cultivated lines, or in other plant species which are able to defend themselves adequately against the pest species. In the first case, conventional breeding techniques can often be used to transfer resistance into the agricultural crop; in the second case, genetic engineering methods are necessary to transfer resistance.

4.2 Defensive methods used by plants against insect attack

4.2.1 *Defensive mechanisms and plant breeding*

Physical defensive mechanisms employed by plants have been the subject of considerable study for use in conventional breeding programmes, where they have been successfully employed. For example, some varieties of bean plants possess hooked trichomes, effective in controlling certain leaf hoppers and aphids, which become impaled on them. Some species of wild potato show a similar mechanism of resistance, although in this case contact with the trichome by the aphid (*Myzus persicae*) causes it to rupture and release an exudate. On oxidation of the exudate, it sets into an insoluble black substance which hardens on the aphid's legs and immobilises it. This form of resistance is being exploited by transfer to modern cultivars (Simons, 1981).

Such mechanisms are in general not suitable for transfer between species, since they generally involve significant changes in phenotype, and even where a single genetic locus can be identified as responsible for the resistance the nature and mode of action of the gene and gene product are unknown. The 'transferability' of such resistance mechanisms is therefore not predictable, even if carrying out the transfer were practical.

Chemical resistance mechanisms have received less attention from conventional plant breeders. Partly this is a result of the poor state of knowledge of resistance mechanisms. Most, if not all, the types of secondary metabolites produced by plants have been implicated as being involved, in one case or another, in the resistance of plants to insect attack. In some cases, such as certain alkaloids, very clear evidence of the effectiveness of secondary metabolites as insecticides has been obtained, by extraction of the compounds and bioassay. In most cases, however, although the plant material may contain many secondary metabolites, these metabolites have only been studied in terms of their phytochemistry, and their role(s) remains a subject of speculation. In other cases plants contain compounds such as ecdysones (insect steroidal hormones) that can be predicted to interfere with insect growth and metabolism, although direct evidence for insecticidal properties may be lacking. The exploitation of chemical resistance mechanisms therefore requires that potential insecticidal compounds can be tested for effectiveness in some kind of bioassay. It is also necessary to be able to assay the compound, and possibly to purify it also; this has proved a further drawback in exploiting chemical resistance mechanisms in field-based breeding programmes.

Fundamental problems with the use of chemical resistance mechanisms for conventional breeding purposes are: lack of variation within a (plant) species, so that the insect pests of that species may have evolved to overcome the chemical defence, or even to use it for their own advantage; toxicity to humans or animals of the chemical defence; and the inability to transfer between species. Genetic engineering can be used to overcome the third of these problems, and thus to address the first.

4.2.2 *Insecticidal plant metabolites*

A classification of potentially useful plant insecticidal compounds may be made based on their chemical nature. Some examples of these compounds are summarised in Table 4.2.

4.2.2.1 *Terpenoids.* Although the volatile terpenoids responsible for scents and some flavours in plants may act as deterrents for insects, most herbivorous insects are attracted to host plants by these metabolites, and actually use them

Table 4.2. Examples of plant secondary compounds from legume seeds with demonstrated insecticidal properties.

Compounds	Bruchid	Legume	References
Non-protein antimetabolites			
Alkaloids			
2,5-dihydroxymethyl-3,4-dihydroxypyrrolidine (DMDP)	*Callosobruchus maculatus*	*Lonchocarpus*	Evans *et al.* (1985)
Castanospermine	*Callosobruchus maculatus*	*Castanospermum australe*	Nash *et al.* (1986)
Non-protein amino acids			
p-Aminophenylalanine	*Callosobruchus maculatus*	*Vigna*	Birch *et al.* (1986)
	Zabrotes subfasciatus		
Rotenoids	*Callosobruchus maculatus*	*Lonchocarpus salvadorensis*	Birch *et al.* (1989)
Saponins	*Callosobruchus chinensis*	*Phaseolus vulgaris*	Ishii (1952)
	Callosobruchus chinensis	*Glycine max*	Applebaum *et al.* (1965)
Polysaccharides			
Pectosans	*Callosobruchus chinensis*	*Phaseolus vulgaris*	Ishii (1952)
Heteropolysaccharide	*Callosobruchus chinensis*	*Phaseolus vulgaris*	Applebaum *et al.* (1972)
Heteropolysaccharide	*Acanthoscelides obtectus*	*Phaseolus vulgaris*	Gatehouse *et al.* (1987)
Polysaccharide (?)	*Callosobruchus chinensis*	*Vigna*	Kitamura *et al.* (1987)
Protein antimetabolites			
Lectins (phytohaemagglutinins)	*Callosobruchus maculatus*	*Phaseolus vulgaris*	Janzen *et al.* (1976)
	Callosobruchus maculatus	*Phaseolus vulgaris*	Gatehouse *et al.* (1984)
α-Amylase inhibitors	*Zabrotes subfasciatus*	*Phaseolus vulgaris*	Gatehouse *et al.* (1987)
	Callosobruchus chinensis	*Phaseolus vulgaris*	Ishimoto and Kitamura (1988)
Protease inhibitors	*Callosobruchus maculatus*	*Vigna unguiculata*	Gatehouse *et al.* (1979)
Arcelin/LLP	*Zabrotes subfasciatus*	*Phaseolus vulgaris*	Osborn *et al.* (1986)
	Zabrotes subfasciatus	*Phaseolus vulgaris*	Gatehouse *et al.* (1990)

for the synthesis of their own metabolites; for example, the cotton boll weevil, *Anthonomus grandis*, is attracted to the cotton plant by odiferous mono- and sesquiterpenes (a-pinene, limonene and caryophyllene), which are used to produce aggregation and mating pheromones. However, these terpenes may be used to 'warn off' other species which do not feed on cotton. Some terpenes do, however, act as insecticidal compounds by mimicking insect hormones, e.g. terpenoids present in balsam fir, preventing or interfering with the metamorphosis of larvae of *Pyrrhocoris apterus* into adults. Although this does not prevent initial damage from insect larvae, it does prevent insect populations building up, and thus provides an effective defence. More complex and modified terpenoids may be very effective insecticides, e.g. the pyrethrins.

4.2.2.2 *Steroids.* Many insects are dependent on the presence of certain steroids in their diet, since they are unable to convert saturated steroids to the unsaturated steroidal hormones used in control of moulting (ecdysones). Some plants contain insect steroid hormones, but there is little evidence that these compounds, when ingested, do in fact interfere with insect development. However, the toxicity of steroidal alkaloids and steroid-based cardiac glycosides is well established (Mann, 1987).

4.2.2.3 *Flavanoids.* Flavanoids in many plants play a primary role in attracting insect pollinators to flowers through their colouration, and thus provide a further link in the co-evolution of insects and plants. The insecticidal properties of flavanoids have received less study, and, despite the observation that the bitter taste of most flavanoids is likely to act as a feeding deterrent for predatory animals, few examples of toxicity of these compounds in bioassays have been reported. The toxicity of condensed tannins, derived from flavanoids, from legume seed coats towards some seed-eating beetles has been shown directly, as in the case of the tannins from *Vicia faba*, which inhibit the development of the bruchid *Callosobruchus maculatus* (Boughdad *et al.*, 1986). Despite the observation that anthocyanin pigments have beneficial effects towards mammals, they may be toxic towards insects. For example, cyanidin 3-glucoside has been shown to strongly inhibit larval growth of the tobacco budworm, *Heliothis virescens* (Hedin *et al.*, 1983). Tissue pigmentation, via tannins or anthocyanins, may thus be a plant defence mechanism that requires further study.

4.2.2.4 *Alkaloids.* Many alkaloids are very effective in deterring insect attack on plants containing them; for example, few insects will feed on plant tissues containing nicotine, or more toxic alkaloids such as strychnine. In some cases the insecticidal properties of the compounds have been demonstrated in bioassays. However, the very effectiveness of alkaloids as insecticides has placed a very strong selection pressure on specialised predatory insects to overcome the defence mechanism, so that pest species of alkaloid-containing plants are immune to the toxic effects; for example, the Danaid butterflies are unaffected by pyrrolizidine alkaloids in their food sources. Alkaloids are a very diverse group of molecules, and it is likely that different alkaloids have different toxic effects. One group of alkaloids whose mode of action has been investigated is the sugar analogues such as castanospermine and DMDP, (2,5-dihydroxymethyl-3,4-dihydroxypyrrolidine) which have been shown to be potent inhibitors of insect glycosidases (Evans *et al.*, 1985; Nash *et al.*, 1986).

4.2.2.5 *Non-protein amino acids.* Legume seeds are a rich source of non-protein amino acids, and whereas some of these, such as *S*-methylcysteine (*Vicia faba*), do not seem to have any protective role, others certainly do. Seeds of the Brazilian vine, *Dioclea megacarpa*, contain large amounts of the arginine analogue canavanine, which is toxic to most animals, including many potential insect predators. This amino acid analogue is used as a storage compound by the plant and is utilised on seed germination. A specialised predator, the bruchid beetle *Caryedes brasiliensis*, is not only unaffected by canavanine (being able to distinguish it from arginine and thus avoid the production of inactive proteins containing canavanine), but is also able to use it as a food source in a manner analogous to the plant metabolism; this is another example of strong selection pressures producing a highly efficiently adapted predator (Rosenthal *et al.*, 1976, 1978).

4.2.2.6 *Miscellaneous secondary metabolites.* Insecticidal properties have been demonstrated, or postulated, for a variety of other secondary metabolites in plants; some well-authenticated examples include the phenol gossypol in cotton, glucosinolates in *Brassica* species and cyanogenic glycosides, such as linamarin in clover. Examples of resistance to insect attack caused by heteropolysaccharides in seeds have also been reported (Applebaum and Guez, 1972; Gatehouse *et al.*, 1987).

4.2.2.7 *Proteins.* Plants contain a number of different types of protein with toxic or antimetabolic properties, some of which have been tested for their effects on insect development in bioassays.

(i) Protease inhibitors. The presence of inhibitors of mammalian digestive proteases in seeds, particularly those of legumes, has been known since 1938, when Read and Haas reported that an aqueous extract of soyabean flour inhibited the ability of trypsin to liquefy gelatin (Read and Haas, 1938). The characterisation of these inhibitors as proteins was incontrovertibly demonstrated by the isolation of crystalline trypsin inhibitor from soya beans by Kunitz in 1945. As has become clear, protein protease inhibitors are remarkably widespread in plants (see the review by Richardson, 1977), although they have been most extensively studied in the Leguminosae, Graminae and Solanaceae, presumably because of the large number of species in these families which are important food crops (Richardson, 1981).

The Kunitz type of inhibitor is less common than the low molecular weight, high-cysteine type of inhibitor, sometimes referred to as Bowman–Birk inhibitors, which form a family of homologous proteins of particularly wide distribution. Other types of low-M_r inhibitor are known. Common properties of protease inhibitors are their stability to heat and extremes of pH and their resistance to proteolysis; in the case of the Bowman–Birk inhibitors, this is achieved by a high degree of cysteine cross-linking in the polypeptide chain. The mechanism of enzyme inhibition is usually through the formation of a very strong complex containing a covalent bond between an active site residue on the enzyme and the inhibitor; in the case of the Bowman–Birk inhibitors, one inhibitor molecule usually contains two reactive sites, which may be directed against the same protease or different proteases. The majority of inhibitors isolated are active against bovine trypsin, although inhibitors active against chymotrypsin, elastase, subtilisin, and other proteases have been reported.

Increasing attention is being paid to investigating the action of protease inhibitors on insect proteases, since, in some cases at least, the insect proteases are not affected by inhibitors of bovine trypsin. Interest in the effects of plant proteinase inhibitors on insects was aroused as early as 1947, when Mickel and Standish observed that larvae of certain pests were unable to develop normally on soya bean products (Mickel and Standish, 1947). Lipke *et al.* (1954) subsequently studied the toxicity of soya bean trypsin inhibitors on development of *Tribolium confusum* (flour beetle), a common pest of stored grain. Although the soya bean trypsin inhibitor gave negative results in their bioassay, the presence of a specific inhibitor of *Tribolium* larval proteases was revealed. This inhibitor was later isolated and shown to inhibit completely larval gut proteolysis of both *T. confusum* and *T. castaneum* (Birk *et al.*, 1963) but to be inactive towards bovine trypsin and chymotrypsin (Applebaum and Konijn, 1966). As a further illustration that each case must be considered individually, a midgut protease from mealworm (*Tenebrio molitor*) was inhibited by a wide range of trypsin inhibitors tested, including soya bean trypsin inhibitor, lima bean trypsin inhibitor and ovomucoid. A systematic study of a wide range of different protease inhibitors has been made in an attempt to find a kinetic parameter which would be useful in predicting the potential for any inhibitor to act as a resistance factor to the larvae of grass grubs (*Costelytra zealandica*), the major insect pest in New Zealand (Christeller and Shaw, 1989); the dissociation constant of the inhibitor:protease (i.e. grass grub gut protease) complex was suggested as such a parameter.

The antimetabolic nature of plant protease inhibitors towards insect development has been shown directly by bioassay in artificial diet. For example, Steffens *et al.* (1978) demonstrated that addition of soya bean trypsin inhibitor at levels of 2–5% in the diet of corn borer (*Ostrinia nubilalis*) larvae inhibited larval growth and delayed pupation, whereas comparable levels of maize trypsin inhibitors had no effect, suggesting a protective role for the legume trypsin inhibitors. An example of the toxic effects of a host's trypsin inhibitor on a pest species of that plant was provided by A.M.R. Gatehouse *et al.* (1979), who showed that elevated levels of the cowpea trypsin inhibitor were insecticidal to developing larvae of the bruchid, *Callosobruchus maculatus*; this example is considered in more detail in a later section. More recently, Broadway and Duffy (1986) not only demonstrated *in vitro* inhibition of larval gut proteolysis in the Lepidopterans *Heliothis zea* and *Spodoptera exigua* by soya bean trypsin inhibitor and potato proteinase inhibitor II, but also showed these proteins to be toxic *in vivo*.

Additional evidence for the protective role of proteinase inhibitors has come from the experiments of Ryan and co-workers (e.g. Green and Ryan, 1972). Mechanical damage, such as insect attack, on leaves of a number of plant species (tomato and potato being the best studied) induces the synthesis of protease inhibitors, not only in the attacked leaf but throughout the plant. A factor, or wound hormone, called proteinase inhibitor-inducing factor (subsequently characterised as an oligosaccharide) was found to be released from the damaged leaves and was transported to other leaves where it induced synthesis and accumulation of inhibitors. Within 2–3 days of the first damage, proteinase inhibitors could account for over 1% of the total soluble proteins of leaves, and they remained in the leaves for long periods, stored in the central vacuoles, presumably as a defence against repeated attack. This forms a particularly useful system to study signalling interactions between parts of the plant, and demonstrates an active response of the plant to insect attack. The reader is referred to the excellent reviews by Ryan for further information (Ryan, 1983, 1985).

The mechanism of antimetabolic action of protease inhibitors is not yet completely clear; direct inhibition of digestive enzymes is unlikely to be the only effect, and possibly a situation analogous to their effect on mammals, where the major deleterious effect is loss of nutrients through pancreatic hypertrophy and overproduction of digestive enzymes, holds in insects. Alternatively, the protease inhibitors may affect other proteases vital for insect development, such as those involved in moulting.

(ii) Inhibitors of other digestive enzymes. Protein inhibitors of mammalian α-amylases have been purified and characterised from many plant seeds; they are abundant in cereal grains, and are also found in legume and other seeds (Garcia-Olmedo *et al.*, 1987). These proteins belong to a number of sequence families, but are often of relatively low molecular weight. Some amylase inhibitors are homologous to the Bowman–Birk-type protease inhibitors, and bifunctional trypsin–amylase inhibitors of this type are known (Campos and Richardson, 1983). As for protease inhibitors, most characterisation of amylase inhibitors has been done using mammalian enzymes (salivary or pancreatic amylases) as assay material, and comparatively little work has been carried out with insect amylases. Where such work has been carried out, it has often been the case that inhibitors that are active against mammalian amylases are ineffective against their insect counterparts, and vice versa.

Amylase inhibitors from *Phaseolus vulgaris* and wheat have been tested for antimetabolic effects in artificial diet; in both cases toxicity towards some insect pests was obtained (A.M.R. Gatehouse *et al.*, 1986; Ishimoto and Kitamura, 1988). The importance of testing the insecticidal properties of potential resistance factors in bioassay is illustrated by the finding that, although an α-amylase inhibitor from wheat was found to be equally effective in inhibiting gut α-amylase activities from a pest and non-pest species *in vitro*, feeding trials demonstrated it to be innocuous towards the pest species but toxic to the non-pest species (A.M.R. Gatehouse *et al.*, 1986). Unlike protease inhibitors, induced synthesis of amylase inhibitors by insect attack has not been observed, and the precise role of these proteins remains open to doubt.

(iii) Lectins. Lectins are carbohydrate-binding proteins found in many plant tissues, but are often present in relatively large amounts (usually approximately 1% of total protein, but in some species, e.g. *Phaseolus vulgaris*, up to 30%) in seeds and other storage tissues. Many lectins belong to a homologous family of proteins based on an amino acid chain of approximately 220 residues, typified by soya bean lectin (Lotan *et al.*, 1975), although totally different sequence types have also been shown to have similar functional properties, e.g. the potato lectin, which is a proline-rich protein unrelated in sequence to the typical legume lectin. Lectins clearly have more than one role in the plant, since the hypothesis of Hamblin and Kent (1973) that they mediate binding of *Rhizobia* to legume roots as a first step in the nodulation process has recently been directly demonstrated in transgenic clover plants (Diaz *et al.*, 1989). An involvment of these proteins in resistance to insect attack was first put forward by Janzen *et al.* (1976), who showed that the lectin from *Phaseolus vulgaris* was toxic to developing larvae of the bruchid beetle *Callosobruchus maculatus.* The toxicity of some lectins towards mammals (Evans *et al.*, 1973) and birds (Jayne-Williams and Burgess, 1974) has been well documented, and the studies of Pustzai and co-workers (Pusztai *et al.*, 1979; King *et al.*, 1980a, b) have shown that the toxic effects are consequent on the binding of the protein to glycoprotein receptors on the gut microvilli. Toxicity towards insects may occur through an analogous process, since A.M.R. Gatehouse *et al.* (1984) have shown that the lectins of *Phaseolus vulgaris* will bind to the midgut epithelial cells of *Callosobruchus maculatus*; they do not bind to the gut wall of the related bruchid *Acanthoscelides obtectus*, which develops successfully on *P. vulgaris* seeds (A.M.R. Gatehouse *et al.*, 1989). The insecticidal properties of another legume seed lectin, from soya bean, have been demonstrated by bioassay against developing larvae of the Lepidopteran, *Manduca sexta* (Shukle and Murdoch, 1983). The differing sugar specificities of different lectins, and their differing susceptibility to gut proteolysis, makes prediction of the toxic effects of any given lectin difficult, but many lectins have now been bioassayed and shown to be effective antimetabolites to at least one insect species (unpublished data).

(iv) Ribosome-inactivating proteins. The most well-known ribosome-inactivating protein (RIP) is the highly toxic ricin, found in castor bean seeds;

this exemplifies the B-type toxins which contain a lectin chain linked to a polypeptide that inactivates ribosomes by hydrolysing the sugar–base linkage at one specific position in the ribosomal RNA. The A-type toxins contain the ribosome-inactivating activity only, and are exemplified by saporin, present in tissues of *Saponaria* spp. Little evidence of the toxicity of these proteins towards insects has been reported, although a recent paper (A.M.R. Gatehouse *et al.*, 1990) shows that both ricin and saporin are highly toxic towards Coleoptera such as boll weevil and bruchid beetles; interestingly, the toxins were ineffective towards the Lepidopteran species tested. RIPs are also toxic towards Orthoptera such as locusts (unpublished results). Involvement of these proteins in insect resistance in the field is, however, largely conjectural.

4.3 Insecticidal compounds from other sources

4.3.1 *Chemical insecticides*

Although a full consideration of chemical insecticides is outside the scope of this review, the extensive, and necessary, use of chemical insecticides must be considered in relation to the potential utility of plants carrying inherent resistance.

The earliest exogenously applied insecticidal compounds tended to be derivatives of insect-resistant plants such as pyrethrum, rotenone, nicotine or neem oil, or to be simple chemicals such as sulphur or metallic salts of copper, lead or mercury. The nature of chemical crop protection was radically altered in the 1940s with the introduction of organochlorine insecticides like DDT. These compounds, which were extremely effective and highly persistent, had a revolutionary impact on agricultural practices which has persisted to the present day; usage of chemical insecticides has increased exponentially since the introduction of DDT (see Metcalf, 1986). Despite the undoubted success in the short term of chemical insecticides, a number of problems with this approach soon became apparent. These include:

(1) Toxicity (to non-target organisms, such as pollinating insects, and the natural predators of pest species, and to humans).
(2) Environmental spread (it has been estimated that more than 99.9% of exogenously applied insecticide is wasted, contributing to the problems outlined in (1), and representing a substantial economic cost).
(3) Loss of effectiveness (build-up of populations of pest species resistant to one or more insecticides has been observed to occur in comparatively few generations, due to the high selective pressure imposed on the target population by the toxicity and heavy application of insecticides).
(4) Consumer pressure (the increasing popularity of 'organic' produce in the developed countries, and public concern over pesticide residues in foodstuffs, is making non-use of chemical insecticides less economically unfavourable).

Although there can be no question that chemical insecticides will continue to be extensively used in agriculture, and that improved insecticides that overcome some of the problems listed above will be developed, there is a clear need to develop alternative strategies of plant protection. Such strategies will be considered as assisting, rather than replacing, protection by exogenously applied chemicals. They will allow pesticides to be applied in lower, more economical, dosages, and make the use of less toxic, more ecologically 'safe' pesticides a viable proposition.

4.3.2 *Bacterial toxins*

Although insects are subject to attack by bacterial parasites, the example of a bacterial toxin with insecticidal properties that has received the most attention is the crystal protein produced by strains of the soil bacterium, *Bacillus thuringiensis*. During sporulation, *B. thuringiensis* produces intracellular crystals, which, under the alkaline conditions of the midgut of some insects, are hydrolysed, releasing proteins of 65 000–160 000 M_r; further proteolytic processing of these proteins releases smaller toxic fragments. The mechanism of toxicity seems to depend on disruption of the membranes of cells lining the gut, and this action is only carried out by the proteolytically processed fragments; the intact proteins are not toxic (Sacchi *et al.*, 1986). The toxicity of such compounds towards humans, or other mammals, is theoretically very low, since the conditions necessary for the production of the active toxic fragments are not present in mammalian guts.

The *B. thuringiensis* (Bt) toxins in various formulations (usually as whole, sporulated bacteria) have been employed as insecticidal crop sprays for more than 20 years (Dulmage, 1981), but their use is limited by high costs of production and the instability of the crystal protein under field spraying and exposure conditions. A further drawback to the use of these proteins is their specificity; different strains of *B. thuringiensis* produce different toxins, which are effective against a limited range of insect species. Commercial Bt toxins effective (in total) against more than 50 Lepidopteran pest species have been identified, but comparatively few toxins with activity against Coleopterans have been isolated. To provide protection against a wide range of pests may therefore involve spraying with a mixture of many different strains of *B. thuringiensis*. Despite these drawbacks, the use of Bt toxins as insecticides seems likely to increase, as their specificity and lack of persistence are advantageous when environmental factors are taken into account; research in extending the range and utility of these compounds is being actively pursued (Gaertner, 1989).

The use of toxins from other sources, such as fungi, insects or higher animals, would seem to offer possibilities in the control of insect pests. However, no examples of their use are available because of the practical difficulties of obtaining enough of the compounds to carry out trials.

4.3.3 *Biological control*

The use of predatory or parasitic insects, nematodes and fungi to control insect pests is well established in horticulture, and is an effective means of control in confined areas. To be effective, however, such control requires that a population of the pest has built up, and that the biological control organism is not itself competed out; success in the field is therefore less easy to attain. The reader is referred to reviews such as those of Hill (1983, 1987) for further information.

4.4 Constraints on the genetic engineering of plants for insect resistance

The requirements for producing fully viable insect-resistant transgenic plants may be listed as follows:

(1) The compounds which confer insect resistance, and are intended to be produced in transgenic plants, should be the products of single genes.
(2) The genes to be transferred should have been identified, and their isolation should be readily accomplished.
(3) Transfer methods for the target plant species should be developed, including the availability of suitable vector systems, suitable tissue culture techniques to allow the production of fertile plants and an efficient screening system to allow the selection of transformed plants.
(4) Expression of the transferred gene in the target plant at a level suitable to confer resistance in the desired tissues should be attainable.
(5) The transferred gene should not render the target plant toxic to humans or animals or make it unsuitable for human use.
(6) There should be no yield penalty, i.e. the transferred gene should not have adverse effects on the metabolism of the target plant.
(7) The transgenic plant should show effective resistance to insect pests under a variety of circumstances, including those of the field.
(8) Resistance should be stably inherited in offspring of the transgenic plant.

Perhaps surprisingly, these stringent requirements can be met, as will be detailed in Section (4.5), but it is also worth considering approaches that fail to meet the criteria.

While the production of plants capable of synthesising chemical insecticides is clearly not feasible, as enzymes capable of carrying out the necessary reactions are not known, the possibility of transferring highly effective natural insecticidal compounds such as alkaloids between species is less remote. However, these compounds are the results of multistep biochemical pathways which are often not fully characterised; in any case, the enzymes that carry out the steps of the pathway are often unknown or unpurified. There is thus not only a requirement to transfer multiple genes, which will be difficult to obtain, but the transferred genes will have to be regulated in the transgenic plant in

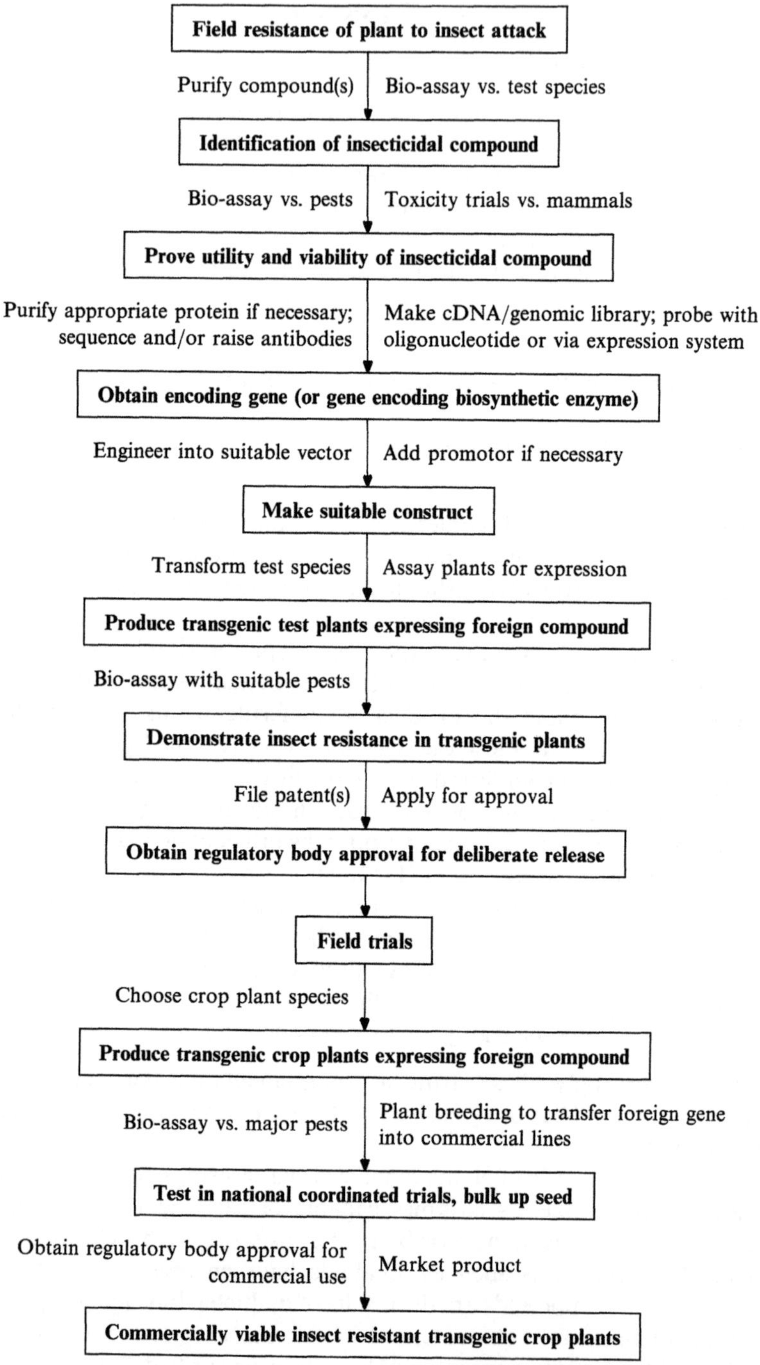

Figure 4.2. Schematic diagram of the stages involved in the production of commercially viable insect-resistant transgenic plants.

such a way that the pathway 'works'. This is well beyond the capability of present techniques. The only possibility which could be feasible would be if the target species lacked only one enzyme in the pathway to make an effective insecticide; it would then be possible to add the missing step from a species that possessed it. An example of this process has been suggested for white clover, which synthesises a precursor of the isoflavanoid phaseollin, an effective insect deterrent; addition of the gene(s) responsible for the final step in the pathway should yield insect resistance (Sutherland, personal communication). However, most promising insecticidal compounds are ruled out for genetic engineering purposes, and the case for using insecticidal proteins, which are direct products of single genes, is made very much stronger.

Not all the potentially useful insecticidal proteins from plants, or other sources, can be considered as candidates for the production of transgenic plants. For example, ribosome-inactivating proteins are too toxic to humans to make their use in crops legislatively acceptable. They may also prove toxic to the transgenic plant itself. Other proteins may be incompatible with the protein biosynthesis system in plants, due to differing glycosylation, targeting, folding and stability requirements (see Chapter 7); this may make the expression of active proteins from non-plant sources difficult, or may inflict yield penalties on the transgenic plant. Achieving levels of expression necessary to cause toxicity does not pose such problems, as promoters from plant, or plant viral, genes actively expressed constitutively (e.g. CaMV promoters) or in a tissue-specific manner (e.g. storage protein genes) are available. However, consideration of codon usage and messenger RNA stability may still be necessary to get sufficient levels of expression of non-plant genes.

While gene transfer technology via *Agrobacterium tumefaciens*-based transfer systems is well established for a limited range of Solanaceae species and some *Brassica* sp., gene transfer to many crop plants is either difficult or impossible. These drawbacks are, however, rapidly being overcome (see Chapter 2), and transformation systems for most crops should soon be established procedures, if still technically difficult. Similarly, rapid advances are being made in the development of systems for the selection and screening of transformants.

An overall scheme for the production of commercially viable insect-resistant transgenic plants is given in Figure 4.2.

4.5 The production of insect-resistant transgenic plants: two case studies

4.5.1 *Transgenic plants expressing Bt toxins*

The use of preparations of *Bacillus thuringiensis* to control Lepidopteran pests in the field has been limited by the high costs of production and the carefully timed, repeated applications necessitated by its instability under field conditions. The potential for increasing the utility of Bt by the genetic engineering of plants to express the Bt toxin was recognised at an early stage; three principal publications report successful attempts to achieve this aim, and show that enhanced resistance to certain insect pests results (Barton *et al.*, 1987; Fischhoff *et al.*, 1987; Vaeck *et al.*, 1987).

The insecticidal properties of sporulating Bt reside in the presence of the δ-endotoxin polypeptides in parasporal crystals (Angus, 1954). These polypeptides vary in structure between different varieties of Bt, and in some cases more than one type may be present within a spore. Each toxin is, however, a single polypeptide, and is thus the product of a single gene. The differences between the polypeptides account for the specificities towards different groups of insects observed between different varieties of Bt.

Bt toxins have been readily isolated and purified, largely on the basis of the unusually low solubility of the crystal protein. Polyclonal and monoclonal antibodies have been raised against the purified proteins, facilitating the identification of the toxins in transgenic organisms. The identification and cloning of the toxin genes was made easier by the observation that in most, though not all, cases the gene resides on a plasmid. It was thus possible to isolate the plasmid from bacterial cultures and to clone restriction fragments into an *E. coli* expression system. *E. coli* expressing the Bt toxin were identified using the antitoxin antibodies. The Bt toxins were expressed at a high level in these engineered *E. coli*, and the toxin produced by them was purified and shown to be as toxic to susceptible insect species as the native protein.

In the case of the Lepidopteran-specific Bt toxins from Bt varieties 'Berliner' and 'Kurstaki', the toxin is produced as a 130 000 M_r precursor, which is cleaved in the insect gut by proteolytic enzymes to produce a smaller (approximately 67 000 M_r) toxic fragment. The cleavage is necessary for toxicity, and only occurs efficiently under the alkaline pH conditions of the Lepidopteran gut. Since only part of the complete Bt toxin gene is necessary for toxicity (corresponding to the *N*-terminal end of the precursor protoxin), various deletions within the cloned Bt toxin gene were constructed. These are referred to as truncated Bt toxins.

Bt toxin-encoding sequences were transferred to plant expression vectors such that the coding sequence was brought under the control of a constitutive plant promoter—either the mannopine synthase gene promoter of plasmid pTiA6 (Velten *et al.*, 1984) or the cauliflower mosaic virus (CaMV) 35S gene promoter. The constructs also included a plant transcription termination/polyadenylation signal. The expression vectors included a chimeric neomycin phosphotransferase (NPT II) gene, with a nopaline synthase promoter, to allow selection of transformed plant tissue on kanamycin. An example of the genetic engineering of a truncated Bt toxin expression vector is shown in Figure 4.3. The constructs were introduced into tomato and tobacco plants via conventional *Agrobacterium* Ti plasmid-mediated gene transfer.

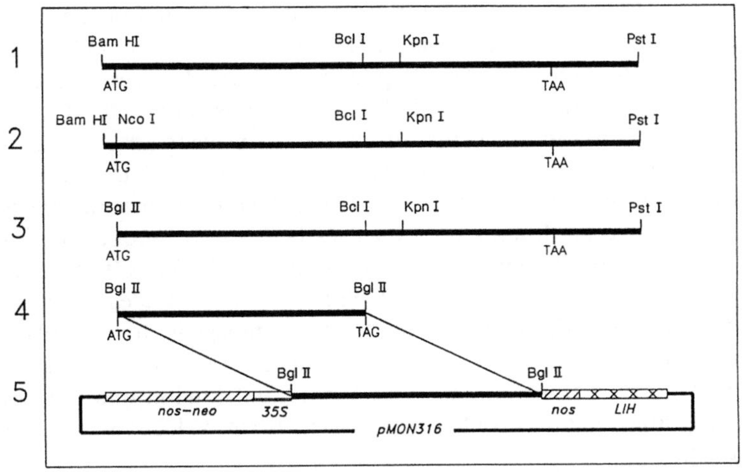

Figure 4.3. Genetic manipulations for the production of a truncated Bt toxin plant expression cassette (Fischhoff *et al.*, 1987). The Bt toxin cDNA plasmid pMAP 4 (1) was subjected to site-directed mutagenesis to introduce an Nco I site adjacent to the translation initiation codon to give pMAP 17 (2). Digestion with Nco I and addition of a synthetic linker containing a Bgl II site completed the manipulations of the 5′ end, giving pMON 294 (3). The C-terminal coding sequences of the toxin were removed by digestion with Bcl I, and an oligonucleotide linker containing a new translation–termination codon and a Bgl II site was added to the 3′ end (4). The resulting Bgl II fragment was ligated into the Bgl II site of the plant expression vector pMON 316 to yield the final construct pMON 9713 (5).

Plants transformed with constructs containing the entire Bt protoxin-coding sequence expressed Bt toxin at very low levels, and the presence of this gene construct appeared to have adverse effects on the plants. Levels of Bt toxin expression were higher, though still very low, in plants transformed with constructs containing truncated Bt toxin-coding sequences; the highest reported level of expression was only approximately 0.02% of total leaf soluble protein. Within any given transformant, NPT II expression was at least an order of magnitude greater than that of the Bt toxin, even when the latter was expressed from the CaMV 35S promoter, which is expected to be about 30 times more active than the *nos* promoter used to drive NPT II synthesis.

Despite these low levels of expression, many of the plants were shown to be insecticidal towards the test insect *Manduca sexta*, the tobacco hornworm. On many plants 100% mortality was recorded within a few days; feeding damage to the plants was drastically reduced in comparison with the untransformed controls, which were reduced to a stalk. These results highlight the very high toxicity which Bt toxins may have towards susceptible insects. *Manduca* was used in these initial trials since, besides being a convenient insect for laboratory studies, it is particularly sensitive to the strains of Bt toxin used. Noctuid Lepidopterans, which constitute a very serious group of insect pests and are responsible for a high use of insecticides, are far less sensitive to Bt toxins. Tomato plants expressing truncated Bt toxin were tested against two species of noctuid pest, *Heliothis virescens*, the tobacco budworm, and *Heliothis zea*, the corn earworm. Despite the low level of expression, resistance to these caterpillars was enhanced over controls, although the level of resistance was low, and probably insufficient to provide adequate control.

In addition to damaging the foliage of tomato plants, *Heliothis* species cause extensive damage by boring into fruits, rendering them unmarketable. Newly hatched larvae begin feeding on the leaves before moving onto the fruit, so foliar expression of Bt might be sufficient to protect the fruits. The CaMV 35S gene promoter is, however, active in all plant tissues, and there is evidence that Bt toxin activity is detectable in the fruits of transgenic tomato plants. *H. virescens* larvae fed on these fruits gained weight at only half the rate of those fed on control (non-transformed) fruits. A major outstanding question remains how acceptable tomatoes containing bacterial toxins will be to their intended consumers.

Various attempts have been, and are being, directed towards increasing the level of expression of Bt toxins in transgenic plants, and thereby potentially increasing the range of insects controlled. Chimeric genes producing a truncated Bt toxin-coding sequence fused to the selectable marker protein NPT II have been introduced into tobacco plants. With these constructs the level of Bt toxin produced, and the insecticidal effectiveness of the plants, was proportional to the level of kanamycin resistance of the transformant. It might therefore be possible to select plants expressing high levels of Bt toxin directly, by exposure to high levels of kanamycin. Unfortunately, the fusion of the Bt toxin sequences to NPT II reduces the level of NPT II expression, so the strategy is not successful. Alternative approaches to increasing the level of Bt toxin expression in transgenic plants, including the use of sequences encoding only the minimum Bt sequence necessary for toxicity, use of enhanced promoters and modifications to the codon usage, have apparently resulted in transgenic plants expressing elevated levels of Bt toxin (Fuchs *et al.*, 1989).

A number of field trials have been conducted with Bt toxin-expressing transgenic plants. Preliminary results with transgenic tomatoes showed a reduction in damage to tomato fruits by *Heliothis zea*. The efficacy of the genetic engineering of these plants was most dramatically demonstrated by a field trial in Florida, when there was a heavy infestation with tomato pinworms (*Keiferia lycopersicella*). Pinworm is a major pest of tomatoes in Florida, California and Mexico, and the absence of an adequate control regimen renders some regions unsuitable for tomato production. The pinworm larvae mine leaves, stems and fruits and are thus protected from topically applied insecticides. However, the presence of Bt toxin in all the tissues of the transgenic tomatoes exposed the insects to this control agent, and the transgenic plants were substantially protected in conditions under which controls were completely defoliated. (Fischhoff, 1989).

The performance of genetically engineered plants expressing Bt toxins in field trials has been very encouraging, although up to the present these trials have, necessarily, been of a limited nature. Large-scale, multisite testing will be necessary for a final assessment of the likely performance of such transgenic plants under normal field conditions.

4.5.2 *Transgenic plants expressing protease inhibitors*

The role of protease inhibitors in legume seeds in conferring resistance to seed-eating bruchid larvae has been, and still is, a disputed area. Although early reports that bruchids did not possess protease activity (Applebaum, 1964) have been shown to be incorrect (A.M.R. Gatehouse *et al.*, 1985; Kitch and Murdoch, 1986), it is the case that any wide-scale correlation of trypsin inhibitor content of seeds with their resistance to insect attack cannot be made. (The failure to obtain this correlation is in the main due to two factors: the common occurrence of multiple mechanisms of resistance in the seeds [see earlier] and the failure to assay inhibitory effects on insect proteases.) Nevertheless, the synthesis of protease inhibitors in the leaves of a number of plant species, including tomato, in response to insect attack clearly suggests that these compounds are of functional significance in some plant–insect interactions (Brown *et al.*, 1985).

In an attempt to explain the enhanced resistance of a collected line of cowpea (*Vigna unguiculata*) to the major storage pest of its seeds, the bruchid beetle *Callosobruchus maculatus*, A.M.R. Gatehouse *et al.* (1979) screened both resistant and susceptible seeds for a variety of antimetabolic compounds. The only significant difference they observed between seeds of the one resistant line and seeds from a selection of susceptible lines was that the resistant seeds contained approximately twice as much trypsin inhibitory activity per gram of seed material as the susceptible. This difference was considered to be a quantitative, rather than qualitative, difference between the cowpea lines, since no significant differences in the isoinhibitor band patterns were observed on isoelectric focusing. Purified cowpea trypsin inhibitor from a susceptible line was then tested by bioassay in artificial diet for its effects on larval development in *C. maculatus*. At low levels (0.1% w/w) the inhibitor had no effects on larval survival, but, at levels comparable to that found in the resistant line (0.8%), no larvae survived. It was suggested that the trypsin inhibitor level in the resistant line was responsible for its resistance to insect attack.

Subsequent work has shown that the cowpea trypsin inhibitor (CpTI) is not the only factor involved in seed resistance to bruchid attack in this example, but that multiple mechanisms of resistance operate; amongst other evidence, it was clear from an examination of progeny resulting from a breeding programme using the resistant line as one parent that the correlation of high inhibitor levels with resistance was not always maintained (Redden *et al.*, 1983). This result was partly explained by subsequent observations that the antimetabolic effects produced by CpTI in artificial diet could be overcome by supplementing the diet with the limiting amino acids (A.M.R. Gatehouse and Boulter, 1983), and that the effect of CpTI on the bruchid gut protease, an acidic thiol protease rather than an alkaline trypsin-like protease, was to give only partial inhibition (A.M.R. Gatehouse *et al.*, 1985); in this case, the effects of CpTI would be dependent on other components present in the seed or diet, and on other factors affecting larval metabolism. On the basis of a co-evolutionary relationship between the cowpea and its predator, it is perhaps not surprising that the bruchid should have evolved to attempt to evade the major defensive mechanism of its host.

CpTI, nevertheless, clearly played a major role in the resistance of a line of cowpea seeds to their insect predator, and it was therefore of interest to determine whether it was effective against other insect species. The purified protein was tested in artificial diets against a wide range of both field and storage insect pests, including members of the Lepidoptera and Coleoptera. As shown in Table 4.3, many insect pests of economic importance, including *Heliothis* spp., *Anthonomus grandis* (cotton boll weevil) and *Diabrotica* spp. (corn root worm), are affected by this antimetabolite. In bioassays against a limited range of insects, CpTI was shown to be more effective as an antimetabolite than certain other legume trypsin inhibitors (A.M.R. Gatehouse and Boulter, 1983). Despite the toxicity of CpTI to insects and its effective inhibition of bovine trypsin, cowpea seeds are not toxic to humans and can be eaten raw (Peterson, 1984). Feeding trials have shown that the inhibitor in untreated cowpea meal did not adversely affect the growth of rats (un-

Table 4.3. Insecticidal efficacy of cowpea trypsin inhibitors (CpTI).

Insects killed by CpTI in artificial diets	CpTI transgenic tobacco plants resistant to the following insects
Lepidoptera	
Heliothis virescens	*Heliothis virescens*
Heliothis zea	*Heliothis zea*
Spodoptera littoralis	*Spodoptera littoralis*
Chilo partellus	
	Manduca sexta
	Autographa gamma
Coleoptera	
Callosobruchus maculatus	*
Anthonomus grandis	*
Diabrotica undecimpunctata	*
Tribolium confusum	*
Costelytra zealandica	*
Orthoptera	
Locusta migratoria	*

*Insects unable to attack control tobacco plants.

published data). This protein was thus judged to be a suitable candidate for transfer to other species via genetic engineering.

CpTI contains a number of isoinhibitors, which can be divided into trypsin–trypsin and trypsin–chymotrypsin double-headed types. The proteins are all small polypeptides of about 80 amino acids, and belong to the Bowman–Birk inhibitor family. They are encoded by a repetitive gene family on the cowpea genome. A full-length coding sequence for a CpTI protein was obtained by screening a cDNA library prepared from poly(A)$^+$ RNA isolated from developing cowpea seeds. (A standard commercial cowpea line was used; see above.) An oligonucleotide derived from the amino acid sequence of the protein was used as a probe for the library. The clone isolated, pUSSRc3/2, contained a 66 amino acid putative 'leader' sequence, and most of the 3′ untranslated sequence of the mRNA, as well as the mature protein-coding region of a trypsin–trypsin double-headed inhibitor (Hilder *et al.*, 1989).

The CpTI sequence was placed under the control of a CaMV 35S promoter in the final construct produced for transfer to plants (shown in Figure 4.4; Hilder *et al.*, 1987). The construct employed the *Agrobacterium tumefaciens* Ti plasmid binary vector pROK 2 (Baulcombe *et al.*, 1986); a terminator from the *nos* gene was placed 3′ to the coding sequence, and the construct also contained a *nos-neo* gene to allow transformants to be selected on the basis of kanamycin resistance. The vector was mobilised into *Agrobacterium*, and the bacteria were used to infect tobacco leaf discs, by standard protocols; subsequent production of rooted plants after selection of regenerating shoots on kanamycin-containing media also followed normal procedures. By taking cuttings from the original transformants, and rooting them, numbers of clonal plants sufficient for insect bioassay could be produced from each of the original transformants. The transformed plants were shown to be expressing CpTI in the leaves at levels varying from undetectable to approximately 1% of total protein by a dot blot immunoassay, i.e. in the range of values expected for the CaMV promoter. In control plants transformed with a construct where the coding sequence of CpTI had been inserted in the incorrect (i.e. 3′–5′) direction relative to the CaMV promoter, no expression was detected. The expression of CpTI was confirmed in the 'correct' transformants by western blotting, and by a direct *in vitro* assay for inhibition of bovine trypsin. The former technique showed that tobacco was capable of processing the precursor CpTI polypeptide encoded by the inserted coding sequence to a polypeptide resembling native CpTI on SDS–PAGE; other plant proteins have been shown to be correctly processed in transgenic tobacco plants (Ellis *et al.*, 1988). The latter technique showed that the CpTI synthesised in transgenic tobacco possessed its normal functional integrity. The lack of complication in obtaining high levels of expression of functional CpTI in tobacco illustrates the advantage of expressing plant proteins in transgenic plants; there are no problems with codon usage, mRNA stability, protein processing, etc. that are liable to occur if attempts to express proteins from other kingdoms are made. It also shows that

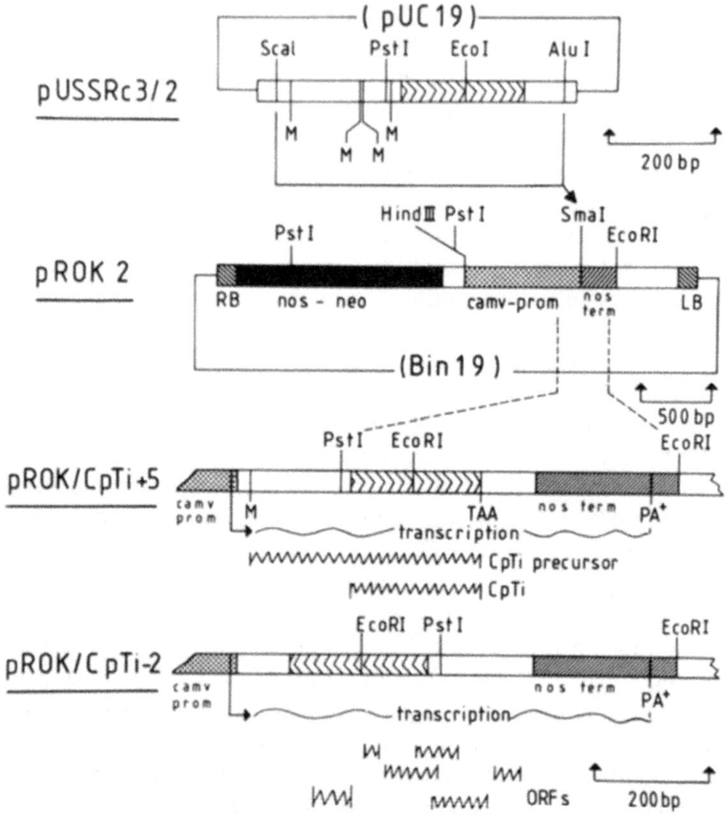

Figure 4.4 Construction of a CpTI expression vector for plant transformation (Hilder *et al.*, 1987). The CpTI cDNA pUSSRc3/2, containing a complete mature CpTI-coding sequence (indicated as > > >; in frame, initiator codons are marked M), was restricted with Alu I and Sca I and ligated into the Sma I site of the expression vector pROK 2. Clones with the coding sequence in the correct orientation relative to the promoter (pROK/CpTI + 5) and in the incorrect orientation (pROK/CpTI-2) were generated. Transcripts generated by the clone with the CpTI-coding sequence in the correct orientation will be translated to produce a CpTI precursor polypeptide; transcripts from the clone with the CpTI-coding sequence in the incorrect orientation contain six short open reading frames.

although much higher levels of CpTI expression (approximately 1% of total protein) are required for protection compared to Bt toxin, the necessary level of expression is easily obtained.

Bioassay of clones of selected transformants was initially carried out by infesting plants with first instar larvae of the Lepidopteran pest, *Heliothis virescens* (tobacco budworm), a serious pest of tobacco, cotton and maize. A high degree of protection against insect attack was observed, with CpTI-expressing plants showing only minor damage, whereas control transformed plants containing the 'reverse' CpTI construct and not expressing CpTI were

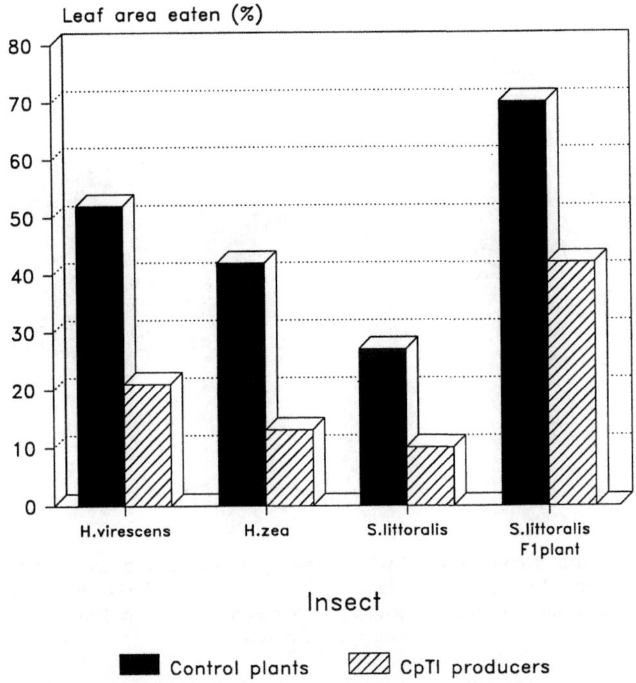

Figure 4.5. Bioassay data for 7-day feeding trials of Lepidopteran pests on transgenic tobacco plants expressing CpTI. Leaf area eaten was measured by computer-aided image analysis of harvested leaves. Figure courtesy of Agricultural Genetics Company Ltd.

reduced to stalks. On the CpTI-producing plants, although the larvae begin to feed and do some very limited damage to the leaves, they either die or fail to develop as they would on control plants. These observations are consistent with the mechanism of CpTI toxicity proposed by Gatehouse and Boulter (1983), in which the predator is subject to a finely controlled balance between availability and non-availability of nutrients; inhibition of proteolysis, or sequestration of essential amino acids in an unavailable source, leads to larval starvation at an early stage. Statistical analysis of the bioassay in terms of insect survival, biomass and plant damage by leaf area confirmed the highly significant protection afforded by CpTI (Hilder *et al.*, 1987). This protection has been subsequently confirmed for the other Lepidopteran pests shown in Table 4.3 and Figure 4.5; on the basis of the protection afforded against *Heliothis zea* (corn earworm), *Spodoptera littoralis* (armyworm) and *Manduca sexta* (tobacco hornworm) (Figure 4.6) it is reasonable to suppose that effectiveness against a wide range of Lepidoptera is provided by CpTI. It has not yet been possible to use the transgenic plants to confirm the antimetabolic properties of CpTI in artificial diet against Coleoptera, due to the lack of a

Figure 4.6. Exposure of control and CpTI-expressing transgenic tobacco plants to larvae of tobacco hornworm (*Manduca sexta*). Plant on the left is a control, showing almost complete destruction; plant on the right is a transgenic CpTI expresser. Reproduced from Hilder *et al.* (1989) *Pestic. Sci.* **27**: 165.

suitable test organism. The wide range of organisms that are affected by CpTI is advantageous for the use of this gene as a broad-spectrum protectant in transgenic plants.

The bioassays on transgenic plants expressing CpTI have confirmed the previous suggestions of the role of proteinase inhibitors in insect resistance, and supplied the direct evidence, previously lacking, of their effectiveness as antimetabolites against specific insect species. A proteinase inhibitor, on its own, can significantly reduce insect attack, although recent evidence suggests that the inhibitors form part of a complex interaction between the plant's nutritional value and the insect's digestive physiology (Broadway *et al.*, 1986). Although CpTI was the first insecticidal protein of plant origin to be expressed in transgenic plants, a recent report (Johnson *et al.*, 1990) demonstrates that expression of proteinase inhibitors from both potato and tomato in transgenic tobacco plants resulted in enhanced resistance to *Manduca sexta* (tobacco hornworm). The general utility of proteinase inhibitor genes is thus demonstrated, although, as discussed earlier, different inhibitors may prove to be more useful against certain specific pest species.

Experiments designed to test whether the CpTI-expressing transgenic plants suffer a 'yield penalty' from having to express comparatively large amounts of a foreign protein have shown that, although a very small penalty results from transformation, no additional yield loss, loss of vigour or resistance to stress results from expression of CpTI (Hilder *et al.* 1990). This result once again emphasises the observation that plants have excess metabolic capacity for synthesis of secondary metabolites. The CpTI genes in transgenic plants were stably inherited, and despite being present as multiple copies behaved for the most part as single genetic loci, suggesting that the insertion of tandem repeats had occurred; this conclusion was confirmed by Southern blotting of the genomic DNA (Hilder and Gatehouse, 1990).

The utility of CpTI as a protectant against insect attack in transgenic plants has thus been established, and, like the Bt toxin, most of the conditions necessary to make its use a practical proposition have been met. Again like the Bt toxin gene, extensive field trials will be necessary before this product can be brought to the marketplace.

4.6 Future prospects

Two independent methods have been used to show that the genetic engineering of plants for insect resistance is not only feasible but is likely to lead to commercially viable products. The two methods should not be considered as competing with each other; the specificity and high activity of Bt toxin is complementary to the broad-spectrum protection afforded by protease inhibitors, and an obvious development is to produce transgenic plants expressing both insecticides, and to look for protection superior to either method on its own. Unfortunately, to the knowledge of the authors, this has yet to be done. The omission is not due to any scientific reason but is a necessary consequence of commercial interests in the various genes employed, and will no doubt be made good in time. Work is actively proceeding to engineer both Bt and CpTI genes into a wider range of crop plants, with the limiting factor being the development of viable transformation systems. The availability of more transgenic plant species will allow a wider range of insect pest species to be tested, and give a clearer picture of the utility of the genes.

Some indications have been given above of the improvements that are being made to Bt toxin genes intended for transfer to plants to give insect resistance; genes with different specificities against insect pests will also be tested and developed. Major research efforts are being put into Bt toxins specific for Coleopteran pests. Sections 4.2 and 4.3 of this review will suggest to the reader other sources of genes that could be used to protect transgenic plants against insect attack; a limiting factor will be the biology of the systems, i.e. demonstrating that a particular candidate is in fact responsible for observed resistance. A particular need at present is for a compound that will give some measure of protection against Hemiptera (sucking pests such as aphids and plant hoppers) and is suitable for introduction into plants via genetic engineering methods.

An example of a different type of plant gene that has been shown to give transgenic plants a degree of insect resistance is *lec*A, the pea (*Pisum sativum* L.)

seed lectin gene (Gatehouse et al., 1987). This gene was transferred to tobacco as a construct where the coding sequence was driven by the CaMV 35S promoter; plants expressing the lectin alone, or in combination with CpTI, were produced. Insect bioassay against *Heliothis virescens* showed that the lectin, expressed at a level of 1.0–1.5% of total protein, afforded significant protection against larval damage but was not as effective a protectant as similar levels of CpTI. However, both antimetabolites together gave greater protection than either separately (Boulter *et al.* 1990). This is the first demonstration of 'pyramiding' resistance genes in a transgenic plant. Further developments in this area are to be expected.

In the longer term, economic and consumer pressure for decreased use of pesticides seems likely to lead to increasing use of genetic engineering as a means of protecting crops. Although it is impossible to predict at present which genes, or combinations of genes, will eventually gain the major shares of market acceptance, there is a high probability that transgenic, insect-resistant plants will be of increasing importance as an agricultural resource.

Acknowledgements

The authors would like to thank the Agricultural Genetics Company, Cambridge, England, for their financial support and commitment to this area, and Professor D. Boulter for his encouragement, advice and help, and for the provision of departmental facilities. The assistance of all our co-workers and collaborators, and those authors who sent preprints of their published work, is gratefully acknowledged.

References

Angus, T.A. (1954) A bacterial toxin paralysing silkworm larvae. *Nature* **173**: 545.

Applebaum, S.W. (1964) Physiological aspects of host specificity in the bruchidae. I. General considerations of developmental compatibility. *J. Insect Physiol.* **10**: 783.

Applebaum, S.W. and Guez, M. (1972) Comparative resistance of *Phaseolus vulgaris* beans to *Callsobruchus chinesis* and *Acanthoscelides obtectus* (Col. Bruchidae): the differential digestion of soluble heteropolysaccharide *Entomol. Exp. App.* **25**: 64.

Applebaum, S.W. and Konijn, A.M. (1966) The presence of a *Tribolium*-protease inhibitor in wheat. *J. Insect Physiol.* **12**: 665.

Barton, K.A., Whiteley, H.R. and Yang, N-S. (1987) *Bacillus thuringiensis*-endotoxin expressed in *Nicotiano tabacum* provides resistance to lepidopteran insects. *Plant Physiol.* **85**: 1103.

Baulcombe, D.C., Saunders, G.R., Bevan, M., Mayo, M.A. and Harrison, B.D. (1986) Expression of biologically active viral satellite RNA from the nuclear genome of transformed plants. *Nature* **321**: 446.

Berardi, L.C. and Goldblatt, L.A. (1980) Gossypol. In '*Toxic Constituents of Plant Foodstuffs*, 2nd edn., Liener, I.E. Academic Press, New York, 183.

Birk, Y., Gertler, A. and Khalef, S. (1963) Separation of a *Tribolium*-protease inhibitor from soybeans on a calcium phosphate column. *Biochim. Biophys. Acta* **67**: 326.

Boughdad, A., Gillon, Y. and Gagnepain, C. (1986) Influence des tannins condenses due tegument de feves (*Vicia faba*) sur le developpement larvaire de *Callosobruchus maculatus. Entomol. Exp. Appl.* **42**: 125.

Boulter, D., Edwards, G.A., Gatehouse, A.M.R., Gatehouse, J.A. and Hilder, V.A. (1990) Additive protective effects of incorporating two different higher plant derived insect resistance genes in transgenic tobacco plants *Crop Protection*, (in press).

Broadway, R.M. and Duffy, S.S. (1986) Plant proteinase inhibitors: mechanism of action and effect on the growth and digestive physiology of larval *Heliothis zea* and *Spodoptera exigua*. *J. Insect Physiol.* **32**: 827.

Broadway, R.M., Duffy, S.S., Pearce, G. and Ryan, C.A. (1986) Plant proteinase inhibitors: A defense against herbivorous insects? *Ent. Exp. Appl.* **41**: 33.

Brown, W.E., Takio, K., Titani, K. and Ryan, C.A. (1985) Wound-induced trypsin inhibitor in alfalfa leaves: Identity as a member of the Bowman–Birk inhibitor family. *Biochemistry* **24**: 2105.

Campos, F.A.P. and Richardson, M. (1983) The complete amino acid sequence of the bifunctional amylase/trypsin inhibitor from seeds of ragi (Indian finger millet, *Eleusine coracana* Gaertn.) *FEBS Lett.* **152**: 300.

Christeller, J.T. and Shaw, B.D. (1989) The interaction of a range of serine proteinase inhibitors with bovine trypsin and *Costelytra zealandica* trypsin. *Insect Biochem.* **19**: 233.

Cromartie, W.J. (1981) The environmental control of insects using crop diversity. In *Handbook of Pest Management in Agriculture*, ed. Pimental, D. CRC Press, Boca Raton, FL, 223.

Diaz, C.L., Melchers, L.S., Hooykaas, P.J.J. Lungtenberg, B.J.J. and Kijne, J.W. (1989) Root lectin as a determinant of host-plant specificity in the *Rhizobium*–legume symbiosis. *Nature* **338**: 579.

Dulmage, H.T. (1981) Insecticidal activity of isolates of *Bacillus thuringiensis* and their potential for pest control. In '*Microbial Control of Pests and Plant Diseases* 1970–1980, ed. Burges, H.D. Academic Press, New York, 193.

Ellis, J.R., Shirsat, A.H., Hepher, A., Yarwood, J.N., Gatehouse, J.A., Croy, R.R.D. and Boulter, D. (1988) Tissue specific expression of a pea legumin gene in seeds of *Nicotiana plumbaginifolia* *Plant Mol. Biol.* **10**: 203.

Evans, R.J., Pusztai, A., Watt, W.B. and Bauer, D.H. (1973) Isolation and properties of protein fractions from navy beans (*Phaseolus vulgaris*) which inhibit growth of rats, *Biochim. Biophys. Acta* **303**: 175.

Evans, S.V., Gatehouse, A.M.R. and Fellows, L.E. (1985) Detrimental effects of 2,5-dihydroxymethyl-3,4-dihydroxypyrrolidine in some tropical legume seeds on larvae of the bruchid *Callosobruchus maculatus*. *Entomol. Exp. Appl.* **37**: 257.

Fischhoff, D.A. (1989) Plants as delivery systems for biopesticides. *Agbiotech '89*, 373.

Fischhoff, D.A., Bowdish K.S., Perlak, F.J., Marrone, P.G., McCormick, S.M., Niedermeyer, J.G., Dean, D.A., Kusano-Kretzmer, K., Mayer, E.J., Rochester, D.E., Rogers, S.G. and Fraley, R.T. (1987) Insect tolerant transgenic tomato plants. *Biotechnology* **5**: 807.

Fuchs, R., MacIntosh, S., Kishore, G., Perlak, F., Dean, D., Stone, T., Sims, S., Hoffmann, N., Greenplate, J., Marrone, P. and Fischhoff, D.A. (1989) Enhanced expression/efficacy of transgenic plants which express the *Bacillus thuringiensis* insect control protein. *Agbiotech '89*, 210.

Garcia-Olmedo, F., Salcedo, G., Sanchez-Monge, R., Gomez, L., Royo, J. and Carbonero, P. (1987) Plant proteinaceous inhibitors of proteinases and alpha-amylases. In *Oxford Surveys of Plant Molecular and Cell Biology*, vol. IV, ed. Miflin, B.J. Oxford, University Press, 275.

Gaertner, F. (1989) *Bacillus thuringiensis*: Utility. *Agbiotech '89*, 354.

Gatehouse, A.M.R. and Boulter, D. (1983) Assessment of the antimetabolic effects of trypsin inhibitors from cowpea (*Vigna unguiculata*) and other legumes on development of the bruchid beetle *Callosobruchus maculatus*. *J. Sci. Food Agric.* **34**: 345.

Gatehouse, A.M.R., Gatehouse, J.A., Dobie, P., Kilminster, A.M. and Boulter, D. (1979) Biochemical basis of insect resistance in *Vigna unguiculata*. *J. Sci. Food Agric.* **30**: 949.

Gatehouse, A.M.R., Dewey, F.M., Dove, J., Fenton, K.A. and Pusztai, A. (1984) Effect of seed lectin from *Phaeolus vulgaris* on the development of larvae of *Callosobruchus maculatus*: mechanism of toxicity. *J. Sci. Food Agric.* **35**: 373.

Gatehouse, A.M.R., Butler, K.J., Fenton, K.A. and Gatehouse, J.A. (1985) Presence and partial characterisation of a major proteolytic enzyme in the larval gut of *Callosobruchus maculatus*. *Entomol. Exp. Appl.* **39**: 279.

Gatehouse, A.M.R., Fenton, K.A., Jepson, I. and Pavey, D.J. (1986) The effects of α-amylase inhibitors on insect storage pests; inhibition of α-amylase *in vitro* and effects on development *in vivo*. *J. Sci. Food Agric.* **37**: 727.

Gatehouse, A.M.R., Dobie, P., Hodges, R.J., Meik, J., Pusztai, A. and Boulter, D. (1987) Role of carbohydrates in insect resistance in *Phaseolus vulgaris*. *J. Insect Physiol.* **33**: 843.

Gatehouse, A.M.R., Shackley, S.J., Fenton, K.A. Bryden, J. and Pusztai, A. (1989) Mechanism of seed lectin tolerance by a major insect storage pest of *Phaseolus vulgaris, Acanthoscelides obtectus. J. Sci. Food Agric.* **47**: 269.

Gatehouse, A.M.R., Barbieri, L., Stirpe, F. and Croy, R.R.D. (1990) Effects of ribosome inactivating proteins on insect development—differences between Lepidoptera and Coleoptera. *Entomol. Exp. Appl.* (in press).

Gatehouse, J.A., Bown, D., Evans, I.M., Gatehouse, L.N. Jobes, D., Preston, P., Croy, R.R.D. (1987) Sequence of the seed lectin gene from pea (*Pisum sativum* L.) *Nucl. Acids Res.* **15**: 7642.

Green, T.R. and Ryan, C.A. (1972) Wound-induced proteinase inhibitor in plant leaves: a possible defense mechanism against insects. *Science* **175**: 776.

Hamblin, J. and Kent, S.P. (1973) Possible role of phytohaemagglutinin in *Phaseolus vulgaris* L. *Nature New Biol.* **245**: 28.

Hedin, P.A., Jenkins, J.N., Collum, D.H., White, W.H. and Parrot, W.L. (1983) In *Plant Resistance to Insects*, ed. Hedin, P.A. American Chemical Society, Washington DC, 347.

Heinrichs, E.A. and Mochida, O. (1983) From secondary to major pest status: The case of insecticide-induced rice brown planthopper, *Nilaparvata lugens* resurgence. Proc. XV Pacific Science Congress, New Zealand.

Hilder, V.A. and Gatehouse, A.M.R. (1990) The phenotypic cost to plants of an 'extra' gene (submitted).

Hilder, V.A., Gatehouse, A.M.R., Sheerman, S.E., Barker, F. and Boulter, D. (1987) A novel mechanism of insect resistance engineered into tobacco. *Nature* **330**: 160.

Hilder, V.A., Barker, R.F., Samour, R.A., Gatehouse, A.M.R., Gatehouse, J.A. and Boulter, D. (1989) Protein and cDNA sequences of Bowman–Birk protease inhibitors from the cowpea (*Vigna unguiculata* Walp.) *Plant Mol. Biol.* **13**, 701.

Hilder, V.A., Gatehouse, A.M.R. and Hughes, S. (1990) Characterisation of transgenic tobacco plants containing cowpea trypsin inhibitor genes (submitted).

Hill, D.S. (1983) In *Agricultural Insect pests of the Tropics and their Control*. Cambridge, University Press, 746.

Hill, D.S. (1987) In *Agricultural Insect Pests of Temperature Regions and their Control*. Cambridge, University Press, 659.

Ishimoto, M. and Kitamura, K. (1988) Identification of the growth inhibitor on azuki bean weevil in kidney bean (*Phaseolus vulgaris* L.). *Japan J. Breed.* **38**, 367.

Janzen, D.H., Juster, H.B. and Liener, I.E. (1976) Insecticidal action of the phytohemagglutinin in black bean on a bruchid beetle. *Science* **192**: 795.

Jayne-Williams, D.J. and Burgess, C.D. (1974) Further observations on the toxicity of navy beans (*Phaseolus vulgaris*) for Japanese quail (*Coturnix coturnix japonica*). *J. Appl. Bacteriol.* **37**: 149.

Johnson, R., Narvaez, J., An, G. and Ryan, C.A. (1990) Expression of proteinase inhibitors I and II in transgenic tobacco plants: effects on natural defense against *Manduca sexta* larvae. *Proc. Natl Acad. Sci. USA* **86**, 9871.

King, T.P., Pusztai, A. and Clarke, E.M.W. (1980*a*) Immunocytochemical localisation of injested kidney bean (*Phaseolus vulgaris*) lectins in rat gut. *Histochem. J.* **12**: 201.

King, T.P., Pusztai, A. and Clarke, E.M.W. (1980*b*) Kidney bean (*Phaseolus vulgaris*) lectin-induced lesions in rat small intestine. 3. Ultrastructural studies. *J. Comp. Pathol.* **92**: 357.

Kitch, L.W. and Murdoch, L.L. (1986) Partial characterization of a major midgut thiol proteinase from larvae of *Callosobruchus maculatus* (F.). *Arch. Insect Biochem. Physiol.* **3**: 561.

Kunitz, M. (1945) Crystallization of a trypsin inhibitor from soybean. *Science* **101**: 668.

Lipke, H., Fraenkel, G.S. and Liener, I.E. (1954) Effect of soybean inhibitors on growth of *Tribolium confusum. J. Agric. Food Chem.* **2**: 410.

Lotan, R., Cacan, R., Cacan, M., Debray, H., Carter, W.C. and Sharon, N. (1975) On the presence of two types of subunit in soybean agglutinin. *FEBS Lett.* **57**: 100.

Mann, J. (1987) *Secondary Metabolism*, 2nd edn. Oxford, University Press.

Marchal, P. (1908) The utilization of auxiliary entomophagus insects in the struggle against insects injurious to agriculture. *Pop. Sci. Mon.* **72**: 352.

Metcalf, R.L. (1986) The ecology of insecticides and the chemical control of insects. In *Ecological Theory and Integrated Pest Management Practice*, ed. Kogan, M. John Wiley & Sons, New York, 251.

Mickel, C.E. and Standish, J. (1947) Susceptibility of processed soy flour and soy grits in storage to attack by *Tribolium castaneum* (Herbst). *Univ. Minn. Agr. Exp. Sta. Tech. Bull.* **178**: 1.

Nash, R.J., Fenton, K.A., Gatehouse, A.M.R. and Bell, E.A. (1986) Effects of the plant alkaloid castanospermine as an antimetabolite of storage pests. *Entomol. Exp. Appl.* **42**: 71.

Norris, D.M. and Kogan, M. (1980) Biochemical and morphological bases of resistance. In *Breeding Plants Resistant to Insects*, eds. Maxwell, F.G. and Jennings, P.R. John Wiley & Sons, London.

Oka, I.N. and Pimentel, D. (1976) Herbicide (2,4-D) increases insect pathogen pests on corn. *Science* **193**: 239.

Peterson, V. (1984) In *The Natural Food Catalogue* **69**. McDonald, London.

Pimental, D., Shoemaker, C., LaDue, E.L., Rovinsky, R.B. and Russell, N.P. (1977) Alternatives for reducing insecticides on cotton and corn: economic and environmental impact. *Environ. Res. Lab. Off. Res. Develop.*, EPA, Athens, G.A., 145.

Pusztai, A., Clarke, E.M.W. and King, T.P. (1979) The nutritional toxicity of *Phaseolous vulgaris. Nutr. Soc.* **38**: 115.

Read, J.W. and Haas, L.W. (1938) Studies on the baking quality of flour as affected by certain enzyme actions V. Further studies concerning potassium bromate and enzyme activity. *Cereal Chem.* **15**: 59.

Redden, R.J., Dobie, P. and Gatehouse, A.M.R. (1983) The inheritance of seed resistance to *Callosobruchus maculatus* F. in cowpea (*Vigna unguiculata* L. Walp.). I. Analysis of parental, F_1, F_2, F_3 and backcross seed generations. *Aust. J. Agric. Res* **34**: 681.

Richardson, M. (1977) The proteinase inhibitors of plants and microorganisms. *Phytochemistry* **16**: 159.

Richardson, M. (1981) Protein inhibitors of enzymes. *Food Chem.* **6**: 235.

Rosenthal, G.A., Dahlman, D.L. and Janzen, D.H. (1976) A novel means for dealing with L-canavanine, a toxic metabolite, *Science* **192**: 256.

Rosenthal, G.A., Dahlman, D.L. and Janzen, D.H. (1978) L-canaline detoxification: a seed predator's biochemical mechanism, *Science* **202**: 528.

Ryan, C.A. (1983) In *Variable Plants and Herbivores in Natural and Managed Systems*, eds. Denno, R.F. and McClure, M.S. Academic Press, New York, 43.

Ryan, C.A. (1985) In *The Biochemistry of Plants: a Comprehensive Treatise*, vol. 6, ed. Marcus, A. Academic Press, New York 351.

Ryan, C.A. (1989) Proteinase inhibitor gene families: strategies for transformation to improve plant defenses against herbivores. *Bio Essays* **10**: 21.

Sacchi, V.F., Parenti, P., Hanozet, G.M., Giordanda, B., Lutly, P. and Wolfersberger, M.G. (1986) *Bacillus thuringiensis* toxin inhibits K^+-gradient-dependent amino acid transport across the brush border membrane of *Pieris brassicae* midgut cells. *FEBS Lett.* **204**: 213.

Shukle, R.H. and Murdoch, L.L. (1983) Lipoxygenase, trypsin inhibitor, and lectin from soybeans: effects on larval growth of *Manduca sexta* (Lepidoptera: Sphingidae). *Environ. Entomol.* **12**: 787.

Simons, P. (1981) The potato bites back. *New Scientist* **1267**: 470.

Singh, S.R. (1978) Resistance to insect pests of cowpea in Nigeria. In: *Pests of Grain Legumes: Ecology and Control*, eds. Singh, S.R., Van Emden, H.F. and Taylor, T.A. Academic Press, New York, 267.

Steffens, R., Fox, F.R. and Kassel, B. (1978) Effect of trypsin inhibitors on growth and metamorphosis of corn borer larvae *Ostrinia nubilalis* (Hubner). *J. Agric. Food Chem.* **26**: 170.

Strong, D.R. (1979) Biogeographic dynamics of insect–host plant communities. *Ann. Rev. Entomol.* **24**: 89.

Vaeck, M., Reynaerts, A., Hofte, H., Jansens, S., De Beuckeleer, M.D., Dean, C., Zabeau, M., Van Montagu, M.V. and Leemans, J. (1987) Transgenic plants protected from insect attack. *Nature* **328**: 33.

Velten, J., Velten, L., Hains, R. and Schell, J. (1984) Isolation of a dual plant promoter fragment from the Ti plasmid of *Agrobacterium tumefaciens. EMBO J.* **3**: 2723.

5 Virus-resistant plants

K.W. BUCK

5.1 Introduction

About 10% of world crop production is lost annually because of plant diseases caused by bacteria, fungi and viruses (Fraser, 1985). Crop losses as a result of virus diseases can be especially serious in developing countries in tropical and subtropical regions. For example, in Africa losses of cassava crops caused by infection with African cassava mosaic virus are valued at more than two hundred million pounds per annum (Harrison, 1987). There are three main types of control measure for preventing virus infections in crops. The first of these involves removal of virus reservoirs, such as infected plants or propagules remaining from previous crops, and producing virus-free planting stocks. The second method aims to prevent plant-to-plant spread of virus, usually by killing vectors that transmit the virus. The third control measure is to produce and grow virus-resistant cultivars of crops.

Virus-resistant plants, which are the subject of this chapter, are economical for the farmer and are likely to become increasingly important in the future for several reasons. Firstly, public pressure to limit the use of environmentally damaging vector-killing pesticides is growing. Secondly, pesticide-resistant vector organisms have become common in some areas, hence reducing the scope of these pesticides (Chung *et al.*, 1982). Thirdly, plant transformation methods, which are becoming available for an increasing number of crop plants, should enable resistance genes that occur naturally in plants to be introduced into plants unrelated to their source plants, which is not possible by conventional plant breeding methods. Finally, increase in our understanding of the molecular biology of virus infections and resistance has not only suggested new ways of manipulating naturally occurring resistance genes, but has created opportunities to create novel types of resistance genes that are not known to occur in nature.

5.2 Basic concepts of resistance to plant virus infection

Inoculation of a susceptible plant with a virus results in replication of the virus in the cells initially entered, followed by cell-to-cell transport, systemic spread of the virus through the plant's vascular system and replication of the virus in non-inoculated tissue. Resistance of a plant to infection by a particular virus may be defined as the ability of the plant, when inoculated with the virus, to reduce substantially the amount of virus replication and spread in the plant compared with that which occurs after infection of a related susceptible plant with the same virus. In nature, host-encoded resistance is common, and most plants are resistant to most viruses (Ponz and Bruening, 1986).

The degree of reduction in virus accumulation in a resistant plant depends on the virus strain, the host resistance gene(s) and environmental factors. Resistance has been considered to operate at three levels (Zaitlin and Hull, 1987).

(1) In extreme cases of resistance, termed total immunity, no virus replication takes place, even in the cells initially entered.
(2) Some resistant hosts may allow virus to replicate in the initially infected cells, but not to spread from cell to cell.
(3) In other hosts some initial replication and spread of a virus may induce a hypersensitive response which restricts the virus to a region around the point of entry.

If resistance in (1), (2) or (3) is partial, a low level of systemic infection of the plant may occur.

Even in a susceptible host, virus replication may be restricted within a plant. For example, bean golden mosaic virus is largely restricted to phloem tissue (Goodman, 1981). Systemic infection of tobacco plants by tobacco mosaic virus (TMV) results in mosaic symptoms on the infected leaves; the dark-green regions or islands contain fewer virus particles than do the yellow-green areas (Atkinson and Matthews, 1970). Plants often show recovery from virus infection and older tissue is frequently more resistant to infection than younger tissue (Ponz and Bruening, 1986).

Host-encoded resistance results from interference by the host with one or more stages of the replication cycle of a virus and its systemic spread through the host. Viruses are dependent on a number of host gene products for their replication and spread; resistance could stem either from the absence of such 'essential' gene products in a particular host or from incompatibility of one or more such gene products with the replication complex or transport machinery of a particular virus. Alternatively a resistant host may encode a product that is not required by the virus but which inhibits some stage of virus replication or transport; such a host gene product could be expressed constitutively or could be induced. When virus infection triggers a hypersensitive response in the host, resistance is often induced not only locally, but also in non-inoculated parts of the plant (induced systemic resistance).

If there is insufficient genetic variation in a crop species to include resistance to a particular virus, a search for resistance may be made in a related non-crop species. For example, the Ry gene for extreme resistance (immunity) to potato virus Y (PVY) (Ross, 1960), identified in the wild species *Solanum stoloniferum*, has been transferred by plant breeders to the cultivated potato *Solanum tubersum*. Resistances transferred in this way have often been shown to be due to single dominant genes (Fraser, 1985, 1986, 1987a, b) and can sometimes be overcome by the appearance of virus mutants (resistance-breaking strains). There are relatively few examples of oligogenic resistance in crop plants, perhaps because breeders have selected for single gene effects. However, it could well occur in plants in different taxonomic groups to known hosts of a virus, for example resistance in monocotyledonous plants to viruses whose host range is confined to dicotyledonous plants. Such resistance could be due, for example, to differences in codon usage (Murray *et al.*, 1989) and absence of the required spectrum of transfer RNAs (tRNAs) in a resistant plant.

Generally the degree of damage to a plant is related to the amount of virus produced (reviewed by Ponz and Bruening, 1986). However low levels of some viruses can cause severe damage, and the tissues in the plant that are infected can be more important than the amount of virus that accumulates. Some plant cultivars may contain genes that suppress symptom formation without reducing virus accumulation. Although such tolerant cultivars can be used for plant protection, they have the potential for creating reservoirs of virus infection. Hence, although it is clear that virus replication and symptom induction can be 'decoupled', it is considered more prudent to control symptoms using resistance genes that reduce symptoms by reducing virus accumulation.

Conversely there are virus mutants that are able to replicate in normally susceptible hosts without causing symptoms. Plants infected by such 'masked' or 'mild' strains of viruses are often resistant or tolerant to superinfection by pathogenic strains of the same or closely related viruses, a phenomenon designated as 'cross-protection'. The replication of, and/or symptoms induced by, some viruses can also be reduced by certain satellite RNAs. Both cross-protection (Fulton, 1986) and satellite RNAs (Baulcombe, 1989) form the basis of a number of methods of protecting plants from virus diseases.

5.3 Key features of the infection cycles of positive-strand RNA plant viruses

Although their mechanisms may in some cases differ, the potential targets for naturally occurring resistance genes and for artificial resistance genes designed by the molecular biologist are the same. Theoretically any stage of the virus infection process could be blocked, starting with plant-to-plant transfer, followed by uptake into cells, uncoating, early gene expression, nucleic acid replication, late gene-expression, assembly of progeny particles and movement of the virus from cell to cell and through the plant's vascular system leading to a systemic infection. To identify potential targets, some key features of the molecular biology of plant virus infections are given.

Most studies of resistance have focused on viruses with genomes of positive (messenger)-sense single-stranded RNA, which comprise the majority (485 or 77%) of known plant viruses (Zaitlin and Hull, 1987). There is considerable diversity among the positive-strand RNA viruses which have been arranged in 27 taxonomic groups (Knudson *et al.*, 1990). Genome sizes vary from about 5 kb up to about 20 kb with similar variations in the number of proteins encoded. Nevertheless, unifying features based on sequences and modes of gene expression have suggested evolutionary relationships between viruses in different groups, including some animal viruses, leading to suggestions for formation of a small number of 'supergroups' (Goldbach, 1986; Zimmern, 1987; Strauss and Strauss, 1988; Gorbalenya and Koonin, 1989; Habili and Symons, 1989). In this section, basic features of a few well-studied positive-strand RNA viruses, selected to illustrate general principles, are presented. A summary of the genome types of these viruses is given in Table 5.1. More extensive accounts of a wider range of plant viruses including those with genomes of negative-stranded RNA, double-stranded RNA, single-stranded DNA or double-stranded DNA may be found in several recent texts (Davies, 1985; Francki, 1985a; Francki *et al.*, 1985, Van Regenmortel and Fraenkel-Conrat, 1987; Rowlands *et al.*, 1987; Koenig, 1988; Milne, 1988; Matthews, 1990).

5.3.1 *Plant-to-plant spread of viruses*

Although some plant viruses are transmitted from plant to plant in the field via pollen or directly through accidental damage to the plant, many viruses are transmitted by insect, fungus or nematode vectors. The plants to which a virus can be transmitted by a vector will depend on the feeding preferences of the vector. An indirect way of reducing virus transmission by vectors would be to make the plant resistant to the vector. For example, leaf hairs of the wild potato *Solanum berthaultii* exude an aphid alarm pheromone which is sufficient to cause aphids to avoid the plant and to reduce the acquisition and transmission of beet yellows virus and PVY (Gibson and Pickett, 1983). The production of transgenic plants resistant to insects is discussed by Gatehouse (this volume, Chapter 4).

Generally the association between a virus and its vector is specific and may involve the capsid protein or other gene products. For example, a 56 kDa protein encoded by PVY is required for successful virus transmission from plant to plant by aphids, probably by a mechanism that involves interaction of the virus and this 'helper component' with the aphid stylet (Berger and Pirone, 1986). Suppression of expression of this gene in a host plant would not make that plant resistant to the virus, but would prevent aphid transmission of the virus to other plants.

5.3.2 *Virus entry, uncoating and early translation*

To enter a cell, a virus particle must penetrate the cuticle, cell wall and plasma membrane. This appears to be a passive process, involving transient damage

Table 5.1. Genome types of positive-strand RNA plant viruses referred to in the text*.

Virus	Abbreviation	Number of genome segments	Particle structure	Structure of genome: 5′ end	Structure of genome: 3′ end	Mode of gene[†] expression
Supergroup A (sindbisvirus-like)[‡]						
Potato virus X (potexvirus)[§]	PVX	1	Flexuous rod	Cap	Poly-A	S
Tobacco mosaic virus (tobamovirus)	TMV	1	Rigid rod	Cap	tRNA-like	R, S
Turnip yellow mosaic virus (tymovirus)	TYMV	1	Isometric	Cap	tRNA-like	P, S
Tobacco rattle virus (tobravirus)	TRV	2	Rigid rod	Cap	tRNA-like	M, R, S
Alfalfa mosaic virus	AlMV	3	Bacilliform	Cap	X_{OH}	M, S
Brome mosaic virus (bromovirus)	BMV	3	Isometric	Cap	tRNA-like	M, S
Cucumber mosaic virus (cucumovirus)	CMV	3	Isometric	Cap	tRNA-like	M, S
Tobacco streak virus (ilarvirus)	TSV	3	Quasi-isometric	Cap	X_{OH}	M, S
Supergroup B (luteovirus-like)						
Barley yellow dwarf virus (luteovirus)	BYDV	1	Isometric	VPg	X_{OH}	F, S
Red clover necrotic mosaic virus (dianthovirus)[‡]	RCNMV	2	Isometric	Cap	X_{OH}	M, F, S
Supergroup C (picornavirus-like)						
Potato virus Y (potyvirus)	PVY	1	Flexuous rod	VPg	Poly-A	P
Tobacco etch virus (potyvirus)	TEV	1	Flexuous rod	VPg	Poly-A	P
Cowpea mosaic virus (comovirus)	CPMV	2	Isometric	VPg	Poly-A	M, P
Tobacco ringspot virus (nepovirus)	TobRV	2	Isometric	Vpg	Poly-A	M, P

*For more comprehensive listings, see Goldbach (1986), Habibi and Symons (1989), Knudson *et al.* (1990).

[‡]Supergroups are those of Habibi and Symons (1989); RCNMV is placed tentatively in supergroup B on the basis of sequence homologies with BYDV.

[†]F, ribosomal frameshift; M, multicomponent genome; P, polyprotein; R, translational read-through; S, subgenomic mRNA.

[§]Taxomomic groups (Knudson *et al.*, 1990) are given in parenthesis after the virus name.

to the cell by the vector or by mechanical means (Plaskitt *et al.*, 1988). There is no evidence for the requirement for specific virus receptors on the surface of plant cells, as there is on bacterial and animal cells.

The majority of plant viruses have a relatively simple particle structure, consisting of the nucleic acid genome surrounded by a coat protein (CP) composed of a single polypeptide species arranged in helical or icosahedral symmetry. There is convincing evidence (reviewed by Wilson, 1988) that uncoating of TMV (tobacco mosaic virus) occurs in a 5′ to 3′ direction and is aided by ribosomes in a process called co-translational disassembly (Wilson, 1984). Removal of a few protein subunits from the concave end of the rod is necessary to expose the 5′ end of the RNA to enable ribosomes to bind to the cap structure and scan the RNA for the initiation codon (Kozak, 1984; see Chapter 1). Protein subunits may be bound less firmly to the 5′ leader sequence which lacks G-residues. *In vitro* a few subunits at the 5′ end are dissociated at pH 8.0 (Wilson, 1984); *in vivo* hydrophobic interactions with membranes or the low intracellular Ca^{2+} concentration may contribute to this dissociation. Once a ribosome has bound to the 5′ end of the RNA, it is envisaged that it translates the 5′ cistrons (to produce that 126 kDa protein (P126) and the 183 kDa protein (P183); Figure 5.1), removing coat protein subunits as it translocates along the RNA in a 5′ to 3′ direction. This mechanism ensures that free RNA is never exposed and therefore is protected from degradation.

Co-translational disassembly would be predicted to stop after synthesis of

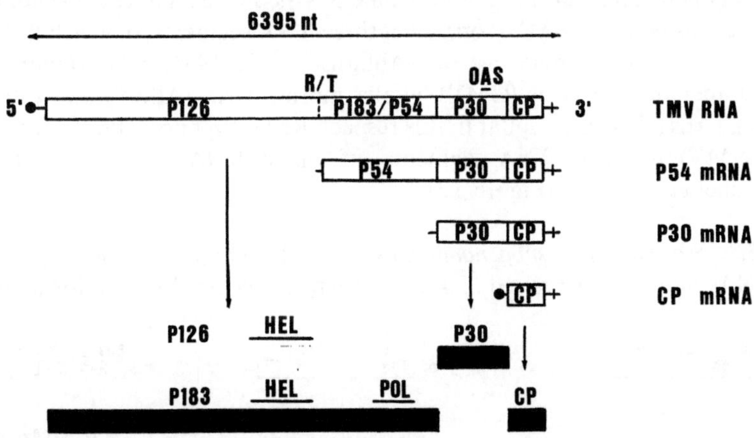

Figure 5.1. Genome organisation and expression of tobacco mosaic virus (TMV). Only the 5′ cistrons of the genomic and subgenomic RNAs are translated. P54 mRNA, but not P54, has been isolated from TMV-infected plants. Open reading frames in RNA molecules are shown as open boxes. Proteins are shown as closed boxes. ●, 7MeGpppN cap structure; +, tRNA-like structure; R/T, translational read-through of UAG stop codon; OAS, origin-of-assembly sequence; CP, coat protein. The positions of conserved sequence motifs in the putative helicase (HEL) and polymerase (POL) regions (Habili and Symons, 1989) are indicated. Other data from Palukaitis and Zaitlin (1986).

P183. P126 and P183 are postulated to be components of an RNA-dependent RNA polymerase needed to replicate the genomic RNA of the virus (Quadt and Jaspars, 1989). It is possible that translation of the 5′ cistrons occurs repeatedly during the early stages of infection to build up a supply of polymerase, while the 3′ cistrons of the RNA remain protected by the coat protein. RNA replication starts by binding of the RNA polymerase to the 3′ end of the RNA and copying the template in a 3′ to 5′ direction. An attractive hypothesis (Wilson, 1988) is that protein subunits may be dissociated from the extreme 3′ end of the RNA *in vivo*, allowing binding of RNA polymerase. Movement of polymerase along the template could then remove the remaining coat protein subunits by 'co-transcriptional disassembly'.

There is evidence that at least some other positive-strand RNA viruses, including icosahedral viruses, can be uncoated by co-translational disassembly (Wilson, 1988), though doubtless variations on the TMV model exist.

5.3.3 *Modes of gene expression*

Eukaryotic ribosomes usually start translation only at the initiation codon of the cistron closest to the 5′ end of the RNA (Kozak, 1984). Plant RNA viruses use one or more of the following strategies to enable all their cistrons to be translated.

5.3.3.1 *Segmentation of the genome.* Division of the genome into more than one segment can result in monocistronic RNAs. For example, the genome of brome mosaic virus (BMV) consists of three RNA segments, of which RNAs 1 and 2 contain a single cistron (Ahlquist *et al.*, 1981, 1984) (Figure 5.2). Cucumber mosaic virus (CMV), alfalfa mosaic virus (AlMV) and tobacco streak virus (TSV) are similar in this respect. Red clover necrotic mosaic virus (RCNMV) has two RNA segments, of which RNA 2 is monocistronic (Lommel *et al.*, 1988) (Figure 5.3).

5.3.3.2 *Formation of subgenomic messenger RNAs.* A series of subgenomic mRNAs may be synthesised *in vivo* from which cistrons located downstream

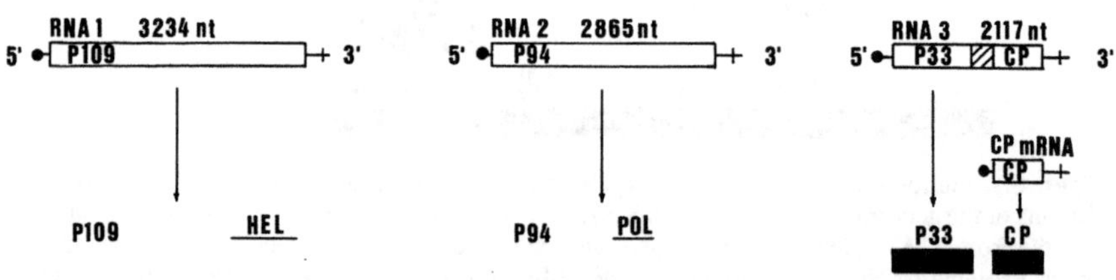

Figure 5.2. Genome organisation and expression of brome mosaic virus (BMV). Notations as in Figure 5.1. The shaded region on RNA 3 is an intergenic region, the complement of which in the negative strand contains the subgenomic promoter for the CP mRNA. Based on Ahlquist *et al.* (1984), Marsh *et al.* (1988).

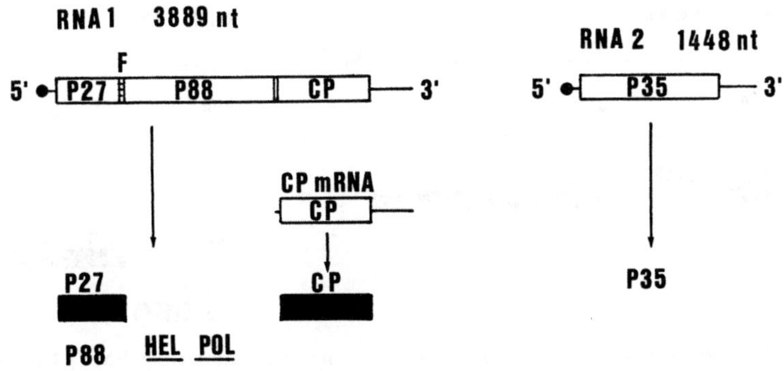

Figure 5.3. Genome organisation and model for expression of red clover necrotic mosaic virus (RCNMV). The sequences of the first 234 amino acids of P27 and P88 are identical. Ribosomes are postulated to either continue in the same reading frame to produce P27 or undergo a frameshift in the striped region marked F, resulting in the synthesis of the P88 fusion protein. P27 and P88 have been detected as *in vitro* translation products, but have not yet been detected *in vivo*. Notations as in Figure 5.1. Based on Lommel *et al.* (1988), Xiong and Lommel (1989).

on the genomic RNA can be translated. For example, the CPs of TMV and RCNMV and the 30 kDa protein (P30) of TMV are translated from subgenomic mRNAs (Morris-Krsinich *et al.*, 1983; Watanabe *et al.*, 1984; Xiong and Lommel, 1989) (Figures 5.1 and 5.3).

5.3.3.3 *Read-through translation.* The first cistron in the genomic RNA may have a 'leaky' termination codon that can be suppressed by a host transfer RNA (tRNA), thereby permitting some of the ribosomes to read through into a downstream cistron. For example in TMV, read-through of a UAG termination codon at the end of the cistron for P126 results in formation of P183 (Beier *et al.*, 1984a) (Figure 5.1). It is conceivable that resistance to infection could result from lack of an appropriate suppressor tRNA in a plant. It is known that wheat-germ does not have sufficient quantities of suppressor tRNA to allow synthesis of P183 (Beier *et al.*, 1984b).

5.3.3.4 *Ribosomal frameshifting.* The first cistron in the genomic RNA may have a sequence prior to its termination codon which allows a proportion of the ribosomes to shift into a different reading frame, thereby allowing expression of an overlapping downstream cistron. For example, there is evidence that in RNA 1 of RCNMV, ribosomal frameshifting can occur near the end of the cistron for the 27 kDa protein (P27), leading to synthesis of the 88 kDa protein (P88) (Xiong and Lommel, 1989) (Figure 5.3). Similar frameshift is postulated to occur in the synthesis of a 99 kDa protein encoded by barley yellow dwarf virus RNA (Miller *et al.*, 1988).

5.3.3.5 *Polyprotein processing.* The genomic RNA contains a long open

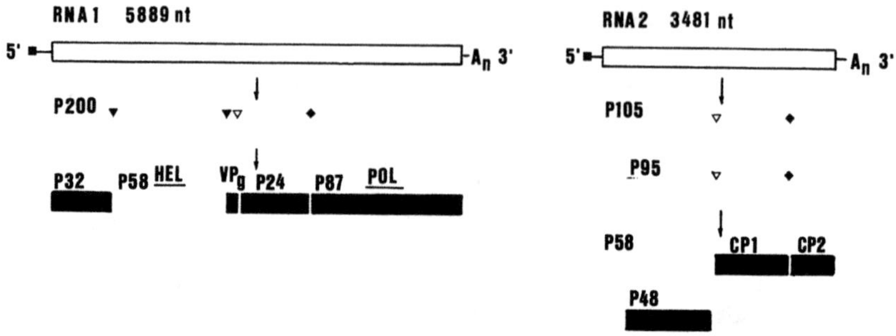

Figure 5.4. Genome organisation and expression of cowpea mosaic virus (CPMV). Notations as in Figure 5.1. ■, VPg. Proteinase (P24) cleavage sites: ▼, Q/S; ▽, Q/M; ◆, Q/G. CP1 (VP37) and CP2 (VP23) are the two capsid polypeptides. P105 and P95 arise from two alternative translational initiation sites in the same reading frame; the sequence of P95 is therefore contained completely within P105. Based on Goldbach (1986).

reading frame which is translated and then cleaved into smaller, functional proteins by viral proteinases. Since these proteinases form part of the polyprotein, initial cleavages must be autocatalytic. Examples include PVY and tobacco etch virus (TEV), each of which has a single RNA with one translational unit sufficient to direct synthesis of a polyprotein of *ca* 350 kDa and cowpea mosaic virus (CPMV) and tobacco ringspot virus (TobRV), which have two RNAs, each of which encode a polyprotein (Van Kammen *et al.*, 1987). The cleavage of the CPMV polyproteins by a virus-encoded proteinase is shown in Figure 5.4.

Clearly, different mechanisms must exist for the control of gene expression, e.g. synthesis of early and late proteins, depending on the mode of gene expression.

5.3.4 *Replication of virus RNA*

Replication of virus RNA takes place in two stages, both catalysed by an RNA-dependent RNA polymerase.

(1) Synthesis of a full-length complementary (negative) RNA strand using the genomic (positive) RNA strand as a template.
(2) Synthesis of progeny genomic RNA using the negative-strand RNA as a template.

There are, as yet, no reports of the isolation of template-dependent plant virus RNA polymerases capable of carrying out both stages together. The virus-encoded proteins required for RNA replication have been deduced from the composition of purified polymerases capable of copying genomic RNA to produce a negative strand, from the use of mutants, for divided genomic viruses from the minimum number of RNA segments needed to infect protoplasts and from the presence of conserved sequence motifs found in polymerases in other systems (Quadt and Jaspars, 1989). Two conserved regions have been identified in those positive-strand RNA plant virus genomes which have been sequenced: (1) a putative helicase region with several conserved sequence motifs, including an NTP (nucleoside triphosphate) binding site, and (2) a putative polymerase region with conserved sequence motifs, including a glycine–aspartic acid–aspartic acid sequence (GDD box) flanked by hydrophobic amino acids (Argos, 1988; Habili and Symons, 1989; Poch *et al.*, 1989). In the case of TMV, the putative helicase elements are located in the region common to P126 and P183, and the GDD box and associated sequences are located in the read-through portion of P183 (Figure 5.1). For BMV, P104 encoded by RNA 1 has the putative helicase elements and P94 encoded by RNA 2 has the GDD box (Figure 5.3). In CPMV, both putative helicase (P58) and polymerase (P87) are encoded as part of the RNA 1 polyprotein (Figure 5.4).

The RNA-dependent RNA polymerase of bacteriophage Qβ, which has been purified to homogeneity, consists of one virus-encoded subunit and three host-encoded subunits (Blumenthal, 1979). In this case a putative helicase function was not identified in the virus-encoded subunit, suggesting that one of the host proteins may serve the function of unwinding the base-paired regions that are formed during RNA synthesis. However it remains likely that replication of plant virus RNA requires at least one host protein. The RNA polymerase of turnip yellow mosaic virus (TYMV), which has been purified to a higher degree than any other plant virus RNA polymerase, is associated with a virus-encoded subunit of 115 kDa and a host-encoded subunit of 45 kDa (Candresse *et al.*, 1986); however it remains to be proved that the host protein is essential for polymerase activity.

Initiation of the synthesis of a negative strand on a positive-strand RNA template requires binding of the polymerase to a recognition site at the 3′ end of the template. The 3′ end of the RNA of many viruses can be folded into a characteristic secondary or tertiary structure which includes the RNA polymerase binding site. In TMV, BMV, CMV and AlMV this is a tRNA-like structure. Some other viral RNAs, such as CPMV and TobRV, terminate with a 3′ poly-A tail, although sequences adjacent to this tail are also essential for replication (Eggen *et al.*, 1989). The polymerase recognition sites have been defined on BMV RNAs by studying the ability of a template-specific polymerase extract to initiate negative-strand synthesis on templates modified by *in vitro* mutagenesis (Dreher and Hall, 1988).

Sequences at the 5′ end of the genomic RNA are also required for RNA infectivity (French and Ahlquist, 1987) and presumably reflect the requirement for binding of the polymerase to a site at the 3′ end of negative-strand RNA. However binding to positive- and negative-strand RNA templates clearly has different requirements since the terminal sequences are different, the synthesis of positive- and negative-strand RNAs are genetically separate events (Kroner *et al.*, 1989), a much larger amount of positive-strand RNA is produced and a polymerase extract capable of initiating synthesis on a positive-strand RNA template could not initiate synthesis of genomic RNA on a negative-strand RNA template (Dreher and Hall, 1988). In the case of BMV RNA 3, some internal sequences are also required for efficient replication (French and Ahlquist, 1987), but there is no evidence that internal sequences are needed for replication of other plant virus RNAs at present.

Synthesis of subgenomic RNAs is initiated by binding of polymerase to subgenomic promoters located on full-length negative-strand templates upstream of the initiation site for subgenomic RNA synthesis. In BMV RNA 3 negative strand, there is one subgenomic promoter site which controls synthesis of the coat protein mRNA (Marsh *et al.*, 1988). In TMV negative strand there are separate subgenomic promoter sites which control synthesis of the mRNAs for P30 (expressed relatively early) and CP (expressed late) (Palukaitis and Zaitlin, 1986; Dawson and Lehto, 1990).

Replication of plant positive-strand RNA viruses takes place in the cytoplasm of infected cells. RNA polymerases appear to be membrane-bound, and some proteins implicated in replication have membrane-binding domains, e.g. P58 encoded by RNA 1 of CPMV. However the precise sites where RNA replication takes place have not been clearly defined and probably differ for different viruses. Granular inclusion bodies have been invoked as the sites for TMV RNA replication (Saito *et al.*, 1987a; Okamoto *et al.*, 1988).

Some plant viruses (e.g. TMV, CMV, BMV, AlMV) have a cap structure (7MeGpppN) at the 5′ end of their genomic RNA similar to those found in most eukaryotic mRNAs. Capping of plant virus RNAs may be a function of the virus replication complex, since capping of host mRNAs occurs in the nucleus. Indeed a methylase activity has recently been found in a replication protein of Sindbis virus, a virus that is evolutionary related to TMV, CMV, BMV and AlMV (Mi *et al.*, 1989). Other plant virus RNAs (e.g. PVY, CPMV, TobRV) have a small virus-encoded protein (genome-linked virus protein or VPg) covalently linked by a phosphodiester linkage to the 5′ terminal nucleotide. Recent investigations with poliovirus, a virus which is evolutionarily related to PVY, CPMV and TobRV (Goldbach, 1986; Strauss and Strauss, 1988), suggest that linkage of VPg to virus RNA occurs by a self-catalysed reaction which may be essential for RNA replication (Tobin *et al.*, 1989).

5.3.5 *Virus assembly*

Particles of TMV and probably many other plant viruses are assembled spontaneously. In TMV RNA, assembly is initiated by interaction of protein discs with a specific region of the RNA, called the origin-of-assembly sequence (OAS), located within the gene for P30. Assembly is bidirectional, with the speed of assembly being faster in the 3′ to 5′ direction (Bloomer and Butler, 1986; Zimmern, 1977).

5.3.6 *Cell-to-cell long-distance movement*

To establish a systemic infection, a plant virus must move from the cell infected initially through a range of cell types, eventually to reach the vascular system where long-distance movement through the plant can occur in the phloem or xylem (reviewed by Hull, 1989). To move to an adjacent cell, the virus particles (or a ribonucleoprotein complex or free nucleic acid) must move through the protoplasmic bridges, the plasmodesmata. This is not a passive process, since the molecular mass exclusion limits of plasmodesmata fall within 700–800 daltons (Oparka and Prior, 1988). There is good evidence that plant viruses encode specific movement proteins which modify plasmodesmata to allow passage of virus particles or nucleic acids. Deom *et al.* (1987) provided unequivocal evidence that P30 of TMV is involved in cell-to-cell movement of the virus. Transgenic plants expressing P30 complemented the Lsl mutant of TMV, a mutant that is temperature-sensitive in cell-to-cell movement due to a single amino acid change in its P30, and potentiated cell-to-cell movement of the Lsl virus at the non-permissive temperature in both inoculated and upper systemic leaves.

TMV P30 accumulates in the plasmodesmata (Tomenius *et al.*, 1987), suggesting that it functions by modifying the structure of plasmodesmata. This suggestion is supported by the recent demonstration, using dextran polymers of different masses, that the molecular exclusion limit of plasmodesmata in transgenic plants expressing P30 is increased to more than 9400 daltons (but less than 17 200 daltons) (Wolfe *et al.*, 1989). However this increase was calculated to be insufficient to allow movement of either TMV particles or free virus RNA. Hence it is possible that an active transport mechanism is involved.

Other plant viruses possess movement proteins analogous to that of TMV, e.g. P35 encoded by RNA 2 of RCNMV (Figure 5.3), P33 encoded by RNA 3 of BMV (Figure 5.2) and P48 encoded by RNA 2 of CPMV (Figure 5.4). Some sequence homologies have been observed between movement proteins of different viruses, e.g. between those of TMV and tobacco rattle virus (TRV) (Boccara *et al.*, 1986), but not between others, e.g. there is little sequence homology between movement proteins of TMV and BMV. Movement proteins of some viruses are able to complement those of others, but this is not universal, i.e. there appears to be some specificity involved in movement. A number of different groups of virus movement proteins has been suggested (Atabekov and Dorokhov, 1984; Hull, 1989).

RNA of TMV mutants that cannot produce CP can infect cells and move from cell to cell in a leaf, but establish a systemic infection in a plant only after a prolonged period (Culver and Dawson, 1989a), indicating that CP is not needed for cell-to-cell movement but is required for rapid vascular transport. However, efficient assembly into particles is not required for long distance movement. T. Meshi and Y. Okada (quoted in Baulcombe and Hull, 1989) constructed a mutant of TMV in which the origin of assembly sequence was disrupted but the CP gene was retained. This mutant, which was defective in particle assembly but expressed coat protein normally, spread systemically through plants almost as well as wild type.

5.4 Molecular bases and exploitation of naturally occurring virus resistance genes

Although there is a considerable amount of information available on the genetics of resistance genes in plants, and many speculative models have been proposed (reviewed by Fraser, 1985, 1986, 1987a, b), there is comparatively little known about the molecular mechanisms of resistance. In this section a few systems have been selected where some progress has been made in recent years to define the molecular basis of resistance.

5.4.1 *Resistance operating by inhibition of virus replication in single cells*

5.4.1.1 *Inhibition of polyprotein processing.* A survey of 1000 cowpea lines revealed one line, Arlington, in which both plants and plant protoplasts were resistant to infection by CPMV. Arlington cowpea seedlings remained free of symptoms and no virus or capsid antigen could be detected after being inoculated with a concentration of virus 100 times that needed to infect susceptible cowpeas. Arlington cowpea protoplasts accumulated only about 10% of the amount of virus accumulated by susceptible lines when inoculated with virions at a concentration of 1–4 μg/ml. Genetic analysis indicated that Arlington cowpea seedling immunity is inherited as a simple dominant character, suggesting that immunity is mediated by an inhibitor of virus replication rather than by the absence of a factor required by the virus.

Extracts of Arlington cowpeas showed three activities that were less apparent or undetected in extracts of susceptible cowpeas and hence might restrict the replication of CPMV (Bruening *et al.*, 1987; Ponz *et al.*, 1988). The first was a proteinase activity that degraded the virus proteins. However the activity was not co-inherited with immunity against the virus in the progeny of cowpea crosses and was unlikely to be the immunity factor.

The second activity interfered with the translation of CPMV RNAs. The inheritance of this activity was complex and probably governed by several loci. At least one of the activities was co-inherited with the immunity gene and could therefore contribute to the immunity.

The third activity was inhibition of CPMV polyprotein processing. Cleavage is catalysed by a proteinase (P24) encoded by RNA 1 (Figures 5.4). Three types of proteolytic cleavage occur: glutamine/glycine (Q/G); glutamine/serine (Q/S); glutamine/methionine (Q/M). The *in vitro* translation products of RNA 2 from two in-phase AUG initiation codons are P105 and

P95. Addition of a source of CPMV proteinase (from extracts of infected protoplasts) results initially in a Q/M cleavage to produce P58 and P48 (from the N-terminal portions of the P105 and P95 respectively) and P60 (the precursor of the two capsid polypeptides, CP1 and CP2). It was demonstrated that this Q/M cleavage was inhibited by extracts of Arlington cowpeas. The activity was co-inherited with immunity in cowpea genetic crosses and hence the inhibitor could be the product of the immunity gene. It is noteworthy that this Q/M cleavage requires the P24 proteinase and a co-factor, P32, also encoded by RNA 1. Hence the inhibitor could prevent the action of either one, or both, of these proteins. It was not determined if the inhibitor interfered with cleavage of any of the other sites in the polyproteins encoded by RNA 1 or RNA 2.

RNAs of two other viruses, cowpea severe mosaic virus and TobRV, are translated to polyproteins which are cleaved in a manner similar to those of cowpea mosaic virus. However the Arlington cowpea inhibitor did not inhibit polyprotein processing of these two viruses. Arlington cowpeas are susceptible to infection by cowpea severe mosaic virus. They are immune to TobRV, but the immunity has been shown to have a different genetic basis to that of immunity to CPMV. Hence the Arlington proteinase inhibitor is specific to CPMV.

5.4.1.2 *Inhibition of virus RNA replication.* Tm-1 is a dominant gene that confers resistance in tomatoes to TMV (reviewed by Fraser, 1985). Inhibition of virus multiplication is greater in Tm-1/Tm-1 tomatoes (homozygous for the Tm-1 gene) than in Tm-1/+ tomatoes (heterozygous for the Tm-1 gene). Resistance is expressed in protoplasts, even when RNA is used as inoculum; synthesis of all virus-specific RNAs (genome and subgenomic) and virus-encoded proteins (P183, P126, P30, CP) (Figure 5.1) is completely blocked (Tm-1/Tm-1) or greatly reduced (Tm-1/+) (Watanabe *et al.*, 1987a). These results suggest that the Tm-1 gene product is expressed constitutively and operates after penetration and uncoating, at an early stage of viral RNA synthesis.

The Tm-1 resistance gene proved not to be durable, being widely overcome by resistance-breaking strains of the virus within a year of its introduction into commercial cultivars (Pelham *et al.*, 1970). Identification of sequence alterations in resistance-breaking strains of viruses is one approach to understanding resistance mechanisms, i.e. to identifying which viral component interacts with a host resistance factor and how the host and viral components interact. Ltal is a resistance-breaking mutant which arose spontaneously from a Tm-1/Tm-1 tomato plant that had been inoculated with a Japanese tomato strain of TMV, L. Unlike L, Ltal spread systemically and caused mosaic symptoms in the Tm-1/Tm-1 tomatoes (Watanabe *et al.*, 1987a). Comparison of the nucleotide sequences of Lta and L revealed five nucleotide changes (Meshi *et al.*, 1988). Two of these (in the P30 and CP cistrons) did not change the amino acid coded and a third, insertion of a U in the 5′ non-coding region, was found in other Tm-1-sensitive isolates; hence none of these was considered significant. The remaining two changes, a C-to-G transversion at nucleotide 3006 and a C-to-U transition at nucleotide 3021, caused amino acid substitutions in the common region of P126 and P183 (glutamine to glutamic acid at position 979 and histidine to tyrosine at position 984). By *in vitro* mutagenesis of cloned full-length cDNA and using an *in vitro* transcription system that allows synthesis of infectious TMV RNA from the cloned cDNA, a strain, designated T1, was constructed which differed from L in only these two amino acids. T1 behaved similarly to Ltal in establishing a systemic infection and mosaic symptoms in Tm-1/Tm-1 plants. The nucleotide sequence around the mutagenised portion of the RNA of progeny virus was unchanged. The same two changes were found in two other independently isolated resistance-breaking strains. Hence these two changes are sufficient for resistance breaking.

To determine if only one amino acid change could cause resistance breaking, two new strains, T2 and T3, were constructed (Meshi *et al.*, 1988). T2 had only the histidine-to-tyrosine change at position 984, and T3 had only the glutamine-to-glutamic acid change at position 979. Both strains replicated in tobacco and tomato (+/+) as efficiently as L, and the introduced mutations were recovered in the progeny virus.

When inoculated onto tomato (Tm-1/Tm-1), T3 induced mosaic symptoms and amounts of progeny virus similar to those induced by Ltal and T1. T2 gave no or only weak mosaic symptoms and variable, but lower, amounts of progeny virus. Hence the glutamine-to-glutamic acid change appeared to be the more important in resistance breaking. However RNA sequence analysis of eight preparations of progeny virus (designated T2p nos. 1–4 for T2 progeny, and T3p nos. 1–4 for T3 progeny) revealed additional base changes in all the progeny except one (T2p no. 4). In three of the progeny (T2p no. 1, T3p no. 1, T3p no. 2) the change resulted in the same sequence as T1. However, substitutions at other residues were found in five of the progeny (Figure 5.5).

In the region sequenced (amino acid residues 947–997), all of the amino acid changes in the naturally occurring resistance-breaking strains or in the progeny of T2 and T3 infection on Tm-1/Tm-1 plants were between residues 975 and 992. This region would therefore be expected to be the target for a resistance factor. The Tm-1 gene is known to be effective against several TMV strains, including strains distantly related to L such as U2 and a cowpea strain. Surprisingly, the sequence of the putative target region is rather variable, there being many base substitutions even between the closely related *vulgare* (U1) and tomato strains (Figure 5.5). The only unifying feature appears to be that the net positive charge of this region in Tm-1-sensitive strains (*L*, *vulgare*, OM) is higher than that in the resistance-breaking strains, suggesting that an electrostatic interaction is involved between the resistance factor and P130 and/or P180 (Meshi *et al.*, 1988). However, alteration of the secondary and

	965					970					975					980					985					990				
L	T	T	L	R	C	P	A	D	V	T	H	F	L	N	Q	R	Y	E	G	H	V	M	C	T	S	S	E	K	K	S
Ltal, T1															E					Y										
T2																				Y										
T2p # 1															E					Y										
T2p # 2											Y									Y										
T2p # 3											Y								E	Y										
T3															E															
T3p # 1															E					Y										
T3p # 2															E					Y								M		
T3p # 3															E													M		
T3p # 4															E					E										

Figure 5.5. Amino acid sequences of P126 and P183 of TMV strains between residues 965 and 994. The sequence of the L strain is shown. For Tm-1 resistance-breaking strains, only residues different from those of the L strain are shown (Meshi *et al.*, 1988). Helicase sequence motif IV (Habili and Symons, 1989) is underlined.

tertiary structure of the proteins caused by the mutations, affecting binding at another site, is another possibility.

There are at least two possible ways in which the Tm-1 resistance gene factor could inhibit virus multiplication.

(1) It could bind to P126 and P183, which are the putative RNA replicase components, thereby inactivating them. The amino acid changes in these proteins in the resistance-breaking strain could prevent this binding or could inactivate the resistance factor.
(2) The resistance gene could be an allele of a gene encoding a host protein which is an essential component of the RNA replicase. The product of the resistance gene would be a mutant of this host protein; it may bind to P126 and P183, but produce a non-functional replicase. In this case, changes in P126 and P183 in resistance-breaking strains would compensate for the mutations in the resistance gene product to allow production of an active replicase.

The temperature sensitivity of the Tm-1 resistance gene for virus multiplication (Fraser and Loughlin, 1982) would favour the first model but does not exclude the second one.

It is noteworthy that the amino acid changes found in the resistance-breaking strains fall within a part of P216 and P183 which contains a series of seven sequence motifs with homology to sequences of a superfamily of nucleic acid helicases of prokaryotic and eukaryotic organisms (Habili and Symons, 1989) (Figure 5.1). Indeed the region containing the mutations is immediately adjacent to sequence motif IV (Figure 5.5), which remains invariant in the Tm-1-sensitive and resistance-breaking strains. It is tempting to speculate that the resistance factor acts by inactivating the helicase function of the virus replicase.

5.4.2 *Resistance restricting virus cell-to-cell movement*

Two dominant, allelic genes, Tm-2 and Tm-2^2, for resistance to TMV have been introduced into cultivated tomatoes (Hall, 1980). The resistance is expressed in whole plants and in leaf discs but not in isolated protoplasts and therefore differs from the resistance conferred by the Tm-1 gene. When Tm-2 or Tm-2^2 plants were inoculated with a tomato strain of TMV, L, only single cells in the epidermis were observed to be stained with a virus-specific fluorescent antibody (Nishiguchi and Motoyoshi, 1987). Tm-2 resistance is specific for TMV and Tm-2 plants can be infected with potato virus X (PVX), a member of another virus group. When Tm-2 plants were preinoculated with PVX, followed by inoculation with TMV strain L, spread of TMV occurred, suggesting that PVX broke the Tm-2 resistance by providing a cell-to-cell movement function (Atabekov and Dorokhov, 1984). These observations suggest that Tm-2 and Tm-2^2 resistances act by restricting the cell-to-cell movement of TMV.

Further information on the molecular basis of Tm-2 resistance has been obtained by studies of resistance-breaking strains of the virus. Strain Ltb1 is a spontaneous mutant isolated from a Tm-2/Tm-2 tomato plant inoculated with strain L. Unlike L, Ltb1 propagates in Tm-2/Tm-2 tomatoes, producing mosaic symptoms in systemically infected leaves. Sequence analysis of Ltb1 RNA, combined with *in vitro* mutagenesis of cDNA clones and synthesis of infectious RNA by *in vitro* transcription, showed that the resistance-breaking properties of Ltb1 were caused by two mutations in the movement protein (P30) of the virus (Meshi *et al.*, 1989). One of these caused a cysteine-to-phenylalanine substitution at position 68, whereas the other caused a glutamic acid-to-lysine substitution at position 133. Reconstructed strains with either one of these mutations induced increased levels of virus multiplication compared with strain L. The cysteine-to-phenylalanine mutation appeared to be more efficient at overcoming resistance than did the glutamic acid-to-lysine mutation. However, neither mutation alone overcame resistance as efficiently as both mutations together.

Analysis of another independently isolated resistance-breaking strain, C32, which arose spontaneously from a different wild-type tomato strain, CH3, revealed two mutations in its movement protein, both of which caused glutamic acid-to-lysine substitutions. One of these, at position 133, was the same as that found in Ltb1. The second occurred at position 52, close to the position of the cysteine-to-phenylalanine substitution in Ltb1. C32 and Ltb1 induced similar symptoms in Tm-2/Tm-2 plants. Hence the ability to overcome Tm-2 resistance can be conferred by two different pairs of mutations in the movement protein of TMV.

There are two regions in the middle position of the movement protein that are relatively well conserved in different TMV strains and which may be important for the structure and function of the protein (Saito *et al.*, 1988).

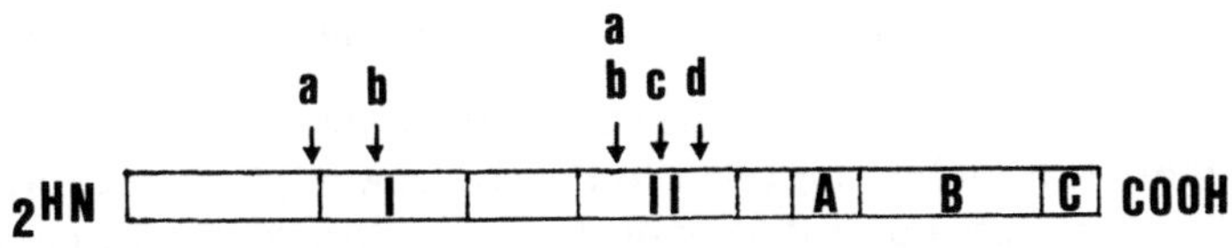

Figure 5.6. Location of mutations in movement proteins of Tm-2 resistance-breaking and temperature-sensitive mutants of TMV. I and II denote relatively well-conserved regions in the movement proteins of different tobamoviruses. Regions A and C are rich in acidic amino acids, while region B is rich in basic amino acids. Arrows indicate the positions of mutations in Tm-2 resistance-breaking strains (a, C32; b, Ltbl) and temperature-sensitive mutants (c, Ni2519; d, Ls1) (data from Saito *et al.*, 1988; Meshi *et al.*, 1989, and references quoted therein).

Temperature-sensitive mutations for cell-to-cell movement have been located in one of these regions. It is noteworthy that the amino acid substitutions responsible for Tm-2 resistance-breaking are located within or close to the well-conserved regions (Figure 5.6).

Meshi *et al.*, (1989) have suggested two possible mechanisms by which the Tm-2 resistance could prevent virus cell-to-cell movement.

(1) A resistance factor could be a mutant form of a host factor that is normally required for virus cell-to-cell movement. The molecular interactions between this host factor and the virus movement protein may be altered, resulting in defective movement. A functional interaction could be restored by the mutations in the movement protein that occur in the resistance breaking strains.
(2) The resistance factor may be unrelated to the normal movement function and may act solely as an inhibitor of the function of the wild-type movement protein. Mutations in the movement protein could evade or repress the resistance reactions.

Like Tm-1 resistance, Tm-2 resistance did not prove to be durable. However the Tm-2^2 resistance gene, which is allelic with the Tm-2 gene, has given durable resistance, and resistance-breaking strains have only rarely been found. Determination of sequence changes in Tm-2^2 resistance breaking strains may suggest reasons why they have not become widespread.

Blocking of cell-to-cell movement may be a common mechanism of resistance. Cowpeas are resistant to infection by tobacco strains of TMV (e.g. U1) but can be infected systemically with cowpea strains of the virus. Sulzinski and Zaitlin (1982) found that only about 1 in 50 000 cells of cowpea leaves inoculated with TMV U1 became infected, although the level of infection in individually infected cells was the same as that in cells of similarly inoculated tobacco cells. This suggests that TMV U1 can replicate in cowpea cells, but cannot spread from cell to cell. It is noteworthy that the amino acids in the movement protein which were altered in Tm-2 resistance-breaking strains were different in the cowpea strain. Hence movement proteins may need to be adapted to allow movement in different hosts.

5.4.3 *Resistance mediated by induction of a host hypersensitive response*

Another type of resistance allows the virus to replicate in initially infected cells and to spread to neighbouring cells in much the same way as in a sensitive host. However, after a period of time, induction of a host resistance mechanism leads to localisation of the virus to a region around the initial site of infection and is often accompanied by necrosis of cells in this region. Hence a common outcome of the inoculation of such resistant hosts with virus is the formation of necrotic local lesions on the inoculated leaves. The host reaction, often termed a hypersensitive response, prevents systemic infection of the plant, and non-inoculated leaves show resistance to virus infection. This resistance is shown not only to the inducing virus, but also to unrelated viruses and even different pathogens, such as bacteria and fungi. Whether this 'induced systemic-induced resistance' is caused by the same mechanism that localises the virus infection on inoculated leaves, 'induced localised resistance', is not clear. The hypersensitive response is important in agriculture as the basis of many examples of field resistance, but has a disadvantage that the response is often temperature-sensitive.

The recognition event that triggers a hypersensitive response is a direct or indirect interaction between a virus gene product and the product of a host resistance gene. In *Nicotiana* species, the N gene, originally from *N. glutinosa*, confers hypersensitive-type resistance to all strains of TMV. The N′ gene from *N. sylvestris* confers hypersensitive-type resistance to some, but not all, strains of TMV.

Recently it has been shown that the CP gene of TMV induces a hypersensitive response to the N′ gene in *N. sylvestris* (Saito *et al.*, 1987b). This was shown by production of CP hybrids of strain L, which induces a hypersensitive response in *N sylvestris*, and strain OM, which induces a systemic infection in *N. sylvestris*. When the coat protein gene of strain L was replaced by the coat protein gene of strain OM, a hypersensitive response was not induced. Furthermore, mutations which caused premature termination of coat protein synthesis after 3, 67, 96 or 110 amino acids abolished the ability of the strain L to induce a hypersensitive response (Saito *et al.*, 1989). Single amino acid changes at several positions in the CP gene can convert a strain which induces a systemic response into one which induces a hypersensitive response (Knorr and Dawson, 1988; Culver and Dawson, 1989b).

A gene other than the CP gene must be responsible for induction of the hypersensitive response in *N. tabacum* cultivars carrying the N gene, since mutants with the entire CP gene deleted, or replaced by a reporter gene, can still induce this response (Dawson *et al.*, 1988, 1989). It has been suggested that the movement protein of strain L may be responsible for induction of a hypersensitive response that sometimes occurs in heterozygous Tm-2/+ tomato plants (Meshi *et al.*, 1989).

A wide range of recognition events can trigger a hypersensitive response, involving a direct or indirect interaction of a product of a gene, called an avirulence gene, in a pathogen (bacterium, fungus or virus) with the product of a specific resistance gene in the host plant. Although little is known of the diversity of molecular interactions that occur between avirulence gene products and resistance gene products, a common set of plant genes is activated in a general defence reaction producing a broad resistance against further infection by viruses, fungi or bacteria. Hence the gene products that confer resistance to virus infection would be expected to be a subset of all those induced in a hypersensitive response.

Many of the gene products induced in a hypersensitive response are also induced as a result of wounding or treatment with chemicals that induce systemic resistance, such as salicylic acid or polyacrylic acid. Candidates for intercellular stress signals that may be common to different types of induction include ethylene, glutathione and oligosaccharides released from plant cell walls. Defence genes have also been shown to be under developmental regulation in healthy plants (Lamb *et al.*, 1989).

The large number of enzymes and metabolites whose synthesis is induced in plants showing a hypersensitive response has been reviewed extensively (Ponz and Bruening, 1986; Fritig *et al.*, 1987; Bol and Van Kan, 1988; Lamb *et al.*, 1989). Some of these, e.g. phytoalexins, inhibitors of polygalacturonases and those pathogenesis-related (PR) proteins that are chitinases, beta $1 \rightarrow 3$ glucanases or proteinase/α-amylase inhibitors, may primarily confer resistance to bacteria, fungi or insects (see Chapter 4). Increased synthesis of cell wall polymers, such as hydroxyproline-rich glycoproteins, lignin, suberin or callose, could confer resistance to viruses as well as other pathogens. Some of these polymers could affect the localisation of virus movement proteins in the plasmodesmata and hence block cell-to-cell movement. Indeed, reduced levels of the TMV movement protein in a cell wall fraction following a hypersensitive reaction have been reported (Moser *et al.*, 1988).

Other induced proteins for which no function has been assigned, e.g. PR-1 proteins and a glycine-rich protein (GRP), have been considered as candidates for antiviral activity. However constitutive expression of PR-1a, PR-1b or GRP in transgenic tobacco plants did not confer resistance to TMV or AlMV (Cutt *et al.*, 1989; Linthorst *et al.*, 1989).

Two other proteins associated with the hypersensitive response are the antiviral factor (AVF) and inhibitor of virus replication (IVR). AVF is a family of proteins whose activity is increased after infection of *N. tabacum* (NN) with TMV. AVF has been reported to act in plants analogously to the action of interferon in animal cells, by increasing the activity of an enzyme which catalyses the synthesis of antivirally active oligoisoadenylates (Sela *et al.*, 1987). IVR, which is serologically unrelated to AVF (Gera and Loebenstein, 1989), is released from protoplasts prepared from TMV-infected *N. tabacum* (NN). The role of AVF and/or IVR in localisation of virus infection after a hypersensitive response remains to be proven.

5.4.4 *Prospects for cloning and manipulating naturally occurring virus-resistance genes*

There are still many crops for which a source of naturally occurring virus resistance genes, which could be introduced by conventional genetic techniques, is not available. Therefore cloning resistance genes to enable them to be introduced into unrelated crop species is a worthwhile goal. Furthermore, knowledge of the nature of resistance gene products and their mode of action may enable us to modify naturally occurring resistance genes to make them more effective and durable. RNA viruses have a high mutation rate (Steinhauer and Holland, 1987) and some resistance genes, e.g. Tm-1, Tm-2, have been overcome easily by only a few base changes. Other resistance genes have proved durable; for example, the N gene in tobacco described by Holmes (1938) has only recently been overcome (Csillery *et al.*, 1983), and such resistance-breaking strains are still very rare. Resistance might be made permanent if it involved inhibition of an essential, highly conserved region of a functional virus gene.

There are several possible approaches to cloning naturally occurring virus resistance genes.

(1) If the resistance gene product can be identified, it can be cloned by conventional means, e.g. the proteinase inhibitor in Arlington cowpeas. Identification of the product might be possible in some other cases by means of a deduced biological activity, e.g. the product of the Tm-1 resistance gene might be identified by testing extracts of Tm-1/Tm-1 tomatoes for inhibition of RNA-dependent RNA polymerase activity extracted from virus-infected +/+ tomatoes. In other cases, *in vitro* translation of mRNA from a resistant plant, prehybridised with cDNA from a sensitive and isogenic plant, could identify proteins specific to the resistant host. Some of the proteins whose synthesis is induced in a hypersensitive response, e.g. AVF, IVR, are additional sources of genes that could confer resistance.

(2) Differential hybridisation and enrichment procedures can be used to screen a cDNA library from a resistant plant to find clones specific for resistance, provided an isogenic susceptible plant is available.

(3) A susceptible host could be transformed with a genomic library of a resistant host and transformants tested for virus resistance. While this shotgun approach may be feasible for plants with comparatively small genomes, such as *Arabidopsis*, Simoens *et al.* (1986) calculated that it would be necessary to examine 1.7×10^7 transformants of tobacco to ensure a 95% chance of a given gene being present. Restriction fragment length polymorphism (RFLP) maps of the genomes of a number of crop plants have been constructed, on which virus resistance genes have been located. With the availability of pulse field gel electrophoresis to isolate large fragments of chromosomes, and YAC cloning technology (see

Chapter 2), it should be possible to produce a regional library containing a particular resistance gene. Use of such a library for plant transformation should reduce the number of transformants which need to be screened to practicable levels.

(4) Transposon tagging, which has been used to isolate a number of plant genes, could potentially be used to isolate virus resistance genes. The maize transposon Ac has been shown to be capable of transposition in a range of plants, and methods have been devised to select cells in which transposition has taken place (Baker *et al.*, 1987; Jones *et al.*, 1989). T-DNA insertions can also be used for gene tagging, and selection for cells in which T-DNA has inserted into a gene (Konz *et al.*, 1989) would reduce considerably the number of plants which must be screened (in this case for loss of resistance).

All of these methods are currently being used in different laboratories, and isolation of several virus resistance genes is anticipated in the next few years.

5.5 Construction and expression of artificial resistance genes in transgenic plants

5.5.1 *The cross-protection phenomenon as a source of potential resistance genes*

Cross-protection is a phenomenon in which a plant that has been infected with a mild strain of a virus that produces few or no symptoms is protected from superinfection by a severe strain of a related virus. Cross-protection differs from induced systemic resistance in that the host is not resistant to either the mild (protecting) or severe (challenging) virus alone and the protection is specific to related viruses. The plant is systemically infected with the protecting virus and the protection is caused by the presence of the protecting virus in the same cell as the challenging virus. Cross-protection can be demonstrated in protoplasts.

Cross-protection has been used in agriculture to protect crops for which no other source of resistance or control measures were available (Fulton, 1986). An outstandingly successful example is the control of tristeza virus disease of citrus trees, which almost wiped out the citrus fruit industry in parts of South America in the 1920s (Costa and Muller, 1980). The mild virus strains used for protection were spontaneous mutants which were naturally protecting a few trees in otherwise heavily diseased orchards. For other crops, mild strains have been obtained from heat-treated infected tissue or by mutagenesis.

There are several potential drawbacks to the widespread use of cross-protection.

(1) The protecting strain might mutate to a more severe form, leading to extensive crop losses.

(2) There is a possibility of conditioning an entire crop for a severe synergistic reaction by an unrelated virus. Such a reaction occurs when PVY infects potato carrying a mild strain of PVX.
(3) A protecting virus might spread to other crops on which its effects may be more severe.
(4) The protecting strain may cause small but significant crop losses.

Powell *et al.* (1986) suggested that these objections might be overcome if cross-protection could be obtained as the result of expression of a single virus gene in a transgenic plant.

Some of the mechanisms that have been proposed to explain cross-protection (reviewed by Palakaitis and Zaitlin, 1984; Sherwood, 1987) have suggested ways in which virus-resistant transgenic plants might be produced.

(1) The presence in a cell of an excess of coat protein of the protecting virus could prevent uncoating of the challenging virus and hence replication of its nucleic acid.
(2) The challenging (positive-strand RNA) virus could initiate replication by synthesis of a negative-strand RNA. However this negative strand could hybridise to a previously synthesised positive strand of the related protecting virus (negative-strand capture), therefore preventing it from acting as a template for synthesis of positive strands of the challenging virus. This mechanism could explain cross-protection by coatless mutants of RNA viruses, e.g. the DT-IG strain of TMV (Gerber and Sarkar, 1989), and by viroids which do not encode proteins.

5.5.2 *Cross-protection in transgenic plants expressing a mild virus strain*

$L_{11}A$ is an attenuated strain of TMV which has been used in Japan to protect tomatoes against wild severe strains. In protoplasts $L_{11}A$ replicates as well as strain L from which it was derived; however the level of subgenomic mRNA for the movement protein is specifically reduced for $L_{11}A$ (Watanabe *et al.*, 1987b). A defect in virus movement could explain the reduced accumulation of $L_{11}A$ in plants compared with L. $L_{11}A$ differs from L in only three amino acids, all of which are located in the common portion of the P126 proteins and P183 proteins (Nishiguchi *et al.*, 1985). It is likely that structural alterations in the virus replicase caused by these amino acid substitutions specifically decreased its ability to bind to the promoter for the movement protein mRNA.

Transgenic plants have been obtained containing an integrated DNA copy of the entire L or $L_{11}A$ genome, expressed under the control of a cauliflower mosaic virus 35S RNA promoter (35S promoter). Infectious virus could be recovered from both types of plant, but, whereas transgenic L plants showed mosaic symptoms typical of virus infection, transgenic $L_{11}A$ plants had no or only mild mosaic on their leaves (Yamaya *et al.*, 1988a, b). When transgenic $L_{11}A$ plants were inoculated with L particles (1 or 5 μg/ml) no symptoms were observed on most plants six weeks later. Protection was also observed when RNA (10 μg/ml) was used as inoculum, indicating that a mechanism other than suppression of virus uncoating is involved.

The method would have the advantage for crops grown annually from seed, such as tomato, that inoculation of the crop with a mild virus strain is avoided; but it retains all the disadvantages of classical cross-protection. Nevertheless it could be useful for studying mechanisms of cross-protection; for example, the effect of deleting the genes for the movement and coat proteins, neither of which is needed for replication in single cells, could be investigated. This could lead to the production of safer virus strains for this form of genetically engineered cross-protection, i.e. strains incapable of reverting to virulence or producing systemic infection in other crops.

5.5.3 *Resistance in transgenic plants expressing virus coat protein genes*

Transgenic tobacco and tomato plants that express the CP-coding sequence of the common (U1) strain of TMV under the control of a 35S promoter have been constructed and found to be partly resistant to infection with TMV (Powell *et al.*, 1986; Beachy *et al.*, 1987; Nelson *et al.*, 1987, 1988). Resistance was manifest as either failure to develop symptoms or a delay in symptom development and a corresponding reduction in virus multiplication. Increasing the concentration of virus in the inoculum shortened the delay in appearance of symptoms. A good correlation was found between the amount of coat protein synthesised in the transgenic plants and the degree of protection afforded.

Protection was obtained against a variety of TMV isolates, although the degree of protection varied. For example when transgenic tomato plants expressing the U1 CP gene were inoculated with the U1 strain, most plants did not develop symptoms or accumulate virus within a 30 day observation period. However when inoculated with the more aggressive PV-230 strain, half of the plants developed symptoms, although all were delayed in appearance of symptoms. Protection was also afforded against tomato strains of the virus, such as strain L and Tm-1 and Tm-2 resistance-breaking strains. Protection can break down when the CPs of the two viruses become too dissimilar, e.g. particles of the cowpea strain of TMV (Cc-TMV) can overcome the 'protection' in U1 CP-transgenic tobacco plants (Anderson *et al.*, 1989); the CPs of Cc and U1 TMV are only 45% homologous. Similarly, transgenic tobacco plants expressing tobacco streak virus CP did not cross-protect against infection by AlMV (Van Dun *et al.*, 1988). However, interference with disease development of some non-related viruses in transgenic plants expressing CP genes of TMV or AlMV has been reported (Anderson *et al.*, 1989).

Transgenic plants expressing CP genes of viruses in a range of different taxonomic groups (Table 5.1), e.g. AlMV (Loesch-Fries *et al.*, 1987; Tumer *et al.*, 1987), CMV (Cuozzo *et al.*, 1988), TRV (Van Dun and Bol, 1988), TSV (Van Dun *et al.*, 1988), TEV (D.M. Stark and R.N. Beachy, unpublished results quoted in Register and Beachy, 1989) and PVX (Hemenway *et al.*, 1988, Hoekema *et al.*, 1989), also show resistance to infection with the respective viruses, suggesting that this approach may be generally applicable for obtaining resistance, at least for positive-strand RNA viruses. Prospects for agricultural use look good. CP genes in transgenic plants appear to be meiotically stable over several generations. In field trials, CP-expressing tomato plants were resistant to infection with tomato strains of TMV; whereas tomato fruit yields of non-transgenic plants were decreased by 25–35% due to infection, yields from CP-expressing lines were unaffected (Nelson *et al.*, 1988). Yields from one CP-expressing line were the same as those from uninoculated non-transgenic plants, indicating that expression of the CP gene at levels sufficient to confer resistance does not intrinsically cause yield depression. That environmental conditions can affect the resistance was illustrated by the decreased levels of CP and resistance to TMV in transgenic tobacco plants exposed continuously to high temperatures (35°C) (Nejidat and Beachy, 1989). Interestingly, CP levels in transgenic tomato plants also dropped under elevated temperatures, but the plants retained their high resistance to TMV.

In the transgenic plants expressing CP genes discussed above, the mRNA contains the 3′ end of the virus RNA as well as the CP-coding region. Hence protection could be caused by the CP or by its mRNA (as in the negative-strand capture model of Palukaitis and Zaitlin, 1984). Proof that the CP mediated the resistance was shown by the observations that transgenic CP-expressing plants which lacked the 3′ tRNA-like structure of TMV RNA retained protection, whereas those in which the initiation codon for the TMV coat protein-coding had been deleted, or in which the AlMV CP-coding region carried a frame shift mutation, failed to resist infection by the challenge virus, despite the presence of the mutant CP mRNA retaining the 3′ genomic sequences (Powell *et al.*, 1988; Van Dun *et al.*, 1988).

Several lines of evidence suggest that, at least for TMV, expression of CP in transgenic plants confers resistance by interfering with an early stage of virus infection, probably co-translational disassembly of virus particles.

(1) The number of cells infected by TMV in inoculated leaves or protoplast suspensions is reduced. However little resistance to infection by TMV RNA in inoculated leaves or protoplasts was shown (Nelson *et al.*, 1987; Register and Beachy, 1988). This suggests that resistance operates before complete uncoating of particles has taken place.

(2) It has been shown that chimeric RNAs containing the 5′ leader sequence of TMV RNA, a β-glucuronidase (GUS) reporter gene and a TMV origin-of-assembly (OAS) sequence can be packaged *in vitro* in TMV CP to form pseudovirus particles. After electroporation-mediated entry into a variety of cell types, these particles uncoat and express their RNA, often very

efficiently (Gallie *et al.*, 1987). However expression of GUS particles was 100 times lower in CP-transgenic protoplasts than in non-transformed protoplasts, suggesting that uncoating of GUS-OAS mRNA is inhibited in CP-transgenic protoplasts (Osbourn *et al.*, 1989a).

(3) In transgenic plants expressing both CP and a chloramphenicol acetyl transferase (CAT) reporter gene linked to a TMV (OAS) sequence, encapsidation of CAT-OAS mRNA or reduction of CAT activity could not be detected, even though such plants were resistant to TMV infection (Osbourn *et al.* 1989b). It was concluded that substantial re-encapsidation of uncoated RNA from challenge virus by endogenous CP in the cytoplasm of CP-transgenic plants is unlikely to account for resistance to virus infection.

This result is consistent with the observation that pH 8.0-treated TMV particles are able to overcome CP-mediated resistance in CP-transgenic protoplasts in much the same way as TMV RNA (Register and Beachy, 1988). Brief exposure of TMV to pH 8.0 has been proposed to remove one turn of CP subunits from the 5′ end of the virus rod, allowing initiation of co-translational disassembly *in vitro* (Wilson, 1988). It is clear that re-encapsidation of this exposed 5′ end does not occur *in vivo*. Therefore CP protection probably acts by interfering with the removal of the first few CP subunits from the 5′ end of the TMV rods and hence preventing the initiation of co-translational disassembly. Experiments with protoplasts suggest that helical, virus-like, aggregates of CP are more effective inhibitors than small aggregates (Register and Beachy, 1989).

Inhibition of nucleocapsid disassembly is probably not the only inhibiting effect of CP. Osbourn *et al.* (1989a) obtained evidence suggesting that a later stage of the replication cycle in protoplasts may be inhibited, to block the replication of the small amount of RNA that does become uncoated. Watanabe *et al.* (1987b) have suggested that CP could block virus cell-to-cell movement in plants by binding to the OAS sequence in the movement protein mRNA, thereby preventing synthesis of the movement protein. There is also evidence that CP expressed in transgenic plants can interfere with long-distance spread of the virus (Register *et al.*, 1989). The action of CP at multiple sites in the virus infection cycle suggests that CP resistance should be an effective and durable trait.

It has been reported that free RNA of PVX does not overcome resistance in PVX CP-transgenic plants (Hemenway *et al.*, 1988). This may reflect differences in the assembly and disassembly of PVX particles. The OAS sequence on PVX RNA is close to the 5′ end of the RNA and co-translational disassembly has not been demonstrated. The possibility that binding of PVX CP to its OAS sequence prevents RNA translation cannot be discounted. Hence there may be different mechanisms involved in CP protection for different viruses.

5.5.4 *Disease attenuation in transgenic plants expressing satellite RNAs*

Isolates of some viruses contain, in addition to the genomic RNA, an extra RNA species termed a satellite RNA (reviewed by Francki, 1985b). Satellite RNAs are totally dependent on their helper viruses for their replication. The association between a satellite RNA and its helper virus is generally specific; only closely related viruses can act as helpers for a given satellite. Satellite RNAs are not required for the replication of their helper viruses and indeed, in many instances, can inhibit the replication of their helper viruses. Satellite RNAs contain sequences unrelated to the nucleic acid sequences of their helper viruses, but do not encode a CP; they are encapsidated by CP synthesised by their helper viruses.

Francki (1985b) distinguished *satellite RNAs* from two other types of RNA that occur in some, but not all, isolates of a virus. RNAs of *satellite viruses*, like satellite RNAs, depend on helper viruses for their replication and contain sequences unrelated to their helper viruses; however, they encode their own coat protein in which they are encapsidated. *Defective-interfering RNAs*, like satellite RNAs, depend on helper viruses for their replication and are encapsidated in proteins encoded by the helper virus; however, they are derived from their helper virus genomes by deletions and sequence rearrangements.

The presence of a satellite RNA can alter the symptoms caused by its helper virus. Generally satellite RNAs cause attenuation of symptoms but in some cases can induce more severe and different symptoms. The type of symptom modification produced depends on the satellite RNA, the strain of helper virus and the host plant. Inoculation of crops with helper virus/satellite RNA combinations that produce symptomless infections has been employed in China to obtain protection against the effects of infection by wild strains of the virus (Tien *et al.*, 1987). However such practices suffer from the disadvantages described for classical cross-protection. The need for preinoculation of plants with helper virus would be obviated if the satellite RNA sequences were expressed from the plant nuclear genome. This has been achieved with satellite RNAs of TobRV and CMV.

5.5.4.1 *Satellite RNA of tobacco ringspot virus.* TobRV infects a wide range of dicotyledonous plants, including soyabean, in which it causes budlight disease. It has a genome of two positive-strand RNA molecules (7.2 kb and 4.8 kb) and its proteins are produced by cleavage of polyprotein precursors in an analogous way to those of CPMV (Figure 5.4). Satellite RNAs (STobRV) with the ability to ameliorate disease symptoms have been found associated with many isolates of TobRV. STobRV is a linear RNA of 359 nucleotides which does not appear to encode functional polypeptide products. The encapsidated strand of STobRV is designated the (+) strand. Replication of STobRV, which presumably utilises the RNA-dependent RNA polymerase encoded by RNA 1 of TobRV, takes place via (−) strand intermediates. Rolling circle mechanisms have been proposed to account for the concatameric (+) and (−) strand transcripts isolated from infected tissues. Concatamers of either (+) or (−) strands undergo autocatalytic ribonucleolytic cleavage at specific sites *in vitro*. Cleavage of (+) strand concatamers occurs between nucleotides 359 and 1 to generate an RNA with the same terminal sequences as SToRV. Cleavage of (−) strand concatamers occurs between nucleotides 48 and 49 (Buzayan *et al.*, 1986a, b; Forster and Symons, 1987; Haseloff and Gerlach, 1989).

Gerlach *et al.* (1987) produced transgenic tobacco plants containing integrated trimers of the cloned permuted monomer of STobRV cDNA flanked by a 35S promoter and *nos* termination sequence. Transgenic ST+ and ST− plants contained the trimers in orientation which allowed transcription of (+) and (−) STobRV sequences respectively. RNA from the transgenic plants was analysed by northern blotting with probes specific for (+) or (−) STobRV sequences. The major transcript from both ST+ and ST− was about SToRV monomer length (359 nucleotides), consistent with autocatalytic cleavage of the ST+ and ST− transcripts. When ST+ and ST− plants were infected with TobRV, after three weeks there was a large increase in the levels of satellite RNA sequences.

A striking feature of the TobRV-infected ST+ plants was the attenuation of symptoms compared with infected control plants. Lesions on inoculated leaves of ST+ plants developed one to two days later and lacked the necrotic centres of lesions on leaves of infected control plants. After three weeks new leaves on infected ST+ plants were symptomless, and by six weeks the growth of these plants was similar to that of uninfected plants. In contrast, after six weeks infected control plants were stunted and newly developed leaves showed severe mottling and were reduced in size.

In the ST+ plants, the attenuated symptoms correlated with a lower level of virus replication than in infected control plants. In fact in upper symptomless leaves, no virus could be detected. This suggests that replication of STobRV to high levels may depress TobRV replication to such an extent that insufficient movement protein is produced to potentiate cell-to-cell movement of the virus.

ST− plants appeared to be as susceptible as control plants in the initial stages of infection. However after six weeks the plants had developed a resistant phenotype and had completely recovered. The difference could be due to the lower level of STobRV transcripts observed in ST− plants than in ST+ plants, or to differences in the mechanism of initiating replication on (+) or (−) transcripts. Previously, (+) but not (−) strand transcripts had been found to initiate replication when inoculated onto plants together with TobRV.

How STobRV inhibits TobRV replication is not known. However one possibility is that STobRV binds the TobRV replicase with high affinity. When TobRV initiates infection, after or concurrently with particle disassembly, the RNA 1 must be translated to produce *inter alia* the replicase and may be engaged repeatedly in this process for some time before replication can be initiated. In contrast STobRV, already present in multiple copies by transcription of the ST+ gene, is probably not translated and is free to bind the first replicase molecules synthesised. Hence the levels of STobRV could be amplified even before replication of TobRV has commenced. Further replication competition would be in favour of the small satellite RNA. A scheme depicting stages in the replication of STobRV in ST+ plants is shown in Figure 5.7.

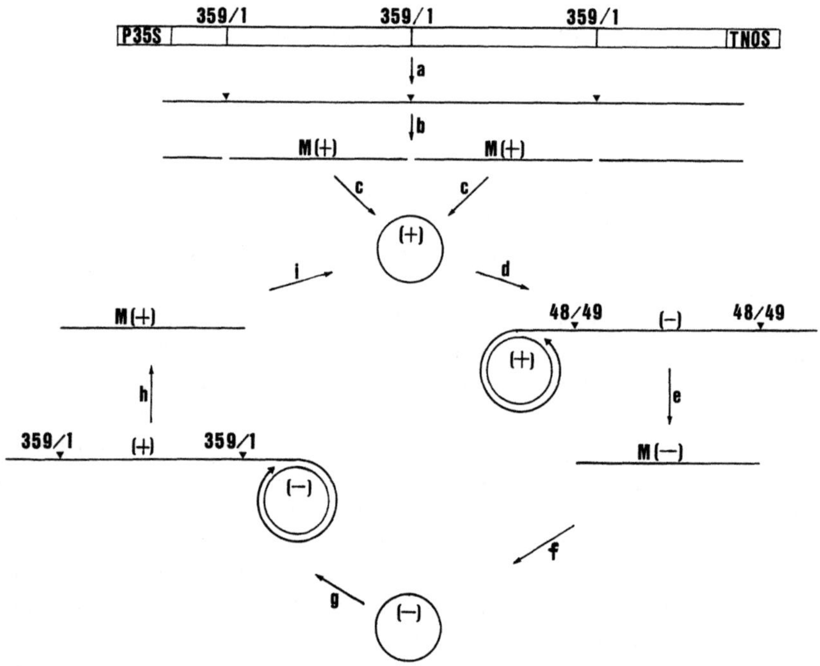

Figure 5.7. Model for the synthesis of satellite tobacco ringspot virus (STobRV) in uninfected transgenic plants and its replication after infection of the plants by TobRV. Arrowheads indicate the sites for self-cleavage of STobRV (+) and (−) RNA strands. (a) Transcription of integrated trimers of the cloned permuted monomer of STobRV cDNA in ST+ transgenic plants under the transcriptional control of a 35S promoter (P35S) and a terminator (polyadenylation signal) from a nopaline synthase gene (TNOS) to produce permuted trimeric STobRV RNA molecules (+strands). (b) Self-cleavage of trimeric RNA molecules between nucleotides 359 and 1 to produce monomeric STobRV (+) strands (M+). (c) Circularisation of monomer units, probably catalysed by an RNA ligase. (d) Rolling circle synthesis of multimeric STobRV (−) strands, using the circular (+) strand template. (e) Self-cleavage of multimers between nucleotides 48 and 49 to generate monomeric STobRV (−) strands (M−). (f) Circularisation of (−) strand monomer units. (g) Rolling circle synthesis of multimeric STobRV (+) strands, using the circular (−) strand template. (h) Self-cleavage of multimers to generate monomeric STobRV (+) strands. (i) Circularisation of monomeric (+) strands. Stages a, b and probably c occur in uninfected ST+ transgenic plants. The subsequent stages occur only after infection of ST+ transgenic plants with TobRV. Based on Foster and Symons (1987), Gerlach *et al.* (1987).

5.5.4.2 *Satellite RNA of cucumber mosaic virus.* Essentially similar results have been obtained in protecting tobacco plants from the effects of infection by CMV by expression of an integrated benign satellite RNA (Baulcombe *et al.*, 1986; Harrison *et al.*, 1987). Transgenic tobacco plants, containing DNA copies of either 1.3 or 2.3 units of satellite RNA under the control of a 35S promoter, were obtained. In these plants transcripts containing 1.3 or 2.3 units of satellite RNA were synthesised, but processing to monomer length did not occur until after infection with CMV. When these transgenic plants were challenged with CMV, symptoms in systemically infected leaves were greatly attenuated and the growth of the plants was almost normal. The attenuation of symptoms was correlated with the formation and replication of the monomeric satellite RNA to high levels and depression of virus multiplication to 5–20% of that in control infections. Hence suppression of symptoms in this system also may depend (at least in part) on the inhibition of virus replication.

When the transgenic satellite RNA-expressing plants were challenged with other viruses some interesting results emerged. As expected, there was no resistance to infection by a range of viruses from different taxonomic groups, which did not support the replication of CMV satellite RNA. Infection of the plants with a closely related virus in the same (cucumovirus) group, tomato aspermy virus, did lead to replication of monomeric satellite RNA and attenuation of symptoms. However the replication of tomato aspermy virus was not inhibited. Hence suppression of symptom development can be caused by factors other than inhibition of virus replication.

Although good protection is obtained with these genetically engineered satellite systems, which, unlike CP-mediated protection, is not overcome at high inoculum levels or with time, they do have some drawbacks. Even when virus replication is inhibited, a limited amount of replication is necessary to replicate the satellite, which can be encapsidated in helper virus particles and transmitted to other plant species where it may cause more severe disease. Alternatively the satellite may mutate to cause more severe disease in the source plant; virulent and benign satellites can differ in only a few nucleotides (Palukaitis, 1988). By studying the properties of *in vitro* recombinants between virulent and benign satellites, good progress is being made in identifying the domains that control symptom production (Devic *et al.*, 1989; Kurath and Palukaitis, 1989). In the future it may be possible to design benign satellites that are incapable of reverting to virulence and incapable of being encapsidated in helper virus, thus preventing transmission to other plants.

5.5.4.3 *Prospects for the construction of artificial satellites and defective-interfering RNAs.* Satellite viruses and defective-interfering (DI) RNAs, as well as satellite RNAs, are able to interfere with the replication of their helper virus. However, these have been reported for only a small fraction of known plant viruses (Francki, 1985b). Theoretically it should be possible to construct artificial DI-RNAs or satellite viruses and express them from a plant genome.

A DI-RNA would need to contain replicase recognition sites from the 3′ and 5′ end of the helper virus RNA, together with any internal sequences needed for replication. A satellite virus would need additionally to contain a CP gene (this could be from a known satellite virus, e.g. the tobacco necrosis virus satellite virus) and the corresponding OAS sequence. As with satellite RNAs, replication of satellite viruses or DI RNAs in a transgenic plant would occur only after infection with the helper virus. Progress in such constructions may be expected in the next few years.

5.5.5 *The antisense RNA approach*

Antisense RNA is known to function naturally in the regulation of gene expression in prokaryotes and has been used to suppress gene expression in both prokaryotes and eukaryotes, including plants (Green *et al.*, 1986; Robert *et al.*, 1989). The regulatory effect is believed to depend on base pairing between the sense and antisense strands, thereby blocking the normal expression pathway of a mRNA. In a study of the suppression of the expression of a *bar* (biolopholas resistance) gene in plants (Cornelissen, 1989), it was found that antisense RNA resulted in a reduced *bar* transcript level and a reduced synthesis of *bar* translation product per transcript. Duplex RNA could not be detected in the nucleus or cytoplasm, but approximately equal quantitities of *bar* and anti-*bar* RNA were detected in the cytoplasm. The results suggested both nuclear and cytoplasmic control by antisense RNA. In the nucleus RNA duplexes formed between *bar* and anti-*bar* RNA may be largely degraded, with only small amounts of *bar* and anti-*bar* RNA escaping to be transported to the cytoplasm. In the cytoplasm the translational efficiency of the *bar* mRNA might be reduced by unstable association with anti-*bar* RNA or by unwinding of a duplex by a helicase that modifies the RNA (Wagner *et al.*, 1989).

Transgenic plants expressing antisense RNA to the 3′ region, including the CP gene, of TMV (Powell *et al.*, 1989), PVX (Hemenway *et al.*, 1988) or CMV (Cuozzo *et al.*, 1988) RNAs showed some protection against infection with the respective viruses or viral RNA. However the protection was much lower than that obtained in transgenic plants expressing virus CP genes and was only observed at very low inoculum levels.

To determine which region of the antisense RNA caused the protection, transgenic plants expressing antisense RNA to either the TMV coat protein gene or the 3′ untranslated region of TMV RNA were constructed. Only the latter plants conferred protection. It is likely that the protection afforded by the antisense RNA is the result of its interaction with the replicase binding site, thereby interfering with virus RNA replication by reducing the synthesis of the virus negative RNA strand.

The relatively low level of protection afforded by antisense RNA may be due to the intrinsic leakiness of its suppression of gene expression. Expression of antisense RNA in transgenic plants has generally resulted in reductions in gene expression of 90–95% (Robert *et al.*, 1989). This degree of reduction would still allow virus replication to occur, and as levels of virus RNA increased the antisense inhibition would be overcome.

The efficiency of gene suppression could be increased by increasing the promoter strength, e.g. by incorporating additional enhancer elements, or by increasing the template copy number. Geminivirus gene amplification vectors may be suitable for the latter purpose since a master copy can be stably integrated into the plant genome, average copy numbers of 100 per cell can be obtained and gene expression is independent of positional effects (Hayes *et al.*, 1989).

The choice of gene or region of the genome to be suppressed may also be important. Hirashima *et al.* (1986) reported that antisense RNA encompassing the 24-base non-coding region plus 216 bases of the coding region of the maturation protein gene was the most effective in reducing proliferation of phage SB in *E. coli*. This may simply have abolished the infectivity of particles by reducing synthesis of maturation protein, rather than inhibition of RNA replication, since the maturation protein is a minor capsid component needed for attachment of particles to cells. The nearest equivalent in plant viruses is the movement protein, which in the case of TMV is synthesised relatively early in infection in small amounts and hence may be a good target.

In the plant viruses, antisense RNA has been tested so far against only RNA viruses, and hence protection has been limited to cytoplasmic effects of antisense RNA. Antisense RNA may be found to be more effective against DNA viruses, such as the gemini viruses, which replicate in the nucleus and therefore would be susceptible to the suppressive effects of antisense RNA in both the nucleus and cytoplasm.

5.5.6 *Virus-resistant plants expressing sense RNA*

There are several ways in which regions of a virus genome in the sense orientation could suppress virus replication. For example the sense RNA could bind to the negative-strand RNA as in the negative-strand capture model of Palukaitis and Zaitlin (1984), invoked to explain cross-protection. Alternatively, a sense RNA containing the RNA replicase binding site at the 3′ end of the virus RNA might competitively bind the RNA polymerase. Morch *et al.* (1987) found that RNAs bearing the tRNA-like region at the 3′ end of the genomic RNA of TYMV inhibited the virus RNA-dependent RNA polymerase *in vitro*. However it should be noted that transgenic plants expressing the 3′ tRNA-like structures of TMV (Powell *et al.*, 1988) or AlMV (Van Dun *et al.*, 1988) did not show significant resistance to virus infection. Amplification of the RNA, e.g. by a replicating geminivirus vector (Hayes *et al.*, 1989), may improve the efficiency of this method.

In a different approach, synthesis of a truncated or mutated form of a protein could inhibit the activity of the functional protein by competitive binding to the substrate or, in the case of a multisubunit protein, by competitive binding to one of the constituent subunits. Two examples illustrate the potential of this 'molecular blocker' approach.

(1) Mutation of the G-residue of the GDD box in the β-subunit of phage Qβ RNA polymerase to A, S, P, M or V inactivated the polymerase. When any of the mutated polymerase genes was incorporated into a plasmid and expressed in *E. coli*, proliferation of phage Qβ in these cells was severely inhibited as a result of suppression of phage RNA synthesis (Inokuchi and Hirashima, 1987).
(2) A transcriptional activation domain of virion protein VP16 of herpes simplex virus is located within its 78 carboxyl-terminal amino acids. Deletion of this domain resulted in loss of the ability to activate transcription of the viral immediate early genes. Replication of herpes simplex virus in mouse L-cells that constitutively expressed a form of VP16 lacking its acidic activating domain was severely impaired (Friedman *et al.*, 1988). It is clear that molecular blockers could be targeted at a variety of different virus genes.

Recently, transgenic tobacco plants were produced which contained the TMV P54 gene, expressed under the transcriptional control of a 35S promoter. Such plants were completely resistant to infection by TMV and the resistance was not overcome by high inoculum levels (M. Zaitlin, D. Golemboski and G. Lomonosoff, personal communication). P54 is encoded by an ORF within the read-through portion of, and in the same reading frame as, P183. Although a subgenomic RNA corresponding to P54 has been isolated (Figure 5.1), it is not yet clear whether P54 is synthesised in plants infected with TMV. It remains to be determined whether the resistance conferred by the P54 construct is determined at the RNA or protein level. Binding of RNA to the negative strand could block either replication of genomic RNA or synthesis of the subgenomic RNA for the movement protein, the promoter for which lies within the P54 ORF. Alternatively P54, which contains the GDD box of the putative RNA replicase, could act as a molecular blocker of RNA polymerase function.

This exciting new approach could be applicable to a wide range of viruses.

5.5.7 *Potential of ribozymes in the construction of artificial virus resistance genes*

Ribozymes (Kruger *et al.*, 1982) are RNA molecules that act as enzymes or exhibit intramolecular catalysis (e.g. self-cleavage or self-splicing). There are two types of ribozymes that can cleave RNA in a sequence-specific manner. The first of these is based on the self-splicing reaction that occurs to remove an intron from the nuclear preribosomal RNA of *Tetrahymena*. Such ribozymes have a four-base recognition sequence and they are useful for specific cleavages of RNA *in vitro* (Zaug *et al.*, 1988).

The second type of ribozyme depends on the self-cleavage reactions that occur during the replication of certain viroids and satellite RNAs. Self-cleavage is believed to be responsible for conversion of linear RNA multimers, produced by rolling circle replication, into unit-size progeny, as depicted for STobRV in Figure 5.7. *In vitro* cleavage requires divalent cations. e.g. Mg^{2+}, and is thought to involve a base-catalysed nucleophilic attack by a 2′ OH group on the adjacent phosphodiester linkage to generate a 2′3′ cyclic phosphate and a 5′ hydroxyl group.

Forster and Symons (1987) identified conserved sequences and a 'hammer-head' secondary structure that were common to many of these self-cleaving RNAs, including STobRV (+) RNA. By carrying out *in vitro* mutagenesis, Haseloff and Gerlach (1988) identified sequences essential and non-essential for self-cleavage of STobRV (+) RNA and showed that the substrate and nuclease (ribozyme) activities could be on separate molecules. Further analysis showed that most of the required conserved nucleotides could be located on the ribozyme strand. The substrate strand need have very little sequence constraint, providing that it can base pair to the ribozyme strand to maintain the required hammerhead secondary structure. This allowed the design of ribozymes to target any substrate sequence *NNNNNN*GU*XNNNNNNNNN* where *N* can be A, C, G or U, *X* can be A, C or U and cleavage takes place between *X* and *N*. The ribozyme contains a core of conserved nucleotides flanked by nucleotides *N′* which are designed to be complementary to the corresponding nucleotides *N* on the substrate (Figure 5.8).

Ribozymes could clearly be targeted against many different regions of a virus genome. However there are several factors which will be expected to influence the antiviral efficacy of ribozymes expressed in transgenic plants.

(1) The stability of the ribozyme *in vivo*.
(2) For cytoplasmic viruses, the ability of the ribozyme to be transported into the cytoplasm.
(3) The influence of stable secondary structure in the target RNA on the ability of the ribozyme to bind.
(4) The ability of the ribozyme to dissociate from its substrate after reaction, if it is to recycle as a true enzyme. This would be affected by the length and nature of the sequences surrounding the GUXN target sequence. Increasing the length of these flanking sequences would increase the specificity of the ribozyme and the stability of the ribozyme–substrate complex. Conversely it would be expected to reduce the efficiency of ribozyme recycling. The requirement of a relatively high temperature (50°C) for maximum efficiency of ribozyme cleavage *in vitro* may reflect the temperature needed for efficient recycling.
(5) The level of expression of the ribozyme. It has been shown that a 500- to

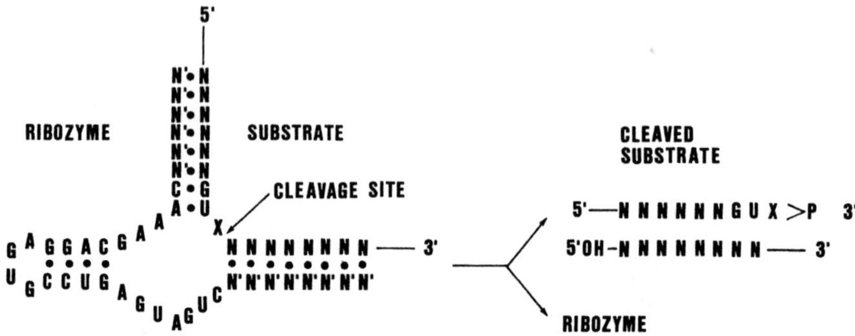

Figure 5.8. Model of the ribozyme designed by Haseloff and Gerlach (1988) showing binding to its substrate in a hammerhead configuration and the cleavage products. Complementary base pairs are indicated by dots.

1000-fold excess of ribozyme over substrate is necessary for efficient cleavage of mRNA in nuclear extracts. Such a high ratio was achieved *in vivo* by incorporating a ribozyme cDNA into a tRNA cDNA expression cassette and injecting it into the nucleus of *Xenopus* oocytes (Cotton and Birnstiel, 1989). Sufficient of the chimeric ribozyme–tRNA was produced and migrated to the cytoplasm to cleave and destroy its substrate (U7snRNA).

Although the full potential of ribozymes in transgenic plants in comparison with other types of virus- resistance gene remains to be evaluated, ribozymes add another mechanism of considerable promise to the repertoire of those available.

5.6 Future prospects

Elucidation of the mechanisms by which naturally occurring resistance genes confer protection to virus infection may be predicted to lead to cloning of these genes, their modification to be less easily overcome by resistance-breaking strains and their transfer to new plant species by genetic engineering techniques. This is a large field of research (Fraser, 1986), the surface of which has only just been scratched. Therefore it can be anticipated that many new mechanisms of resistance will be uncovered and new types of resistance gene will be cloned.

Studies of the virus genome organisation, modes of gene expression and replication, as well as cross-protection and satellite RNAs, has led to the production of 'artificial' resistance genes, such as those based on coat protein, antisense RNA, sense RNA, satellite RNA and ribozymes, and doubtless new ways of selectively suppressing virus gene expression will be found in the future. In transgenic plants the most effective virus resistance has been provided by coat protein, sense RNA and satellite RNAs. Antisense RNA and ribozymes may need to be amplified to high levels to be effective; tRNA expression cassettes or gemini virus vectors may provide the answer to this problem. Where two resistance mechanisms have a different basis, improved resistance might be obtained by combining both forms of resistance in the same plant, e.g. by blocking particle disassembly with coat protein and replication with antisense RNA to the 3′ end of the RNA or by incorporating a ribozyme into an antisense RNA.

This chapter has concentrated on positive-strand RNA viruses with relatively small genomes and numbers of genes. Some of the methods which have been shown to be effective with these viruses, e.g. coat protein, may prove to be ineffective on other types of virus, e.g. some DNA viruses. Conversely, consideration of the mechanisms of replication and gene expression of single-stranded DNA, double-stranded DNA, negative-stranded RNA and double-stranded RNA viruses, and the functions of the additional genes which occur in positive-strand RNA viruses with larger genomes, will doubtless suggest new types of resistance gene that will be effective against these types of virus.

Constitutive expression of resistance genes at high levels could be detrimental to a plant, particularly when multiple resistance to a variety of viruses is required. For transgenic plants expressing cloned resistance genes there would be an advantage in expressing the gene under the control of a promoter which is inducible by virus infection. Most studies of genes induced by virus infection have involved a hypersensitive response. However there is evidence that host genes are also induced during systemic virus infections (Bol and Van Kan, 1988). These could prove to be useful as a source of virus-inducible promoters for controlling the expression of artificial virus resistance genes in transgenic plants.

There seems little doubt that at least some of the novel types of resistance discussed above will find their way into agricultural use within the next decade. The few field trials that have been carried out are promising, although more will be needed to test the efficacy of the resistance under a range of environmental conditions. Safeguards will be needed to ensure that plants containing parts of the genome of pathogenic agents do not pose a threat to the environment. The problems of the use and safeguards needed with satellite RNA have been discussed. With other types of resistance two questions need to be addressed: (1) the possibility of recombination between the nucleic acid of an infecting virus and viral sequences expressed by a resistant host and (2) the possibility that such recombination might generate a more virulent pathogen. Recombination at either the DNA level (homologous or non-homologous recombination) or the RNA level (template switching during replication) is theoretically possible. Although the risks for genes such as coat protein would appear to be low, it would be worthwhile carrying out experiments to assess these risks quantitatively. For food products, checks should be made to ensure that the products of novel resistance genes are not toxic.

References

Ahlquist, P., Luckow, V. and Kaesberg, P. (1981) Complete nucleotide sequence of brome mosaic virus RNA 3. *J. Mol. Biol.* **153**: 23.

Ahlquist, P., Dasgupta, R. and Kaesberg, P. (1984) Nucleotide sequence of the brome mosaic virus genome and its implications for virus replication. *J. Mol. Biol.* **172**: 369.

Anderson, E.J., Stark, D.M., Nelson, R.S., Powell, P.A., Tumer, N.E. and Beachy, R.N. (1989) Transgenic plants that express the coat protein genes of TMV or AlMV interfere with disease development of some non-related viruses. *Phytopathology* **79**: 1284.

Argos, P. (1988) A sequence motif in many polymerases. *Nucl. Acids Res.* **16**: 9909.

Atabekov, J.G. and Dorokhov, Y.L. (1984) Plant virus-specific transport function and resistance of plants to viruses. *Adv. Virus Res.* **29**: 313.

Atkinson, P.H. and Matthews, R.E.F. (1970) On the origin of dark green tissue in tobacco leaves infected with tobacco mosaic virus. *Virology* **40**: 344.

Baker, B., Coupland, G., Fedoroff, N., Starlinger, P. and Schell, J. (1987) Phenotypic assay for excision of the maize controlling element Ac in tobacco. *EMBO J.* **6**: 1547.

Baulcombe, D. (1989) Strategies for virus resistance in plants. *Trends Genet.* **5**: 56.

Baulcombe, D. and Hull, R. (1989) Spreading the word. *Nature* **341**: 189.

Baulcombe, D.C., Saunders, G.R., Bevan, M.W., Mayo, M.A. and Harrison, B.D. (1986) Expression of a biologically active viral satellite RNA from the nuclear genome of transformed plants. *Nature* **321**: 446.

Beier, H., Barciszewska, M., Krupp, G., Mitnacht, R. and Gross, H.J. (1984*a*) UAG readthrough during TMV RNA translation: isolation and sequence of two tRNAs with suppressor activity from tobacco plants. *EMBO J.* **3**: 351.

Beier, H., Barciszewska, M. and Sickinger, H.-D. (1984*b*) The molecular basis for the differential translation of TMV RNA in tobacco protoplasts and wheat germ extracts. *EMBO J.* **3**: 1091.

Berger, P.H. and Pirone, T.P. (1986) The effect of helper component on the uptake and localisation of potyviruses in *Myzus persicae*. *Virology* **153**: 256.

Bloomer, A.C. and Butler, P.J.G. (1986) Tobacco mosaic virus: Structure and self-assembly. In *The Plant Viruses*. Vol. 2, *The Rod-Shaped Plant Viruses*, eds. Van Regenmortel, M.M.V. and Fraenkel-Conrat, Plenum Press, London, 19.

Blumenthal, T. (1979) RNA replication: function and structure of Qβ-replicase. *Ann. Rev. Biochem.* **48**: 525.

Boccara, M., Hamilton, W.D.O. and Baulcombe, D.C. (1986) The organisation and interviral homologies of genes at the 3′ end of tobacco rattle virus RNA 1. *EMBO J.* **5**: 223.

Bol, J.F. and Van Kan, J.A.L. (1988) The synthesis and possible functions of virus-induced proteins in plants. *Microbiol. Sci.* **5**: 47.

Bruening, G., Ponz, F., Glascock, C., Russell, M.L., Rowhani, A. and Chay, C. (1987) Resistance of cowpea mosaic virus and to tobacco ringspot virus. In *Plant Resistance to Viruses Ciba Foundation Symposium 133*, 23.

Buzayan, J.M., Gerlach, W.L. and Bruening, G. (1986*a*) Satellite tobacco ringspot virus RNA: a subset of the RNA sequence is sufficient for autolytic processing. *Proc. Natl Acad. Sci. USA* **83**: 8859.

Buzayan, J.M., Gerlach, W.L. and Bruening, G. (1986*b*) Non-enzymatic cleavage and ligation of RNAs complementary to a plant virus satellite RNA. *Nature* **323**: 349.

Candresse, T., Mouches, C. and Bove, J.M. (1986) Characterization of the virus-encoded subunit of turnip yellow mosaic virus RNA replicase. *Virology* **152**: 322.

Chung, T-C., Sun, C-N. and Hung, C-Y. (1982) Resistance of *Nilaparvata lugens* to six insecticides in Taiwan. *J. Econ. Entomol.* **75**: 199.

Cornelissen, M. (1989) Nuclear and cytoplasmic sites for anti-sense control. *Nucl. Acids Res.* **17**: 7203.

Costa, A.S. and Muller, G.W. (1980) Tristeza control by cross protection. *Plant Dis.* **64**: 538.

Cotten, M. and Birnstiel, M.L. (1989) Ribozyme mediated destruction of RNA *in vivo*. *EMBO J.* **8**: 3861.

Csillery, G., Tobias, I. and Rusko, J. (1983) A new pepper strain of tomato mosaic virus. *Acta Phytopathol. Acad. Sci. Hung.* **18**: 195.

Culver, J.N. and Dawson, W.O. (1989*a*) Tobacco mosaic virus coat protein: an elicitor of the hypersensitive reaction but not required for the development of mosaic symptoms in *Nicotiana sylvestris*. *Virology* **173**: 755.

Culver, J.N. and Dawson, W.O. (1989*b*) Point mutations in the coat protein gene of tobacco mosaic virus induce hypersensitivity in *Nicotiana sylvestris*. *Molecular Plant–Microbe Interactions* **2**: 209.

Cuozzo, M., O'Connell, K.M., Kaniewski, W., Fang, R-X., Chua, N-H. and Turner, N.E. (1988) Viral protection in transgenic tobacco plants expressing the cucumber mosaic virus coat protein or its antisense RNA. *Biotechnology* **6**: 549.

Cutt, J.R., Harpster, M.H., Dixon, D.C., Carr, J.P., Dunsmuir, P. and Klessig, D.F. (1989). Disease response to tobacco mosaic virus in transgenic tobacco plants that constitutively express the pathogenesis-related PRlb gene. *Virology* **173**: 89.

Davies, J.W., ed. (1985) *Molecular Plant Virology*, vols. 1 and 2. CRC Press, Boca Raton, FL.

Dawson, W.O. and Lehto, K.M. (1990) Regulation of tobamovirus gene expression. *Adv. Virus Res.* (in press).

Dawson, W.O., Bubrick, P. and Grantham, G.L. (1988) Modifications of the tobacco mosaic virus coat protein gene affect replication, movement and symptomatology. *Phytopathology* **78**: 783.

Dawson, W.O., Lewandowski, D.J., Hilf, M.E., Bubrick, P., Raffo, A.J., Shaw, J.J., Grantham, G.L. and Dejardins, P.R. (1989) A tobacco mosaic virus-hybrid expresses and loses an added gene. *Virology* **172**: 285.

Deom, C.M., Oliver, M.J. and Beachy, R.N. (1987) The 30-Kilodalton gene product of tobacco mosaic virus potentiates virus movment. *Science* **237**: 389.

Devic, M., Jaegle, M. and Baulcombe, D. (1989) Symptom production on tobacco and tomato is determined by two distinct domains of the satellite RNA of cucumber mosaic virus (strain Y). *J. Gen. Virol.* **70**: 2765.

Dreher, T.W. and Hall, T.C. (1988) RNA replication of brome mosaic virus and related viruses. In *RNA Genetics*. Vol. 1, RNA-Directed Virus Replication. CRC Press, Boca Raton, FL, 91.

Eggen, R., Verver, J., Wellink, J. Pleij, K. and Van Kammen, A. (1989) Analysis of sequences involved in cowpea mosaic virus RNA replication using site-specific mutants. *Virology* **173**: 456.

Forster, A.C. and Symons, R.H. (1987) Self-cleavage of plus and minus RNAs of a virusoid and a structural model for the active sites. *Cell* **49**: 211.

Francki, R.I.B. (ed.) (1985*a*) *The Plant Viruses*. vol. 1, *Polyhedral Virions with Tripartite Genomes*. Plenum Press, London.

Francki, R.I.B. (1985*b*) Plant virus satellites. Ann. Rev. Microbiol. **39**: 151.

Francki, R.I.B., Milne, R. and Hatta, T. (1985) *Atlas of Plant Viruses*, vols. 1 and 2. CRC Press, Boca Raton, FL.

Fraser, R.S.S. (1985) Some basic concepts and definitions in resistance studies. In *Mechanims of Resistance to Plant Diseases*, ed. Fraser, R.S.S. Martinus Nijhoff/Dr W. Junk Publishers, Dordrecht, 1.

Fraser, R.S.S. (1986) Genes for resistance to plant viruses. *CRC Crit. Rev. Plant Sci.* **3**: 257.

Fraser, R.S.S. (1987*a*) Resistance to plant viruses. In *Oxford Surveys of Plant Molecular and Cellular Biology*, ed. Miflin, B.J. Oxford, University Press, 1.

Fraser, R.S.S. (1987*b*) Genetics of plant resistance to viruses. In *Plant Resistance to Viruses, Ciba Foundation Symposium 133*. Wiley, Chichester, 6.

Fraser, R.S.S. and Loughlin, S.A.R. (1982) Effects of temperature on the Tm-1 gene for resistance to tobacco mosaic virus in tomato. *Physiol. Plant Pathol.* **20**: 109.

French, R. and Ahlquist, P. (1987) Intercistronic as well as terminal sequences are required for efficient amplification of brome mosaic virus RNA 3. *J. Virol.* **61**: 1457.

Friedman, A.D., Friezenberg, S.J. and McKnight, S.L. (1988) Expression of a truncated viral *trans*-activator selectively impedes lytic infection by its cognate virus. *Nature* **335**: 452.

Fritig, B., Kauffmann, S., Dumas, B., Geoffroy, P., Kopp, M. and Legrand, M. (1987) Mechanism of the hypersensitivity reaction of plants. In *Plant Resistance to Viruses Ciba Foundation Symposium* 133. Wiley, Chichester 92.

Fulton, R.W. (1986) Practices and precautions in the use of cross protection for plant virus disease control. *Ann. Rev. Phytopathol.* **24**: 67.

Gallie, D.R., Sleat, D.E., Watts, J.W., Turner, P.C. and Wilson, M.A. (1987) *In vivo* uncoating and efficient expression of foreign mRNAs packaged in TMV-like particles. *Science* **236**: 1122.

Gera, A. and Loebenstein, G. (1989) Evaluation of antisera to an inhibitor of virus replication (IVR) *J. Phytopathol.* **124**: 366.

Gerber, M. and Sarkar, S. (1989) The coat protein of tobacco mosaic virus does not play a significant role for cross-protection. *J. Phytopathol.* **124**: 323.

Gerlach, W.L., Llewellyn, D. and Haseloff, J. (1987) Construction of a plant disease resistance gene from the satellite RNA of tobacco ringspot virus. *Nature* **328**: 802.

Gibson, R.W. and Pickett, J.A. (1983) Wild potato repels aphids by release of aphid alarm pheromone. *Nature* **302**: 608.

Goldbach, R.W. (1986) Molecular evolution of plant RNA viruses. *Ann. Rev. Phytopathol.* **24**: 289.

Goodman, R.M. (1981) Geminiviruses. *J. Gen. Virol.* **54**: 9.

Gorbalenya, A.E. and Koonin, E.V. (1989) Viral proteins containing the purine NTP-binding sequence pattern. *Nucl. Acids Res.* **17**: 8413.

Green, P.J., Pines, O. and Inouye, M. (1986) The role of antisense RNA in gene regulation. *Ann. Rev. Biochem.* **55**: 569.

Habili, N. and Symons, R.H. (1989) Evolutionary relationship between luteoviruses and other RNA plant viruses based on sequence motifs in their putative RNA polymerases and nucleic acid helicases. *Nucl. Acids. Res.* **17**: 95443.

Hall, T.J. (1980) Resistance at the Tm-2 locus in the tomato to tomato mosaic virus. *Euphytica* **29**: 189.

Harrison, B.D. (1987) Introduction. In *Plant Resistance to Viruses, Ciba Foundation Symposium 133*. Wiley, Chichester, 1.

Harrison, B.D., Mayo, M.A. and Baulcombe, D.C. (1987) Virus resistance in transgenic plants that express cucumber mosaic satellite RNA. *Nature* **328**: 799.

Haseloff, J. and Gerlach, W.L. (1988) Simple RNA enzymes with new and highly specific endoribonuclease activities. *Nature* **334**: 585.

Haseloff, J. and Gerlach, W.L. (1989) Sequences required for self-cleavage of the satellite RNA of tobacco ringspot virus. *Gene* **82**: 43.

Hayes, R.J., Coutts, R.H.A. and Buck, K.W. (1989) Stability and expression of bacterial genes in replicating geminivirus vectors in plants. *Nucl. Acids Res.* **17**: 2391.

Hemenway, C., Fang, R-X., Kaniewska, W.K., Chua, N-H. and Turner, N.E. (1988) Analysis of the mechanism of protection in transgenic plants expressing the potato virus X coat protein or its antisense RNA. *EMBO J.* **7**: 1273.

Hirashima, A., Sawaki, S., Inokuchi, Y. and Inouye, M. (1986) Engineering of the mRNA-interfering complementary RNA immune system against viral infection. *Proc. Natl Acad. Sci. USA* **83**: 7726.

Hoekema, A., Huisman, M. J., Molendijk, L., Van den Elzen, P.J.M. and Cornelissen, B.J.C. (1989) The genetic engineering of two commercial potato cultivars for resistance to potato virus X. *Biotechnology* **7**: 273.

Holmes, F.O. (1938) Inheritance of resistance to tobacco mosaic disease in tobacco. *Phytopathology* **28**: 553.

Hull, R. (1989) The movement of viruses in plants. *Ann. Rev. Phytopathol.* **27**: 213.

Inokuchi, Y. and Hirashima, A. (1987) Interference with viral infection by defective RNA replicase. *J. Virol* **61**: 3946.

Jones, J.D.G., Carlands, F.M., Maliga, R. and Dooner, H.K. (1989) Visual detection of transposition of the maize element *Activator* (Ac) in tobacco seedlings. *Science* **244**: 204.

Koenig, R. (ed.) (1988) *The Plant Viruses.* Vol. 3, *Polyhedral Virions with Monopartite RNA Genomes.* Plenum Press, London.

Konz, C., Martini, N., Mayerhofer, R., Koncz-Kalman, Z., Korber, H., Redei, G.P. and Schell, J. (1989) High-frequency T-DNA-mediated gene tagging in plants. *Proc. Natl Acad. Sci. USA* **86**: 8467.

Knorr, D.A. and Dawson, W.O. (1988) A point mutation in the tobacco mosaic virus capsid protein gene induces hypersensitivity in *Nicotiana sylvestris*. *Proc. Natl. Acad. Sci. USA* **85**: 170.

Knudson, D.L., Atherton, J.G. and Brown, F. (1990) Classification and nomenclature of viruses. *Fifth Report of the International Committee on Taxonomy of Viruses.* S. Karger, Basle (in press).

Kozak, M. (1984) Compilation and analysis of sequences upstream from the translational start site in eucaryotic mRNAs. *Nucl. Acids Res.* **12**: 857.

Kroner, P., Richards, D., Traynor, P. and Ahlquist, P. (1989) Defined mutations in a small region of the brome mosaic virus 2a gene cause diverse temperature-sensitive RNA replication phenoytpes. *J. Virol.* **63**: 5302.

Kruger, K., Grabowski, P.J., Zaug, A.J., Sands, J., Gottschling, D.E. and Cech, T.R. (1982) Self-splicing RNA: autoexcision and autocyclization of the ribosomal RNA intervening sequence of Tetrahymena. *Cell* **31**: 147.

Kurath G. and Palukaitis, P. (1989) Satellite RNAs of cucumber mosaic virus:recombinants constructed *in vitro* reveal independent functional domains for chlorosis and necrosis in tomato. *Molecular Plant–Microbe Interactions* **2**: 91.

Lamb, C.J., Lawton, M.A., Dron, M. and Dixon, R.A. (1989) Signals and transduction mechanisms for activation of plant defenses against microbial attack. *Cell* **56**: 215.

Linthorst, H.J.M., Meurvissen, R.L.J., Kauffmann, S. and Bol, HJ.F. (1989) Constitutive expression of pathogenesis-related proteins PR-1, GRP and PR-S in tobacco has no effect on virus infection. *Plant Cell* **1**: 285.

Loesch-Fries, L.S. Merlo, D., Zinnen, T., Burhop, L., Hill, K., Kraha, K., Jarvis, N., Nelson, S. and Halk, E. (1987) Expression of alfalfa mosaic virus RNA 4 in intransgenic plants confers virus resistance. *EMBO J.* **6**: 1845.

Lommel, S.A., Weston-Fina, M., Xiong, Z. and Lomonossoff, G.P. (1988) The nucleotide sequence and gene organization of red clover necrotic mosaic virus RNA-2. *Nucl. Acids Res.* **16**: 8587.

Marsh, L.E., Dreher, T.W. and Hall, T.C. (1988) Mutational analysis of the core and modulator sequences of the BMV RNA 3 subgenomic promoter. *Nucl. Acids Res.* **16**: 981.

Matthews, R.E.F. (1990) *Plant Virology*, 3rd edn. Academic Press, New York.

Meshi, T., Motoyoshi, F., Adachi, A., Watanabe, Y., Takamatsu, N. and Okada, Y. (1988) Two concomitant base substitutions in the putative replicase genes of tobacco mosaic virus confer the ability to overcome the effects of a tomato resistance gene, Tm-1. *EMBO J.* **7**: 1575.

Meshi, T., Motoyoshi, F., Maeda, T., Yoshiwoka, S., Watanabe, H. and Okada, Y. (1989) Mutations in the tobacco mosaic virus 30-kd protein gene overcome Tm-2 resistance in tomato. *Plant Cell* **1**: 515.

Mi, S., Durbin, R., Huang, H.V., Rice, C.M. and Stollar, V. (1989) Association of the Sindbis virus RNA methyltransferase activity with the nonstructural protein nsPl. *Virology* **170**: 385.

Miller, W.A., Waterhouse, P.M. and Gerlach, W.L. (1988) Sequence and organization of barley yellow dwarf virus genomic RNA. *Nucl. Acids Res.* **16**: 6097.

Milne, R.G. (ed.) (1988) *The Plant Viruses.* Vol. 4, *The Filamentous Plant Viruses.* Plenum Press, London.

Morch, M.D., Joshi, R.L., Denial, T.M. and Haenni, A.L. (1987) A new 'sense' RNA approach to block viral RNA replication *in vitro. Nucl. Acids Res.* **15**: 4123.

Morris-Krsinich, B.A.M., Forster, R.L.S. and Mossop, D.W. (1983) Translation of red clover necrotic mosaic virus RNA in rabbit reticulocyte lysate: identification of the virus coat protein cistron on the larger RNA strand of the bipartite genome. *Virology* **124**: 349.

Moser, O., Gagly, M.J., Godefroy-Colburn, T., Stussi-Garaud, C., Ellivart-Tschurtz, M., Nitschko, H. and Mundry, K.W. (1988) The fate of the transport protein of tobacco mosaic virus in systematic and hypersensitive tobacco hosts. *J. Gen. Virol.* **69**: 1367.

Murray, E.A., Lotze, J. and Eberle, M. (1989) Codon usage in plants. *Nucl. Acids Res.* **17**: 477.

Nejidat, A., and Beachy, R.N. (1989) Decreased levels of TMV coat protein in transgenic tobacco plants at elevated temperatures reduce resistance to TMV infection. *Virology* **173**: 531.

Nelson, R.S., Powell Abel, P. and Beachy, R.N. (1987) Lesions and virus accumulation in inoculated transgenic tobacco plants expressing the coat protein gene of tobacco mosaic virus. *Virology* **158**: 126.

Nelson, R.S., McCormick, S.M., Delannay, X., Dube, P., Layton, J., Anderson, E.J., Kaniewska, M., Proksch, R.K., Horsch, R.B., Roers, S.G., Fraley, R.T. and Beachy, R.N. (1988) Virus tolerance, plant growth, and field performance of transgenic tomato plants expressing coat protein from tobacco mosaic virus. *Biotechnology* **6**: 403.

Nishiguchi, M. and Motoyoshi, F. (1987) Resistance mechanisms of tobacco mosaic virus strains in tomato and tobacco. In *Plant Resistance to Viruses, Ciba Foundation Symposium 133.* Wiley, Chichester, 38.

Nishiguchi, M., Kichuchi, S., Kiho, Y., Ohno, T., Meshi, T. and Okada, Y. (1985) Molecular basis of plant virulence; the complete nucleotide sequence of an attenuated strain of tobacco mosaic virus. *Nucl. Acids Res.* **13**: 5585.

Okamoto, S., Machida, Y. and Takebe, I. (1988) Subcellular localization of tobacco mosaic virus minus strand RNA in infected protoplasts. *Virology* **167**: 194.

Oparka, K.J. and Prior, D.A.M. (1988) Movement of lucifer yellow CH in potato tuber storage tissues: a comparison of symplastic and apoplastic transport. *Planta* **176**: 533.

Osbourn, J.K., Watts, J.W., Beachy, R.N. and Wilson, T.M.A. (1989*a*) Evidence that nucleocapsid diassembly and a later step in virus replication are inhibited in transgenic tobacco protoplasts expressing TMV coat protein. *Virology* **172**: 370.

Osbourn, J.K., Plaskitt, K.A., Watts, J.W. and Wilson, T.M.A. (1989*b*) Tobacco mosaic virus coat protein and reporter gene transcripts containing the TMV origin-of-assembly sequence do not interact in double-transgenic tobacco plants: implication for coat protein-mediated protection. *Molecular Plant–Microbe Interactions* **2**: 340.

Palukaitis, P. (1988) Pathogenicity regulation by satellite RNAs of cucumber mosaic virus: minor nucleotide sequence changes alter host responses. *Molecular Plant–Microbe Interactions* **4**: 175.

Palukaitis, P. and Zaitlin, M. (1984) A model to explain the 'cross-protection' phenomenon shown by plant viruses and viroids. In *Plant–Microbe Interactions: Molecular and Genetic Perspectives.* eds. Konege, T. and Nester, E.W. Macmillan, New York, 420.

Palukaitis, P. and Zaitlin, M. (1986) Tobacco mosaic virus infectivity and replication. In *The Plant Viruses*, vol. 2, eds. Van Regenmortel, M.H.V. and Fraenkel-Conrat, H. Plenum Press, New York, 105.

Pelham, J., Fletcher, J.T. and Hawkins, J.H. (1970) The establishment of a new strain of tobacco mosaic virus resulting from the use of resistant varieties of tomato. *Ann. Appl. Biol.* **65**: 293.

Plaskitt, K.A., Watkins, P.A.C., Sleat, D.E., Gallie, D.R., Shaw, J.G. and Wilson, T.M.A. (1988) Immunogold labelling locates the site of disassembly and transient gene expression of tobacco mosaic virus-like pseudovirus particles *in vivo*. *Molecular Plant–Microbe Interactions* **1**: 10.

Poch, O., Sauvaget, I., Delarue, M. and Tordo, N. (1989) Identification of four conserved motifs among the RNA-dependent polymerase encoding elements. *EMBO J.* **8**: 3867.

Ponz, F. and Bruening, G. (1986) Mechanisms of resistance to plant viruses. *Ann. Rev. Phytopathol.* **24**: 355.

Ponz, F., Glascock, C.B. and Bruening, G. (1988) An inhibitor of polyprotein processing with the characteristics of a natural virus resistance factor. *Molecular Plant–Microbe Interactions* **1**: 25.

Powell, P.A., Nelson, R.S., De, B., Hoffmann, N., Rogers, S.G., Fraley, R.T. and Beachy, R.N. (1986) Delay of disease development in transgenic plants that express the tobacco mosaic virus coat protein gene. *Science* **232**: 738.

Powell, P.A., Sanders, P.R., Fraley, R.T. and Beachy, R.N. (1988) The involvement of the coat protein mRNA, the coat protein, and the tRNA-like structure in cross protection of transgenic tobacco. In *Molecular Biology of Plant–Pathogen Interactions.* UCLA Colloquia. *J. Cellular Biochem. Suppl.* 12C. A.R. Liss, New York, 272.

Powell, P.A., Stark, D.M., Sanders, P.R. and Beachy, R.N. (1989) Protection against tobacco mosaic virus in transgenic plants that express tobacco mosaic virus antisense RNA. *Proc. Natl Acad. Sci. USA* **86**: 6949.

Quadt, R. and Jaspars, E.M.J. (1989) RNA polymerases of plus-strand RNA viruses of plants. *Molecular Plant–Microbe Interactions* **2**: 219.

Register, J.C. and Beachy, R.N. (1988) Resistance to TMV in transgenic plants results from interference with an early event in infection. *Virology* **166**: 524, 532.

Register, J.C. and Beachy, R.N. (1989) Effect of protein aggregation state on coat protein-mediated protection against tobacco mosaic virus using a transient protoplast assay. *Virology* **173**: 656.

Register, J.C., Powell, P.A., Nelson, R.S. and Beachy, R.N. (1989) Genetically engineered cross protection against TMV interferes with viral infection and long-distance spread of the virus. In *Molecular Biology of Plant–Pathogen Interactions*, eds. Staskawicz, B., Ahlquist, P. and Yoder, O. Alan R. Liss, New York, 269.

Robert, L.S., Donaldson, P.A., Ladaique, C., Altossar, I., Arnison, P.G. and Fabijanski, S.F. (1989) Antisense RNA inhibition of β-glucuronidase gene expression in transgenic tobacco plants. *Plant Mol. Biol.* **13**: 399.

Ross, H. (1960) Der braxis der zuchtung auf infektionsresistenz und extremer resistenz (immunitat) gegen das Y-virus der kartoffel. *Eur. Potato J.* **3**: 296.

Rowlands, D.J., Mayo, M.A. and Mahy, B.W.I. (eds.) (1987) *The Molecular Biology of Positive-Strand RNA Viruses.* Academic Press, London.

Saito, T., Hosokawa, D., Meshi, T. and Okada, Y. (1987*a*) Immunochemical localisation of the 130K and 180K proteins (putative replicase components) of tobacco mosaic virus. *Virology* **160**: 477.

Saito, T., Meshi, T., Takamatsu, N. and Okada, Y. (1987*b*) Coat protein gene sequence of tobacco mosaic virus encodes a host response determinant. *Proc. Natl Acad. Sci. USA* **84**: 6074.

Saito, T., Imai, Y., Meshi, T. and Okada, Y. (1988) Interviral homologies of the 30-kd proteins of tobamoviruses. *Virology* **164**: 653.

Saito, T., Yamanaka, K., Watanabe, Y., Takamatsu, N., Meshi, T. and Okada, Y. (1989) Mutational analysis of the coat protein gene of tobacco mosaic virus in relation to hypersensitive response in tobacco plants with the N′ gene. *Virology* **173**: 11.

Sela, I., Grafi, G., Sher, N., Edelbaum, O., Yager, H. and Gerassi, E. (1987) Resistance systems related to the N gene and their comparison with interferon. In *Plant Resistance to Viruses Ciba Foundation Symposium 133*. Wiley, Chichester, 109.

Sherwood, J.L. (1987) Mechanisms of cross-protection between plant virus strains. In *Plant Resistance to Viruses, Ciba Foundation Symposium 133*. Wiley, Chichester, 136.

Simoens, C., Van Montagu, M. and Schell, J. (1986) A binary vector for transferring genomic libraries to plants. *Nucl. Acids Res.* **14**: 8073.

Steinhauer, D.A. and Holland, J.J. (1987) Rapid evolution of RNA viruses. *Ann. Rev. Microbiol.* **41**: 409.

Strauss, J.H. and Strauss, E.G. (1988) Evolution of RNA viruses. *Ann. Rev. Microbiol.* **42**: 657.

Sulzinski, M.A. and Zaitlin, M. (1982) Tobacco mosaic virus replication in resistant and susceptible plants: in some resistant species virus is confined to a small number of initially infected cells. *Virology* **121**: 12.

Tien, P., Xinhua, Z. and Binsheng, Q. (1987) Satellite RNA as a biological control agent of plant diseases caused by cucumber mosaic virus. *Ann. Appl. Biol.* **111**: 143.

Tobin, G.J., Young, D.C. and Flanegan, J.B. (1989) Self-catalyzed linkage of poliovirus terminal protein VPg to poliovirus RNA. *Cell* **59**: 511.

Tomenius, K., Clapham, D. and Meshi, (1987) Localisation by immunogold cytochemistry of the virus-coded 30K protein in plasmodesmata of leaves infected with tobacco mosaic virus. *Virology* **160**: 363.

Tumer, N.E., O'Connell, K.M., Nelson, R.S., Sanders, P.R., Beachy, R.N., Fraley, R.T. and Shah, D.M. (1987) Expression of alfalfa mosaic virus coat protein gene confers cross-protection in transgenic tobacco and tomato plants. *EMBO J.* **6**: 1181.

Van Dun, C.M.P. and Bol, J.F. (1988) Transgenic tobacco plants accumulating tobacco rattle virus coat protein resist infection with tobacco rattle virus and pea early browning virus. *Virology* **167**: 649.

Van Dun, C.M.P., Overduin, B., Van Vloten-Doting, L. and Bol, J. (1988) Transgenic tobacco expressing tobacco streak virus or mutated alfalfa mosaic virus coat protein does not cross protect against alfalfa mosaic virus infection. *Virology* **164**: 383.

Van Kammen, A., Welink, J., Goldbach, R., Eggen, R. and Vos, P. (1987) The use of full-length DNA copies in the study of the expression and replication of the RNA genome of cowpea mosaic virus. In *Plant Molecular Biology*, eds. Van Weltstein and Chua, N-H. Plenum Press, New York, 433.

Van Regenmortel, M.H.V. and Fraenkel-Conrat, H. (eds.) (1986) *The Plant Viruses*. Vol. 2, *The Rod-Shaped Plant Viruses*. Plenum Press, London.

Wagner, R.W., Smith, J.E., Cooperman, B.S. and Nishikura, K. (1989) A double-stranded RNA unwinding activity introduces structural alterations by means of adenosine to inosine conversions in mammalian cells and *Xenopus* eggs. *Proc. Natl Acad. Sci.* **86**: 2647.

Watanabe, Y., Meshi, T. and Okada, Y. (1984) The initiation site for transcription of the TMV 30-kDa protein messenger RNA. *FEBS Lett.* **173**: 247.

Watanabe, Y., Kishibayashi, N., Motoyoshi, F. and Okada, Y. (1987*a*) Characterization of Tm-1 gene action on replication of common isolates and a resistance breaking isolate of TMV. *Virology* **161**: 527.

Watanabe, Y., Morita, N., Nishiguchi, M. and Okada, Y. (1987*b*) Attenuated strains of tobacco mosaic virus. Reduced synthesis of a viral protein with a cell-to-cell movement function. *J. Mol. Biol.* **194**: 699.

Wilson, T.M.A. (1984) Cotranslational disassembly of tobacco mosaic virus *in vitro*. *Virology* **137**: 255.

Wilson, T.M.A. (1988) Structural interactions between plant RNA viruses and cells. *Oxford Surveys of Plant Molecular and Cell Biology* **5**: 89.

Wolfe, S., Deom, C.M., Beachy, R.N. and Lucas, W.J. (1989) Movement protein of tobacco mosaic virus modifies plasmodesmatal size exclusion limit. *Science* **246**: 377.

Xiong, Z. and Lommel, S.A. (1989) The complete nucleotide sequence and genome organisation of red clover necrotic mosaic virus RNA-1. *Virology* **171**: 543.

Yamaya, J., Yoshioka, M., Meshi, T., Okada, Y. and Ohno, T. (1988*a*) Expression of tobacco mosaic virus RNA in transgenic plants. *Mol. Gen. Genet.* **211**: 520.
Yamaya, J., Yoshioka, M., Meshi, T., Okada, Y. and Ohno, T. (1988*b*) Cross protection in transgenic tobacco plants expressing a mild strain of tobacco mosaic virus. *Mol. Gen. Genet.* **215**: 173.
Zaitlin, M. and Hull, R. (1987) Plant virus–host interactions. *Ann. Rev. Plant Physiol.* **38**: 291.
Zaug, A.J., Crosshans, C.A. and Cech, T.R. (1988) Sequence-specific endoribonuclease activity of the *Tetrahymena* ribozyme. *Biochemistry* **27**: 8924.
Zimmern, D. (1977) The nucleotide sequence at the origin for assembly on tobacco mosaic virus RNA. *Cell* **11**: 463.
Zimmern, D. (1987) Evolution of RNA viruses. In *RNA Genetics*, eds. Holland, J., Domingo, E. and Ahlquist, P. CRC Press, Boca Raton, FL.

6 Targeting of proteins to chloroplasts and mitochondria

C. ROBINSON

6.1 Introduction

All cells transport proteins across membranes, but the complexity of protein traffic in plant cells is particularly striking. Plant cells contain a wide variety of organelles, each specialised to carry out particular functions, and each of which contains a distinct complement of proteins. A number of these organelles form the endomembrane system (endoplasmic reticulum, Golgi apparatus, vacuole, protein bodies and glyoxysomes), and proteins destined for these organelles are almost always transported initially into the endoplasmic reticulum (see Chapter 7). In leaf cells, however, many other proteins are destined for either the chloroplast or the mitochondrion. In terms of protein targeting, the biogenesis of these organelles differs from that of the endomembrane system in two significant respects. Firstly, the biogenesis of both the chloroplast and the mitochondrion requires the activities of two genetic systems, since each of these organelles synthesises a small, but essential, proportion of the organellar proteins. Secondly, each of these semiautonomous organelles is composed of more than one membrane, and intraorganellar protein 'sorting' is therefore a central element in the biogenesis of both organelles. The purpose of this chapter is, firstly, to review our current knowledge of the synthesis and transport of chloroplast and mitochondrial proteins, and, secondly, to describe how this knowledge has been utilised in recent years to target foreign proteins into either organelle.

6.2 Structure and biogenesis of chloroplasts

Although best known for its photosynthetic role, the chloroplast is the site of many important metabolic processes in plant cells. These include the synthesis of amino acids, proteins and lipids, as well as light capture, carbon fixation and photorespiration. It is perhaps not surprising that such a multiplicity of functions is carried out within an organelle which is, in structural terms, particularly complex: the chloroplast consists of three distinct membranes which enclose three discrete soluble phases (Figure 6.1). The organelle is

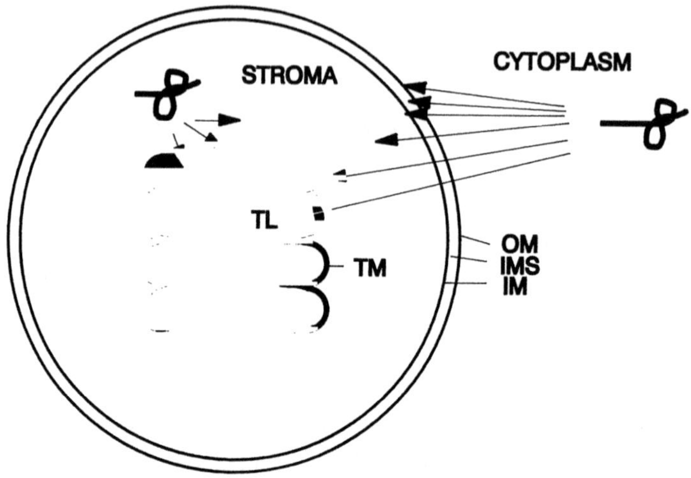

Figure 6.1. Transport of proteins into and within the chloroplast. The chloroplast compartments are the outer and inner envelope membranes (OM, IM) and intermembrane space (IMS), the stroma, thylakoid membrane (TM) and thylakoid lumen (TL). Proteins are synthesised in the cytoplasm and targeted into all six compartments. Chloroplast-encoded proteins either remain in the stroma or are targeted to the thylakoid membrane or lumen. No evidence currently exists for the targeting of chloroplast-encoded proteins to any of the envelope compartments, but this possibility cannot be excluded.

bounded by a double-membrane envelope, which accounts for only about 1% of total chloroplast membrane, but which nevertheless carries out important functions. The inner membrane is a selective permeability barrier which mediates complex interactions between the cell cytoplasm and the chloroplast interior. The outer membrane, on the other hand, is relatively porous to small molecules and functions mainly in biosynthetic reactions (notably lipid synthesis) and protein transport (Douce and Joyard, 1979; Heber and Heldt, 1981). Particularly well characterised is the internal thylakoid membrane, which is primarily involved in light capture, electron transport and photophosphorylation. In most higher plant chloroplasts, the thylakoid network is itself composed of two types of lamellae: stacked, 'granal' lamellae and single, 'stromal' lamellae.

The three chloroplast membranes enclose in turn three soluble phases: the intermembrane space, stroma and thylakoid lumen. Little is known about either the functions or the constituents of the intermembrane space, but studies on the synthesis of stromal and thylakoid lumen proteins have provided a great deal of information on the protein transport and sorting events which are central to chloroplast biogenesis.

The biosynthesis of chloroplast proteins is a complex issue, since proteins functioning in each of the six compartments described above must first be directed to these destinations from their sites of synthesis. Furthermore, chloroplast proteins are synthesised by two distinct genetic systems; a proportion (*ca* 20%) are encoded by chloroplast DNA and synthesised within the organelle, whereas the remainder are synthesised in the cytoplasm and transported into the chloroplast. Studies on the origins of a variety of chloroplast proteins have shown that cytoplasmically synthesised proteins are targeted to all six suborganellar compartments, and that chloroplast-synthesised proteins are targeted (at least) into and across the thylakoid membrane (illustrated in Figure 6.1). The scale and complexity of this protein traffic have attracted considerable interest in recent years, and four areas in particular have been the subject of intensive research:

(1) The pathways taken by imported chloroplast proteins.
(2) The structures and modes of action of the targeting signals which ensure the delivery of newly synthesised chloroplast proteins to their correct destinations.
(3) The structures, specificities and mechanisms of action of the elements of the import machinery which decode chloroplast-targeting signals.
(4) The evolution of the chloroplast protein transport systems.

Within these areas have been varying rates of progress, and, without question, some of the key processes in chloroplast protein transport are presently very poorly understood. On the other hand, some notable success stories have emerged in recent years, especially with respect to the import of thylakoid proteins, and the chloroplast is now beginning to realise its potential as a formidable system for the study of protein transport and sorting.

6.3 Import of stromal proteins

Early studies on chloroplast protein transport concentrated almost exclusively on a stromal protein, ribulose bisphosphate carboxylase–oxygenase (rubisco). This enzyme consists of eight large and eight small subunits which are synthesised in the chloroplast and cytoplasm respectively. The enzyme is extremely abundant, representing up to 50% of soluble leaf protein, and it is no coincidence that the small subunit (SSU) has been a favourite tool in protein transport research. In 1977 it was demonstrated that SSU is initially synthesised as a larger precursor (Dobberstein *et al.*, 1977), and in 1978 two groups showed that this precursor (pre-SSU) could be post-translationally imported by isolated chloroplasts (Chua and Schmidt, 1978; Highfield and Ellis, 1978). Many other imported chloroplast proteins have since been analysed, and all of them, with one possible exception, are also synthesised in precursor form with an amino-terminal presequence or transit peptide. The presequences of stromal proteins are typically between 40 and 60 residues in length, and their structures and functions are considered in greater detail later in this section.

The possible exception to this rule has been proposed by Grim *et al.* (1989).

These workers isolated a cDNA clone encoding a heat shock protein in *Chlamydomonas reinhardii*, and provided evidence that this protein is synthesised without a cleavable presequence but is nevertheless destined for the chloroplast stroma. However, these workers were unable to demonstrate *in vitro* transport of this protein into *Chalmydomonas* chloroplasts (a difficult task unless a cell wall-less mutant is used), and therefore final confirmation of this exception remains to be obtained. If this protein is indeed imported in the absence of proteolytic processing, it will be interesting to determine whether the protein contains an internal targeting signal, as is the case for some mitochondrial and secretory proteins.

In vitro reconstitution systems have been widely used to study the import of cytoplasmically synthesised proteins into chloroplasts, and a number of key events have been identified in the import mechanism.

6.3.1 *Binding to import receptors in the chloroplast envelope*

After synthesis in the cytoplasm, precursors of stromal proteins are believed to bind to specific receptors on the chloroplast surface. The most direct evidence for the existence of such receptors comes from studies in which isolated chrloroplasts were treated with thermolysin, to degrade proteins exposed on the external face of the envelope membranes, prior to incubation with pre-SSU. It was found that this treatment drastically reduced both the rate of import of pre-SSU into chloroplasts and the levels of pre-SSU which are able to bind to the chloroplasts with high affinity (Cline *et al.*, 1985). However, although these observations represent compelling evidence for the presence of receptor proteins in the outer envelope membrane, further details of the structure and specificity of the receptors have been slow to emerge. In part, this has been due to limitations in the *in vitro* import systems which have been available; precursor proteins are usually synthesised in cell-free translation systems based on the wheat-germ or reticulocyte lysate. Both of these systems synthesise very low amounts of protein, and consequently attempts to analyse the chloroplast import receptor, e.g. by saturation or competition studies, have been fraught with difficulties. In spite of these problems, Cornwell and Keegstra (1987) attempted to identify the receptor protein(s) by cross-linking them to *in vitro*-synthesised pre-SSU. They found that the activation of a photoactivatable cross-linker caused prebound pre-SSU to bind to a protein of about 60 kDa (the pre-SSU–protein complex had a molecular weight of approximately 86 000) and speculated that this may be a component of the transport system.

Pain *et al.* (1988) used a more indirect approach to probe the structure of the import receptors. These workers raised an antiserum to a peptide having the sequence of the carboxy-terminal 30 residues of the pea SSU presequence; a second, 'anti-idiotypic' antiserum was then raised against the antipeptide antibodies. The rationale behind this approach is that a certain proportion of the antipeptide antibodies may have a binding site which is structurally similar to that of the pre-SSU receptor, and that consequently the anti-idiotypic antiserum will recognise both types of molecule. Certain properties of the anti-idiotypic antibodies suggested that this approach may be a promising one: the antiserum was shown to inhibit the import of pre-SSU into isolated chloroplasts, and was found to bind to chloroplast envelopes at 'contact sites' between the inner and outer membranes–a likely site of import machinery. However, further work will be required before the properties of the either the antiserum or the putative pre-SSU receptors can be fully understood.

6.3.2 *Translocation across the envelope membranes*

The chloroplast envelope is a major site of protein transport in the plant cell, since several hundred different proteins are transported across the envelope into a developing chloroplast. These proteins vary enormously in size, structure and hydrophobicity, and one of the major problems in the field of chloroplast biogenesis is to unravel the mechanism(s) by which these proteins are specifically transported across the double-membrane envelope, given that the inner membrane is impermeable even to most small ions. Perhaps surprisingly, relatively few studies have so far addressed this problem, and it is therefore rather unfortunate that one of the most interesting aspects of chloroplast biogenesis is also one of the most obscure. Many of our current hypotheses on the mechanism of protein translocation across the envelope have been inspired by much more detailed studies on this aspect of mitochondrial protein import (see Section 6.6).

It is generally believed that protein translocation across the envelope occurs at 'contact sites' between the outer and inner membranes, so that both membranes are traversed in essentially a single step. The existence of such sites in mitochondria has been well documented, and a similar system is thought to be likely to operate in chloroplasts, although it must be emphasised that biochemical evidence in favour of this model is lacking.

Another major gap in our knowledge concerns the conformation adapted by a protein during translocation. Whereas a number of studies have elegantly demonstrated that proteins must unfold during transport across the mitochondrial membranes, only one detailed study has addressed this question in chloroplasts. The subject of this study was the stromal enzyme 5-enolpyruvylshikimate-3-phosphate synthase (EPSP synthase), which is synthesised in the cytoplasm in precursor form (pEPSPS) and imported into the chloroplast (see Chapter 3). Della-Cioppa and Kishore (1988) tested the effects of a tightly binding ligand, the herbicide glyphosate, on the import of pEPSPS. They found that a low (10 μM) concentration of glyphosate almost completely inhibited the import of pEPSPS and suggested that the tightly bound glyphosate had prevented pEPSPS from unfolding to an extent required for translocation across the envelope membranes. Consistent with this possibility, it was found that glyphosate had no effect on the import of other precursors into chloroplasts. These data strongly suggest that precursors must at least partly unfold in order to traverse the envelope membranes, although further studies are required for an in-depth understanding of this topic.

As with all protein-translocating membranes, the transport of proteins across the chloroplast envelope requires energy. This aspect of chloroplast biogenesis has been studied in some detail, and a variety of techniques have been used to analyse the form of energy required and the locations at which energy is utilised (Grossman *et al.*, 1980; Flügge and Hinz, 1986; Olson *et al.*, 1989; Theg *et al.*, 1989). Some of these reports were, to some extent, contradictory, but the emerging consensus seems to be that ATP is required outside the chloroplast for the binding of precursors to the chloroplast, and inside the organelle for protein transport across the envelope. Little is known, however, of the precise means by which ATP promotes either binding or translocation.

6.3.3 *Proteolytic maturation of imported stromal proteins*

During or shortly after transport across the envelope, precursors of stromal proteins are processed to the mature sizes by a stromal processing peptidase (SPP). The enzyme is highly specific for imported precursors; partially purified SPP from pea chloroplasts processes a number of precursors of stromal proteins to the mature sizes, but fails to cleave a wide variety of foreign proteins. The precise structure and composition of the enzyme have yet to be determined, but size fractionation and inhibitor studies indicate that SPP is a metal-dependent (i.e. chelator-sensitive) endopeptidase of about 180 kDa (Robinson and Ellis, 1984a). No data have so far emerged to indicate the existence of more than one variety of SPP, but this possibility cannot yet be ruled out since only a small proportion of imported stromal proteins have been analysed in *in vitro* processing assays (and the enzyme is not yet available in purified form).

In normal circumstances, precursor forms of stromal proteins cannot be detected inside chloroplasts, indicating that precursors are rapidly cleaved after import, or perhaps even during translocation. However, SPP does not appear to be an essential element of the translocation machinery; we have found that carboxymethylation of pre-acyl carrier protein blocks cleavage by SPP, but the modified precursor is nevertheless imported into the stroma (Barnett, Musgrove and Robinson, manuscript submitted for publication).

One aspect of SPP has for some time remained puzzling: the peptidase cleaves some precursors at more that one site, at least *in vitro*. This was first observed in studies on the maturation of pre-SSU; this precursor is processed to the mature size by the isolated enzyme in two steps (Robinson and Ellis, 1984b). Whether this occurs *in vivo* is not known, but there are no compelling reasons to suspect that the enzyme exhibits different specificities *in vivo* and *in vitro*. On the other hand, it is difficult to envisage why some precursors should be processed in more than one step.

Although SPP is clearly highly specific for imported precursors, the basis for this reaction specificity is very difficult to discern. The primary structures of many chloroplast protein presequences have now been deduced, but these sequence data have proved to be of little value for determining the nature of the signal which specifies cleavage by SPP. Different presequences exhibit little or no primary sequence homology, even around the SPP cleavage sites, and we can only assume that conserved (but as yet unknown) structural features are recognised by this enzyme.

6.3.4 *Structure and location of stroma-targeting signals*

In vivo experiments have for some time indicated that the chloroplast protein import mechanism is both efficient and specific. Precursor forms of chloroplast proteins cannot be detected in leaf tissue under normal circumstances, and there is no evidence that chloroplast proteins can be targeted to the wrong organelle. What, then, are the features of chloroplast protein precursors which are recognised so specifically by the import apparatus? Numerous studies have addressed this important question, using approaches which have probed the location of the targeting signals within the precursor molecules.

The role of the amino-terminal presequences in chloroplast protein targeting has been tested by fusing these sequences in front of a variety of 'foreign' proteins. From these studies it is now clear that the presequences of several stromal proteins contain all of the information required to direct the import and processing of foreign proteins. The first reports involved the use of the pea SSU presequence to target a bacterial protein, neomycin phosphotransferase, into the stroma in transgenic plants (Schreier *et al.*, 1985; Van den Broeck *et al.*, 1985). More recently, brome mosaic virus coat protein, *Escherichia coli* β-glucuronidase, yeast mitochondrial superoxide dismutase, and various other proteins, have all been successfully targeted into the stroma either *in vitro* or *in vivo* (Gatenby *et al.*, 1988; Kavanagh *et al.*, 1988; Lubben *et al.*, 1989). On the basis of these findings, it seems likely that almost any protein can be targeted into this location. Another important point to emerge from this type of study is that *chloroplast*-encoded proteins can be efficiently transported into the chloroplast when provided with a suitable presequence. For example, Gatenby *et al.*, (1988) showed that the rubisco SSU presequence can mediate the import of the rubisco large subunit *in vitro*.

These results hold enormous implications for the genetic engineering of plants, for they offer the very real possibility that novel or mutated proteins can be efficiently and selectively targeted into chloroplasts in those species capable of being transformed. By extension, it should also prove possible to target proteins to any other plastid type, since castor bean seed leucoplasts (which are functionally distinct from chloroplasts) have been shown to import chloroplast precursor proteins (Boyle *et al.*, 1986; Halpin *et al.*, 1989a), indicating that the basic import mechanisms are similar.

As yet, there have been few cases of this retargeting technology being used to alter the properties of chloroplasts in transformed plants. However, one example serves to illustrate the potential of this approach. Della-Cioppa *et al.* (1987) isolated a gene from *E. coli* encoding a mutant form of EPSP synthase which is resistant to the herbicide glyphosate, a potent inhibitor of chloroplast EPSP synthases which prevents plant growth by blocking the synthesis of aromatic amino acids (see Chapter 3). The bacterial gene was fused to a 5′ cDNA sequence encoding the presequence of the corresponding petunia chloroplast enzyme, and the translation product from this construct was imported by chloroplasts *in vitro* and *in vivo*, giving rise to a stable, glyphosate-resistant enzyme.

Although a number of studies have pointed to the vital function of the presequence in targeting proteins to the stroma, other work has shown that this signal may not always be sufficient for the *efficient* targeting of foreign proteins. For example, Wasmann *et al.* (1986) compared the *in vitro* import of a 'clean' fusion protein (pea SSU presequence linked to neomycin phosphotransferase II) with a construct in which the same passenger protein was fused behind the pea SSU presequence together with 23 residues of the mature SSU protein. They found that the import efficiency of the clean fusion protein was less than 10% of the construct containing mature rubisco residues. Kavanagh *et al.* (1988) observed a similar phenomenon when using the LHCP presequence to target bacterial β-glucuronidase into chloroplasts in transformed tobacco plants. Fusion proteins containing the presequence and either four or 16 amino acids of mature LHCP were imported very inefficiently (less than 1% in each case). In contrast, fusions containing the presequence plus 24, 53 or 126 residues of mature protein were imported with efficiencies of approximately 98, 96 and 75% respectively. These examples raise the possibility that chloroplast-targeting signals are not in fact located solely in the presequences of imported proteins, and that signals in the mature sequences are important for efficient import. Another, perhaps more likely, explanation is that in these fusion proteins the sequences of mature chloroplast protein serve as convenient linker sequences, projecting the presequences further from the foreign proteins and making them more available for interaction with the import apparatus. Whatever the reason, these findings pose problems for the genetic engineer interested in the efficient targeting of functional proteins into the chloroplast stroma. Clearly, the safest approach is to test a range of constructs, starting from a 'clean' fusion (i.e. presequence plus passenger protein) and then incorporating increasing amounts of mature chloroplast protein into the targeting signal. In the latter cases, however, processing by SPP will generate a passenger protein which still contains extra residues at the amino terminus. This may have adverse effects on the biological activity of some proteins, particularly if amino-terminal residues play a key role in activity. Alternatively, the presence of extra residues may affect the stability or turnover rate of some imported proteins. Clearly, further work is required in this field before any 'rules' can be formulated on how best to target proteins into chloroplasts *in vivo*.

6.4 Transport of proteins into the thylakoid system

6.4.1 *Biogenesis of thylakoid lumen proteins*

Considerable attention has centred on the import of nuclear-encoded thylakoid proteins, since these proteins must traverse both the envelope and the stromal space to reach their target membrane. Of particular interest is the biogenesis of thylakoid lumen proteins, since these proteins undergo a uniquely complex import pathway which involves traversing all three membranes. The first luminal protein to be cloned was plastocyanin, a small, soluble electron carrier (Smeekens *et al.*, 1985), and subsequent work led to the proposal of a two-step import pathway for this protein. In this model, pre-plastocyanin is initially transported into the stroma and processed to an intermediate size by SPP; the intermediate is subsequently transported across the thylakoid membrane and processed to the mature size by a second thylakoidal processing peptidase, TPP (Hageman *et al.*, 1986; Smeekens *et al.*, 1986). The evidence in favour of this model came initially from two critical observations: transient, stromal intermediate forms can be detected during the import of pre-plastocyanin into intact chloroplasts, and pre-plastocyanin is processed to intermediate and mature sizes, respectively, by isolated SPP and TPP. More recently, it has been shown that three other luminal proteins, the

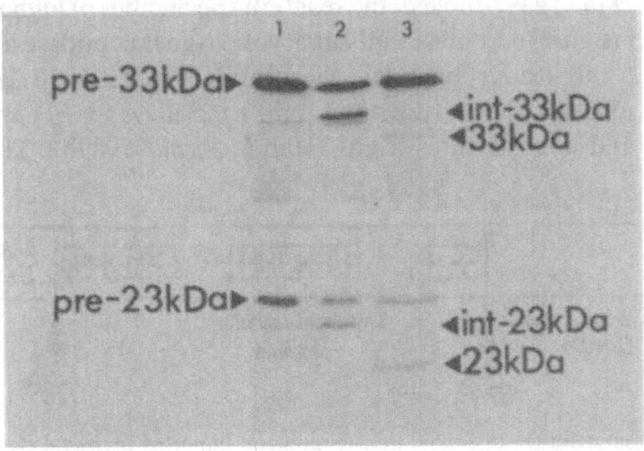

Figure 6.2. Proteolytic processing of precursors to thylakoid lumen proteins. The autoradiogram shows the cleavage of precursors to the 33 kDa and 23 kDa proteins of the oxygen-evolving complex. Lane 1 represents *in vitro*-synthesised precursors (pre-33 kDa, pre-23 kDa) which are processed to intermediate (int-) forms by incubation with partially purified stromal processing peptidase (SPP) (lane 2) or to mature sizes by thylakoidal processing peptidase (TPP) (lane 3).

33 kDa, 23 kDa and 16 kDa proteins of the photosynthetic oxygen-evolving complex, are imported by a similar mechanism (James *et al.*, 1989). Figure 6.2 shows the results of an *in vitro* processing experiment in which wheat pre-23 kDa is cleaved by SPP or TPP to either the intermediate or mature size respectively.

Analysis of partially purified TPP activity has shown that this enzyme, like SPP, catalyses a very specific cleavage reaction. However, unlike the stromal enzyme TPP is highly hydrophobic and can only be released in an active form from the thylakoid membrane using non-ionic detergents (Kirwin *et al.*, 1987). The active site of the peptidase is on the luminal side of the membrane, indicating that processing takes place either during or shortly after translocation (Kirwin *et al.*, 1988).

In keeping with their two-step import pathway, thylakoid lumen proteins are initially synthesised with a bipartite pre-sequence which contains two separate targeting signals: an amino-terminal 'envelope transfer' domain followed by a 'thylakoid transfer' domain. The envelope transfer domains are structurally identical to the presequences of imported stromal proteins (and are likewise removed by SPP). Thus, when regions of the thylakoid transfer domain are deleted, pre-plastocyanin is imported only into the stromal space (Hageman *et al.*, 1990). The structures of the thylakoid transfer domains are markedly different. As a whole, they are much more hydrophobic than envelope transfer domains, and they are characterised by the presence of apolar 'core' regions and short-chain residues (usually alanine) at the -3 and -1 positions (relative to the terminal processing site). These features, which are depicted in Figure 6.3, are also found in 'signal' sequences which direct proteins across the endoplasmic reticulum or the bacterial plasma membrane (von Heijne *et al.*, 1989). Indeed, the reaction specificities of higher plant TPP, *E. coli* leader (signal) peptidase and eukaryotic signal peptidase are essentially identical; *E. coli* leader peptidase accurately processes thylakoid lumen protein precursors to the mature sizes, and TPP cleaves signal peptides of both eukaryotic and prokaryotic origin (Halpin *et al.*, 1989b). These findings

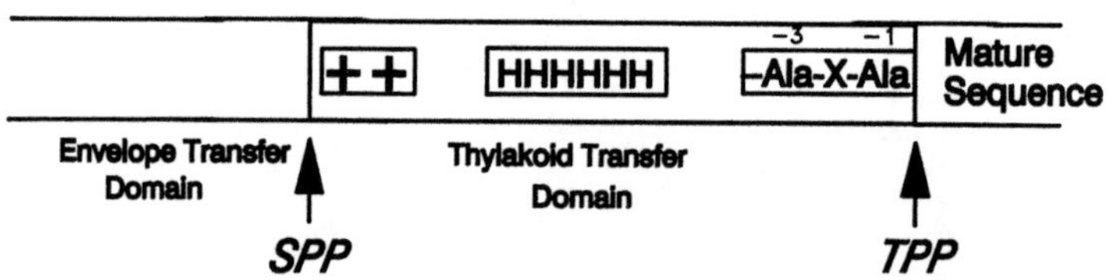

Figure 6.3. Structure of the thylakoid transfer domain. Imported thylakoid lumen proteins are synthesised with a presequence consisting of two domains: an amino-terminal envelope transfer domain followed by a thylakoid transfer domain. The latter signal has features in common with bacterial export signals: an amino-terminal, positively charged region, a hydrophobic central core region (H) and short-chain residues at the -3 and -1 positions (often alanine). The two domains are removed by the stromal and thylakoidal processing peptidases (SPP and TPP).

suggest that the thylakoidal protein import/maturation system may have evolved from a bacterial export system. Possibly, in the ancestral endosymbiotic cyanobacteria (presumed to be the progenitors of the chloroplast), proteins were directed across the plasma and thylakoid membranes by similar mechanisms (or by the same system). Evidence supporting this possibility has come from studies on the synthesis of thylakoid lumen proteins in free-living cyanobacteria. In these photosynthetic bacteria, luminal proteins such as plastocyanin and the 33 kDa PS II protein are synthesised with presequences which closely resemble signal sequences and the thylakoid transfer domains of their higher plant counterparts (Kuwabara *et al.*, 1987; Wallace *et al.*, 1989). It therefore appears that, after the initial endosymbiotic events, these genes were transferred to the nucleus and acquired additional signals to target the proteins from the cytoplasm into the chloroplast stroma. From this point, completion of the import process is mediated by the ancestral, prokaryotic-like thylakoidal system.

Although it is clear that the processing signals within thylakoid transfer domains and signal sequences are remarkably similar, it remains to be determined whether the translocation signals are equally conserved (i.e. whether a bacterial export signal will target a protein across the thylakoid membrane). Similarly, we do not yet know whether the thylakoidal protein transport system bears any resemblance, in terms of composition and mechanism, to bacterial export systems. However, the recent demonstration of ATP-stimulated protein import by isolated thylakoids (Kirwin *et al.*, 1989) means that this system can now be analysed independently of the envelope-based transport system.

As with stroma-targeting sequences, the presequences of thylakoid lumen proteins have been tested for their ability to direct foreign proteins into the thylakoid system. Initial reports were not encouraging; the presequence of *Silene pratensis* plastocyanin did not direct either ferredoxin (a stromal protein), yeast mitochondrial superoxide dismutase or mouse dihydrofolate reductase (DHFR) into the thylakoid lumen of isolated chloroplasts. Instead, stromal intermediate forms accumulated in each case (Smeekens *et al.*, 1986, 1987). More recently, however, it has been shown that the presequence of the wheat 33 kDa PS II protein can direct DHFR into the lumen *in vitro* (Meadows *et al.*, 1989), and the presequence of the same protein from *Arabidopsis* was shown to mediate the transport of DHFR and rubisco small subunit, but not glycollate oxidase, into the lumen (Ko and Cashmore, 1989). It is not clear why DHFR was successfully targeted into the lumen by the two 33 kDa presequences, but not by that of plastocyanin. Possibly some presequences are inherently more efficient than others for these purposes. The failure to target glycollate oxidase into the lumen also suggests that some passenger proteins may be more difficult than others to transfer across the thylakoid membrane.

6.4.2 *Import and integration of the light-harvesting chlorophyll-binding protein*

Situated between the soluble stromal and thylakoid lumen phases are numerous integral thylakoid membrane proteins, approximately half of which are nuclear encoded. The most abundant of these proteins, and by far the best studied, is the light-harvesting chlorophyll-binding protein (LHCP), a hydrophobic component of the photosystem II complex (Thornber, 1975). This protein is synthesised with a presequence which appears only to be a stroma-targeting signal; when the presequence is replaced by that of SSU, the resulting construct is still imported into the chloroplast and targeted into the thylakoid membrane (Lamppa, 1988). This observation strongly suggests that the information specifying integration into the thylakoid membrane is located in the mature LHCP sequence. The LHCP import mechanism thus appears to differ fundamentally from that proposed for thylakoid lumen proteins, in which the targeting of the latter proteins into and across the thylakoid membrane is mediated by a cleavable thylakoid transfer domain. A comparison of the different import pathways is shown in Figure 6.4.

The integration of either pre- or mature-size LHCP into thylakoid

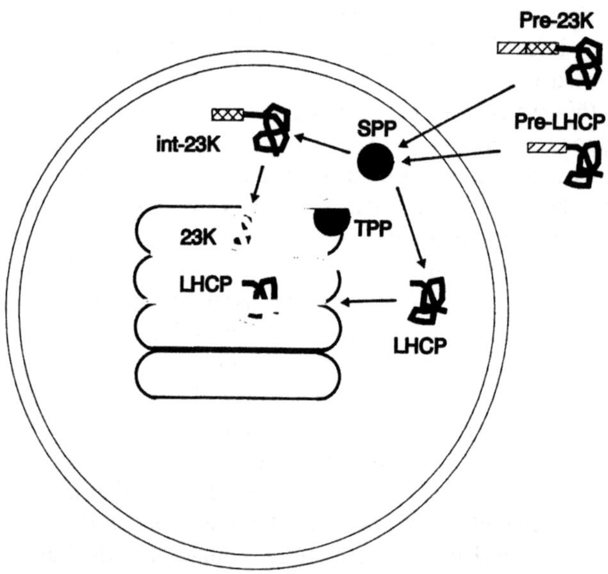

Figure 6.4. Sorting of imported thylakoid membrane and lumen proteins. The 23 kDa oxygen-evolving complex protein (23 K) and LHCP are located in the thylakoid lumen and membrane respectively. Both precursors are imported initially into the stroma and processed by SPP: pre-23 K to an intermediate form (int-23 K) and pre-LHCP to the mature size. Int-23 K is subsequently transferred across the thylakoid membrane and processed to the mature size by TPP. LHCP integrates into the thylakoid membrane by a process which, *in vitro*, requires ATP and a stromal factor.

membranes can be reconstituted *in vitro*. Efficient integration requires ATP and the presence of at least one stromal factor, a protein of about 65 kDa (Cline, 1986; Viitanen *et al.*, 1988; Fulson and Cline, 1989; Kohorn and Tobin, 1989). It has been suggested that the stromal factor may act as an 'unfoldase' which helps the LHCP to assume an integration-competent conformation. Further work is required to test this possibility, and to unravel the role of ATP in this process, but the availability of an *in vitro* integration assay will clearly continue to be of immense value in these studies.

6.5 Import of proteins into the envelope membranes

In recent years there have been a number of significant advances in our understanding of protein import into the stroma and thylakoids, and several distinct import pathways are now apparent. In contrast, we know almost nothing about the import of proteins into the inner or outer envelope membranes, or the intermembrane space, and the absence of even basic information about these import pathways remains a serious obstacle to any comprehensive understanding of the chloroplast as a protein-sorting system. For all of the functional importance of the reactions taking place in the stroma, thylakoid membrane and lumen, it is worth emphasising that these locations represent only half of the total number of chloroplast compartments.

The lack of progess is largely due to technical problems in the isolation and cloning of envelope proteins; these proteins represent only about 1% of total chloroplast membrane protein and are consequently of low abundance. The scale of the problem is such that, to date, only one envelope protein has been cloned: the phosphate translocator in the inner membrane (Flügge *et al.*, 1989). This protein is synthesised in precursor form, and preliminary evidence (U.-I. Flügge and D.L. Willey, personal communication) indicates that the presequence is removed by SPP. This may mean that information in the mature protein dictates localisation into the inner membrane, in which case it will be of interest to determine the mechanism by which this protein inserts into the correct membrane. A simple model in which hydrophobic stretches of residues anchor the protein in the inner membrane (perhaps after removal of the presequence by SPP) seems unlikely. If this were the case, it would be expected that the mature LHCP sequence, which contains several hydrophobic stretches, would likewise prevent transport of pre-LHCP to the thylakoid network. As yet, we do not know how hydrophobic proteins discriminate between the thylakoid membrane and the inner envelope membrane, and this promises to be one of the most interesting areas of research in the near future.

The biogenesis of outer membrane proteins is equally obscure, but at least one protein is known to be initially synthesised as a larger precursor (Flügge and Wessel, 1984). This observation raises intriguing questions about the import pathway taken by such proteins; it will be of particular interest to determine whether these proteins are first targeted into the stroma. If the import mechanism involves cleavage by SPP, subsequent targeting to the outer membrane may turn out to be mediated by a separate export system, as is the case with mitochondrial intermembrane space proteins (see below).

6.4 Transport of proteins into mitochondria

In comparison with chloroplasts, plant mitochondria have been virtually ignored as an experimental system for the study of protein transport, and very little is known of the underlying import mechanisms. In total contrast, protein transport into fungal mitochondria has been investigated with great success, and the biogenesis of yeast and *Neurospora* mitochondria is now understood in some detail. The salient features of these import systems are briefly

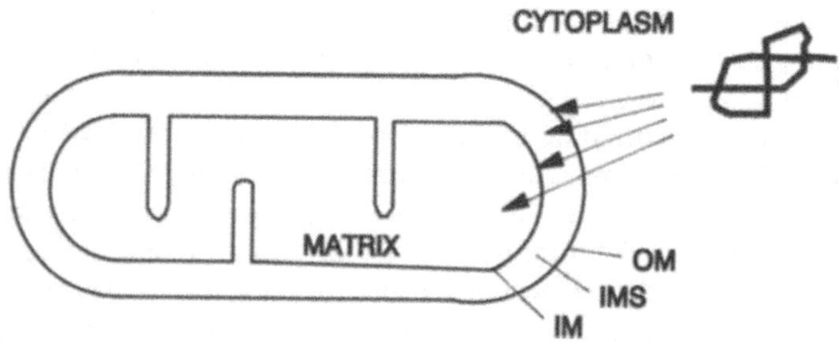

Figure 6.5. Protein transport into the mitochondrion. Cytoplasmically synthesised proteins are transported into all four mitochondrial compartments: the outer membrane (OM), inner membrane (IM), intermembrane space (IMS) and matrix.

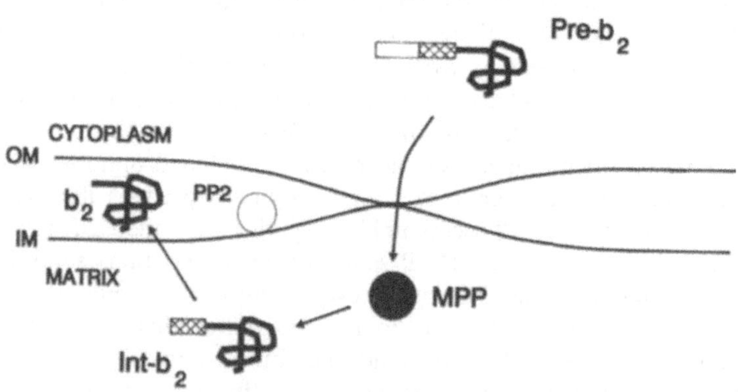

Figure 6.6. Import of intermembrane space proteins into mitochondria. Proteins such as cytochrome b_2 are synthesised as a precursor (pre-b_2) which are imported and processed to an intermediate form (Int-b_2) by the matrix processing peptidase (MPP). Int-b_2 is subsequently exported across the inner membrane and processed to the mature size by a second processing peptidase (denoted PP2) whose precise location is uncertain.

described below; a more detailed description is given by Pfaller *et al.*, (1988). It should, however, be borne in mind that the import of proteins into plant mitochondria may differ in some respects. Figures 6.5 and 6.6 show schematic diagrams of the import of proteins into mitochondria.

6.6.1 *Early stages in the import pathway*

As with imported chloroplast proteins, cytoplasmically synthesised mitochondrial proteins are often initially synthesised with cleavable amino-terminal presequences. However, this is not always the case, and some proteins are imported in the absence of proteolytic processing. These proteins appear to contain internal targeting signals, but will be referred to as 'precursors' in view of evidence that the initial translation products differ in conformation from the imported, functional proteins (Zimmermann and Neupert, 1980; Pfaller and Neupert, 1987). A basic feature common to both organelles is that protein import takes place by a post-translational ATP-dependent mechanism. However, mitochondria differ from chloroplasts in that an energised inner membrane is also required, at least for protein transport into the matrix, inner membrane and intermembrane space (Schleyer *et al.*, 1982).

After synthesis in the cytoplasm, precursor molecules bind to specific receptors in the mitochondrial outer membrane. As with chloroplasts, the presence of these receptors was first inferred from studies in which the protease treatment of isolated organelles was observed to inhibit the binding and import of *in vitro*-synthesised precursors (Zwizinski *et al.*, 1984). More recently though, the receptors have been characterised in much greater detail. Antibodies raised against a 19 kDa mitochondrial outer membrane protein (termed MOM 19) were observed to inhibit protein import by isolated mitochondria, and Söllner *et al.* (1989) have provided evidence that MOM 19 is indeed the receptor for a variety of imported proteins. However, MOM 19 does not appear to mediate the import of the ADP/ATP carrier, a major inner membrane protein, and further studies have indicated that a second protein, termed MOM 72, acts as a receptor for the ADP/ATP carrier (N. Pfanner and W. Neupert, unpublished data).

6.6.2 *Transport across the mitochondrial membranes*

Protein transport into the mitochondrial matrix takes place at 'contact sites' between the inner and outer mitochondrial membranes. This has been clearly demonstrated by the generation of trapped translocation intermediates, in which the transported protein spanned both membranes (Schleyer and Neupert, 1985). These studies also provided evidence that proteins must traverse the mitochondrial membranes in an unfolded conformation, a notion which has been elegantly confirmed by Eilers and Schatz (1986). These workers showed that the presequence of a mitochondrial matrix protein (cytochrome oxidase subunit IV) was capable of targeting DHFR, a cytoplasmic protein, into the matrix. However, import was almost completely blocked in the presence of methotrexate, a folate analogue which binds with high affinity to the DHFR active site. The clear implication from this result was that proteins must unfold prior to or during translocation into the matrix. During or shortly after translocation into the matrix, precursors are processed by a matrix processing peptide (MPP), which carries out a similar function to chloroplast SPP activity. As with the stromal enzyme, MPP is highly specific for imported precursors and is metal-dependent (Böhni *et al.*, 1980; McAda and Douglas, 1982). Interestingly, MPP alone exhibits only low processing activity, and efficient processing requires the presence of a second protein, termed processing-enhancing protein or PEP (Hawlitschek *et al.*, 1988). The requirement for PEP suggests that the processing reaction may represent a novel type of hydrolysis mechanism, but how these components interact is as yet unknown.

6.6.3 *Sorting of imported mitochondrial proteins*

Although the mitochondrion is not as structurally complex as the chloroplast, a number of markedly different import pathways are exhibited by proteins destined for different locations. The biogenesis of intermembrane space proteins, for example cytochromes b_2 and c_1, is particularly interesting. These proteins are synthesised with a two-domain presequence, the first domain being a matrix-targeting signal. Precursors are imported across both mitochondrial membranes and processed to intermediates by MPP; the intermediates are then exported across the inner membrane by a *different* transport system, and processed to the mature sizes by a second processing peptidase (Hartl *et al.*, 1987). This import pathway, which is illustrated in Figure 6.6, has obvious parallels with the pathway utilised by thylakoid lumen proteins. One hypothesis which is currently favoured is that both organellar import systems have evolved to transport all imported proteins into the soluble phase corresponding to the cytoplasm of the ancestral endosymbiotic bacteria (i.e. the matrix or stroma). From this point, the further transport of any proteins destined for other locations within the organelles is mediated by the ancestral transport systems in the thylakoid or mitochondrial inner membrane (and possibly in the chloroplast inner membrane).

Evidence in support of the above 'conservative sorting' hypothesis (Hartl *et al.*, 1987) has come from studies on the import of mitochondrial outer membrane proteins. In several respects, the import of these proteins differs from proteins destined for other compartments. Outer membrane proteins are not synthesised with cleavable presequences, they are not imported at contact sites, and their import does not require an energised inner membrane (Pfanner *et al.*, 1988). However, Hurt *et al.* (1985) showed that the first 12 residues of a yeast 70 kDa outer membrane protein function as a matrix-targeting signal capable of mediating the import of a soluble protein into the matrix. They proposed that a subsequent hydrophobic stretch of amino acids (residues 12–39) acted as a 'stop-transfer' signal anchoring the 70 kDa protein in the outer membrane.

6.6.4 *Transport of proteins into plant mitochondria*

In contrast to chloroplast and fungal mitochondria, plant mitochondria have been largely ignored as an experimental system for studying protein transport, and very little is known of the underlying import mechanisms. Only one report of *in vitro* protein import by plant mitochondria has emerged (Whelan and Harmey, 1988) and detailed biochemical analyses have yet to be carried out. It is likely that plant and fungal mitochondria are fairly similar in this respect, but equally there may be significant differences in some aspects of protein import. For example, it is known that the two types of mitochondrion import a slightly different range of proteins; the α-subunit of the F_1 ATPase is mitochondrion-encoded in maize but nuclear-encoded in all other eukaryotic organisms examined (Hack and Leaver, 1983).

Despite the lack of background information on protein transport into plant mitochondria, the targeting of foreign proteins into this organelle *in vivo* has been achieved by three laboratories. Boutry *et al.* (1987) showed that the amino-terminal 90 residues of the tobacco β-F_1 ATPase precursor target bacterial chloramphenicol acetyltransferase into the matrix in transformed tobacco. More, recently, Schmitz and Lonsdale (1989) targeted β-glucuronidase into tobacco mitochondria using the presequence of a yeast mitochondrial protein. Finally, a cytoplasmic plant glutamine synthetase isoenzyme has been retargeted to the mitochondrial matrix (again in transgenic tobacco) using the presequence of the tobacco β-F_1 ATPase subunit (P., Hemon; M.P., Robbins and J.V., Cullimore, manuscript submitted for publication). In all three cases, the imported proteins were active and the targeting of foreign proteins into this organelle therefore appears to be fairly straightforward.

References

Böhni, P.C., Daum, G. and Schatz, G. (1983) Partial purification of a matrix-located protease involved in cleavage of mitochondrial precursor polypeptides. *J. Biol. Chem.* **258**: 4937.

Boyle, S.A., Hemmingsen, S.M. and Dennis, D.T. (1986) Uptake and processing of the precursor to the small subunit of ribulose 1,5-bisphosphate carboxylase by leucoplasts from the endosperm of developing castor oil seeds. *Plant Physiol.* **81**: 817.

Boutry, M., Nagy, F., Poulsen, C., Aoyagi, K. and Chua, N.-H. (1987) Targeting of bacterial chloramphenicol acetyl transferase to mitochondria in transgenic plants. *Nature* **328**: 340.

Chua, N.-H. and Schmidt, G.W. (1978) Post-translational transport into intact chloroplasts of the precursor of the small subunit of ribulose 1,5-bisphosphate carboxylase. *Proc. Natl Acad. Sci. USA* **75**: 6110.

Cline, K. (1986) Import of proteins into chloroplasts: membrane integration of a thylakoid precursor protein reconstituted in chloroplast lysates. *J. Biol. Chem.* **261**: 14804.
Cline, K., Werner-Washburne, M., Lubben, T.H. and Keegstra, K. (1985) Precursors to two nuclear-encoded chloroplast proteins bind to the outer envelope membrane before being imported into chloroplasts. *J. Biol. Chem.* **260**: 36911.
Cornwell, K.L. and Keegstra, K. (1987) Evidence that a chloroplast surface protein is associated with a specific binding site for the precursor to the small subunit of ribulose bisphosphate carboxylase. *Plant Physiol.* **85**: 750.
Della-Cioppa, G. and Kishore, G.M. (1988) Import of a precursor protein into chloroplasts is inhibited by the herbicide glyphosate. *EMBO J.* **7**: 1299.
Della-Cioppa, G., Baner, S.C., Taylor, M.L., Rochester, D.E., Klein, B.K., Shah, D.M., Fraley, R.T. and Kishore, G.M. (1987) Targeting a herbicide-resistant enzyme from *Escherichia coli* to chloroplasts of higher plants. *Biotechnology* **5**: 579.
Dobberstein, B., Blobel, G. and Chua, N.-H. (1977) *In vitro* synthesis and processing of a putative precursor for the small subunit of ribulose bisphosphate carboxylase of *Chlamydomonas reinhardii*. *Proc. Natl Acad. Sci. USA* **74**: 1082.
Douce, R. and Joyard, J. (1979) Structure and function of the plastid envelope. *Adv. Bot. Res.* **7**: 1.
Eilers, M. and Schatz, G. (1986) Binding of a specific ligand inhibits import of a purified precursor protein into mitochondria. *Nature* **322**: 228.
Flügge, U.-I. and Hinz, G. (1986) Energy depdendence of protein translocation into chloroplasts. *Eur. J. Biochem.* **160**: 563.
Flügge, U.-I. and Wessel, D. (1984) Cell-free synthesis of putative precursors for envelope membrane polypeptides of spinach chloroplasts. *FEBS Lett.* **168**: 255.
Flügge, U.-I., Fischer, K., Gross, A., Sebald, W., Lottspeich, F. and Eckershorn, C. (1989) The triose phosphate-3-phosphoglycerate-phosphate translocator from spinach chloroplasts: nucleotide sequence of a full-length cDNA clone and import of the *in vitro* synthesised precursor protein into chloroplasts. *EMBO J.* **8**: 39.
Fulson, D.R. and Cline, K. (1988) A soluble protein factor is required *in vitro* for membrane insertion of the thylakoid precursor protein pLHCP. *Plant Physiol.* **88**: 1146.
Gatenby, A.A., Lubben, T.H., Ahlquist, P. and Keegstra, K. (1988) Imported large subunits of ribulose bisphosphate carboxylase, but not imported β-ATP synthase subunits, are assembled into holoenzyme in isolated chloroplasts.
Grossman, A.R., Bartlett, S.G. and Chua, N.-H. (1980) Energy-dependent uptake of cytoplasmically synthesised polypeptides by chloroplasts. *Nature* **285**: 625.
Hageman, J., Robinson, C., Smeekens, S. and Weisbeek, P. (1986) A thylakoid processing peptidase is required for complete maturation of the lumen protein plastocyanin. *Nature* **324**: 567.
Halpin, C., Musgrove, J.E., Lord, J.M. and Robinson, C. (1989*a*) Import and processing of proteins by castor bean leucoplasts. *FEBS Lett.* **258**: 32.
Halpin, C., Elderfield, P.D., James, H.E., Zimmermann, R., Dunbar, B. and Robinson, C. (1989*b*) The reaction specificities of the thylakoidal processing peptidase and *Escherichia coli* leader peptidase are identical. *EMBO J.* **8**: 3917.
Hartl, F.-U., Ostermann, J., Guiard, B. and Neupert, W. (1987) Successive translocation into and out of the mitochondrial matrix: targeting of proteins to the intermembrane space by a bipartite signal peptide. *Cell* **51**: 1027.
Hawlitschek, G., Schneider, H., Schmidt, B., Tropshug, M., Hartl, F.-U. and Neupert, W. (1988) Mitochondrial protein import. Identification of processing peptidase and of PEP, a processing enchancing protein. *Cell* **53**: 795.
von Heijne, G., Steppuhn, J. and Herrmann, R.G. (1989) Domain structure of mitochondrial and chloroplast targeting peptides. *Eur. J. Biochem.* **180**: 535.
Heber, U. and Heldt, H.W. (1981) The chloroplast envelope: structure, function and role in leaf metabolism. *Ann. Rev. Plant Physiol.* **32**: 139.
Highfield, P.E. and Ellis, R.J. (1978) Synthesis and transport of the small subunit of chloroplast ribulose bisphosphate carboxylase. *Nature* **271**: 420.
Hurt, E.C., Müller, U. and Schatz, G. (1985) The first twelve amino acids of a yeast mitochondrial outer membrane protein can direct a nuclear-encoded cytochrome oxidase subunit to the mitochondrial inner membrane. *EMBO J.* **4**: 3509.
James, H.E., Bartling, D., Musgrove, J.E., Kirwin, P.M., Herrmann, R.G. and Robinson, C. (1989)

Transport of protein into chloroplasts. Import and maturation of precursors to the 33, 23 and kDa proteins of the oxygen-evolving complex. *J. Biol. Chem.* **264**: 19573.

Kavanagh, T.A., Jefferson, R.A. and Bevan, M.W. (1988) Targeting a foreign protein to chloroplasts using fusions to the transit peptide of a chlorophyll a/b protein. *Mol. Gen. Genet.* **215**: 38.

Kirwin, P.M., Elderfield, P.D. and Robinson, C. (1987) Transport of proteins into chloroplasts. Partial purification of a thylakoidal processing peptidase involved in plastocyanin biogenesis. *J. Biol. Chem.* **262**: 16386.

Kirwin, P.M., Elderfield, P.D., Williams, R.S. and Robinson, C. (1988) Organisation, orientation and lateral distribution of the plastocyanin processing peptidase in the thylakoid membrane. *J. Biol, Chem.* **263**: 18.

Kirwin, P.M., Meadows, J.W., Shackleton, J.B., Musgrove, J.E., Elderfield, P.D., Hay, N.A. and Robinson, C. (1989) ATP-stimulated import of a luminal protein by isolated thylakoid vesicles. *EMBO J.* **8**: 2251.

Ko, K. and Cashmore, A.R. (1989) Targeting of proteins to the thylakoid lumen by the bipartite transit peptide of the 33 kD oxygen-evolving protein. *EMBO J.* **8**: 3187.

Kohorn, B.D. and Tobin, E.M. (1989) A hydrophobic carboxy-proximal region of a light-harvesting chlorophyll a/b protein is necessary for stable integration into thylakoid membranes. *Plant Cell* **1**: 159.

Kuwabara, T., Reddy, K.J. and Sherman, L.A. (1987) Nucleotide sequence of the gene from the cyanobacterium *Anacystis nidulans* R2 encoding the Mn-stabilising protein involved in photosystem II water oxidation. *Proc. Natl. Acad. Sci. USA* **84**: 8230.

Lamppa, G.K. (1988) The chlorophyll a/b binding protein inserts into the thylakoids independent of its cognate peptide. *J. Biol. Chem.* **263**: 14996.

Lubben, T., Gatenby, A.A. Ahlquist, R. and Keegstra, K. (1989) Chloroplast import characteristics of chimeric proteins. *Plant Mol. Biol.* **12**: 13.

McAda, P. and Douglas, M.G. (1982) A neutral metallo endoprotease involved in the processing of an F_1 ATPase subunit precursor in mitchondria. *J. Biol. Chem.* **257**: 3177.

Meadows, J.W., Hulford, A., Shackleton, J.B. and Robinson, C. (1989) Targeting of a foreign protein into the thylakoid lumen of pea chloroplasts. *FEBS Lett.* **253**: 244.

Olson, L.J., Theg, S.M. Selman, B.R. and Keegstra, K. (1989) ATP is required for the binding of precursor proteins of chloroplasts. *J. Biol. Chem.* **264**: 6724.

Pain, D., Kanwar, Y.S. and Blobel, G. (1988) Identification of a receptor for protein import into chloroplasts and its localisation to envelope contact zones. *Nature* **331**: 232.

Pfaller, R. and Neupert, W. (1987) High-affinity binding sites involved in the import of porin into mitochondria. *EMBO J.* **6**: 2635.

Pfanner, N. Hartl, F.-U. and Neupert, W. (1988) Import of proteins into mitochondria: a multistep process. *Eur. J. Biochem.* **175**: 205.

Robinson, C. and Ellis, R.J. (1984a) Transport of proteins into chloroplasts. Partial purification of a chloroplast protease involved in the processing of imported precursor proteins. *Eur. J. Biochem.* **142**: 337.

Robinson, C. and Ellis, R.J. (1984b) Transport of proteins into chloroplasts. The precursor of the small subunit of ribulose bisphosphate carboxylase is processed in two steps. *Eur. J. Biochem.* **142**: 343.

Schleyer, M. and Neupert, W. (1985) Transport of proteins into mitochondria:translocation intermediates spanning contact sites between outer and inner membranes. *Cell* **43**: 339.

Schleyer, M. Schmidt, B. and Neupert, W. (1982) Requirement of a membrane potential for the post-translational transfer of proteins into mitochondria. *Eur. J. Biochem.* **125**: 109.

Schmitz, U.K. and Lonsdale, D.M. (1989) A yeast mitochondrial presequence functions as a signal for targeting to plant mitochondria *in vivo*. *Plant Cell* **1**: 783.

Schreier, P.H., Seftor, E.A., Schell, J. and Bohnert, H.J. (1985) The use of nuclear-encoded sequences to direct the light-regulated synthesis and transport of a foreign protein into plant chloroplasts. *EMBO J.* **4**: 25.

Smeekens, S., Groot, M., Van Binsbergen, J. and Weisbeck, P. (1985) Sequence of the precursor of the chloroplast thylakoid lumen protein plastocyanin. *Nature* **317**: 456.

Smeekens, S. Bauerle, C., Hageman, J., Keegstra, K. and Weisbeck, P. (1986) The role of the transit peptide in the routing precursors toward different chloroplast compartments. *Cell.* **46**: 365.

Smeekens, S., Van Steeg, H., Bauerle, C., Beetenbroeck, H., Keegstra, K. and Weisbeck, P. (1987)

Import into chloroplasts of a yeast mitochondrial protein directed by ferredoxin and plastocyanin transit peptides. *Plant Mol. Biol.* **9**: 377.

Theg, S.M., Bauerle, C., Olsen, L.J., Selman, B.R. and Keegstra, K. (1989) Internal ATP is the only energy requirement for the translocation of precursor protein across chloroplastic membranes. *J. Biol. Chem.* **264**: 6730.

Thornber, J.P. (1975) Chlorophyll-proteins: light-harvesting and reaction centre components of plants. *Ann. Rev. Plant Physiol.* **26**: 127.

Van den Broeck, G., Timko, M.P., Van Montagu, M. and Herrera-Estrella, L. (1985) Targeting of a foreign protein to chloroplasts by fusion to the transit peptide from the small subunit of ribulose 1,5-bisphosphate carboxylase. *Nature* **313**: 358.

Viitanen, P.V., Doran, E.R. and Dunsmuir, P. (1988) What is the role of the transit peptide in thylakoid integration of the light-harvesting chlorophyll a/b protein'? *J. Biol. Chem.* **263**: 15000.

von Heijne, G., Steppuhn, J. and Herrmann, R.G. (1989) Domain structure of mitochondrial and chloroplast targeting peptides. *Eur. J. Biochem.* **180**: 535.

Wallace, T.P., Stewart, A.C. Pappin, D. and Howe, C.J. (1989) Gene sequence for the 9 kDa component of photosystem II from the (cyanobacterium) *Phormidium lamindsum* and comparison with other leader sequences. *Mol. Gen. Genet.* **216**: 334.

Wasmann, C.C., Reiss, B., Bartlett, S.G. and Bohnert, H.J. (1986) The importance of the transit peptide and the transported protein for protein import into chloroplasts. *Mol. Gen. Genet.* **205**: 446.

Zimmermann, R. and Neupert, W. (1980) Transport of proteins into mitochondria. *Eur. J. Biochem.* **109**: 217.

Zwizinski, C., Schleyer, M. and Neupert, W. (1984) Proteinaceous receptors for the import of mitochondrial precursor proteins. *J. Biol. Chem.* **259**: 7850.

7 Protein transport and targeting within the endomembrane system of plants

A.B. BENNETT and K.W. OSTERYOUNG

7.1 Introduction

The subcellular compartmentation of eukaryotic cells provides for spatial separation and specialisation of biochemical pathways and presumably accounts for the functional complexity of eukaryotic organisms. Thus, the mechanisms underlying the biogenesis of subcellular compartments and the sorting of proteins to establish biochemically specialised compartments are of central importance to the eukaryotic way of life.

Subcellular compartments can be conveniently divided into two groups: the semiautonomous organelles (mitochondria and chloroplasts) and the endomembrane organelles (endoplasmic reticulum, nuclear envelope, Golgi apparatus, vacuole, protein bodies and peroxisomes). In many respects, including mechanisms of protein targeting, the mitochondria and chloroplasts differ from the endomembrane organelles, and protein import to these semiautonomous organelles is reviewed elsewhere in this volume (Chapter 6). The organelles of the endomembrane system are naturally related to each other since they all arise, either directly or indirectly, from the endoplasmic reticulum (ER) and are linked by vesicle flow originating in the ER. Proteins entering the endomembrane system are nuclear encoded and are synthesised on ER-bound ribosomes. After entry into the ER those proteins that are not resident there must be sorted and transported to destinations beyond the ER. Thus proteins localised within endomembrane compartments must possess signals directing their primary entry into the ER and secondary signals directing their subsequent transport, sorting and/or retention in specific compartments.

Although the topic of protein transport and sorting has been a central feature of studies in cell biology, only a few laboratories have focused on this problem in plants. With rapid advances in the application of genetic engineering technology, it is now clear that information in this area is needed to design and implement strategies for genetic engineering of biochemical pathways within subcellular compartments of plants. Thus, the molecular basis of protein targeting represents a critical gap in understanding at the interface of plant cell biology and plant biotechnology. In addition to revealing this gap in knowledge of plant cell biology, recent advances in plant genetic engineering have also provided the experimental tools to make rapid progress in improving our understanding of this area. In this chapter, our current understanding of protein transport within the endomembrane system of plant cells is reviewed. However, to fill in gaps in our understanding of plant cell processes, frequent reference to analogous mechanisms in other eukaryotic cells is made, recognising that these 'heterologous' systems may not fully reflect the situation in plants.

7.2 Biogenesis of endomembrane compartments

7.2.1 *Endoplasmic reticulum*

The ER is the primary site of phospholipid biosynthesis in plant cells, possessing a complement of enzymes required for the synthesis of the major membrane phospholipids, phosphatidylcholine and phosphatidylethanolamine, from fatty acid precursors (Moore, 1982). In addition, ribosomes are bound to domains of the ER where they are active in carrying out protein synthesis. Ribosomes interact with specific components of the ER membrane that facilitate translocation of nascent polypeptides into the lumen of the ER or integration into the membrane itself. The capacity to synthesise both membrane lipid as well as soluble and membrane-bound proteins endows the ER with the ability to construct the components of all of the endomembrane organelles.

7.2.2 *Nuclear envelope*

The nuclear envelope is reported to have direct connections to the ER, and in this respect may be considered as a specialised or differentiated portion of the ER (Robinson, 1985). The nuclear envelope has been reported to bind ribosomes, synthesise glycoproteins and bud vesicles destined for the Golgi, and so appears to participate in functions generally thought to belong exclusively to the ER (Robinson, 1985). Thus the nuclear envelope clearly belongs to the endomembrane system of organelles arising from the ER, and proteins destined for the intermembrane space of the nuclear envelope enter the normal endomembrane pathway of protein transport. However, proteins destined for the nucleus itself do not enter the endomembrane pathway of protein transport but are acquired instead by import of cytosolic proteins through nuclear pores (Dingwall and Laskey, 1986). In the context of protein transport then the nuclear envelope and nucleus can be properly considered as distinct organelles with only the envelope and intermembrane space associated with the endomembrane system.

7.2.3 *Golgi apparatus*

Although direct connections between the Golgi apparatus and ER have been reported, these connections only occur in specialised structures and are not the rule. Rather, the association between the Golgi and ER is indirect and results from vesicle flow between the organelles. The evidence for this vesicle flow in plant cells comes predominantly from ultrastructural identification of small vesicles present between the ER and proximal or *cis* face of the Golgi apparatus (Robinson, 1985). Despite several reports of immunocytochemical localisation of vacuolar storage proteins in the ER and Golgi apparatus, localisation of these proteins in ER-to-Golgi transition vesicles has not been reported. Thus, while unequivocal evidence for vesicle flow from the ER to the Golgi is lacking, circumstantial evidence supports this model. In yeast and mammalian cells the evidence for ER-to-Golgi vesicle flow is strong. In addition, retrograde transport of protein from the Golgi to the ER is also well documented. Thus two-way traffic between the ER and Golgi apparatus appears to occur in eukaryotic cells.

Within the Golgi apparatus of mammalian cells the cisternae can be functionally distinguished as three compartments: the *cis* , medial, and *trans* cisternae. The lumina of individual Golgi compartments are not contiguous but are indirectly associated by vesicle flow between cisternae. These different Golgi compartments are characterised by distinct carbohydrate-processing activities that sequentially modify the oligosaccharide moieties of glycoproteins. In animal cells the compartments can be resolved biochemically by density-gradient centrifugation and have been identified *in situ* by immunocytochemical localisation of oligosaccharide-processing enzymes in specific cisternae (Farquhar, 1985). Similar oligosaccharide-processing enzymes are present in the Golgi apparatus of plant cells, yet there is only one report of the biochemical resolution of Golgi cisternae in plants (Ali *et al.*, 1986). Other attempts, however, to resolve the sub-Golgi compartments of plants have been unsuccessful (Sturm *et al.*, 1987a).

7.2.4 *Cell surface*

Although not an organelle, the apoplast or extracellular space can be considered an endomembrane compartment since it is bordered by the plasma membrane, which is derived indirectly from the ER. Vesicle flow from the Golgi to the plasma membrane is well documented both ultrastructurally and by pulse-chase labelling of cell wall components and isolation of secretory vesicles (Griffing and Ray, 1985). Thus Golgi-derived vesicles participate in the delivery of cell wall components, extracellular proteins and plasma membrane components. In addition, a parallel pathway of protein secretion by direct vesicle flow from the ER to the cell surface has been proposed in some tissues (Akazawa and Hara-Nishimura, 1985). Recycling of plasma membrane to the

Golgi or to a partly coated reticulum has also been proposed to occur (Pesacreta and Lucas, 1984). Because of this proposed recycling, membrane flow to the cell surface is bidirectional, providing mechanisms for both export and import of macromolecules. Endocytosis of specific extracellular ligands has been reported, supporting this model of bidirectional membrane flow in plants (Horn *et al.*, 1989).

7.2.5 *Vacuole*

Despite its prominence as the largest subcellular compartment of most plant cells, biogenesis of the vacuole has been largely neglected. Although often interpreted as arising from the swelling and coalescence of ER or Golgi vesicles (Matile and Moor, 1968), other research has suggested a distinct structure, the Golgi-associated endoplasmic reticulum–lysosomal complex (GERL), as the site of vacuole formation. This complex consists of a fenestrated network of cisternae directly connected to both the rough ER and elements of the Golgi apparatus. Such a region has been shown to be the site of packaging of acid phosphatase into provacuoles (Marty, 1978). The GERL network is located at the *trans* face of the Golgi apparatus and appears to be analogous to structures that give rise to lysosomes in animal cells (Novikoff *et al.*, 1971). However, the GERL network has only been characterised in root tips of *Euphorbia*, and the generality of this structure as the site of vacuole origin has not been confirmed.

Transport of proteins to the vacuole has been studied primarily in vacuoles specialised for protein storage (see Section 7.2.6). The cumulative evidence indicates that many, if not all, vacuolar proteins are synthesised on the ER and transported through the ER and Golgi apparatus to the vacuole (Chrispeels, 1985a).

7.2.6 *Protein bodies*

Protein bodies are specialised organelles found primarily in seeds but also in other organs that function in the storage of reserve proteins. In terms of their biogenesis, protein bodies form two distinct classes of organelles, one class arising directly from enlargement of the rough ER and a second class consisting of vacuoles that become specialised for protein storage.

Protein bodies arising directly from enlargement of the rough ER are best characterised in maize but are common in seeds of many monocotyledonous plants (Robinson, 1985). Continuities between ER and protein body membranes are evident at the ultrastructural level, as is the presence of ribosomes on the protein body membrane (Khoo and Wolf, 1970). In addition, cytochrome c reductase, an enzymatic marker of the ER, is associated with isolated protein bodies from maize (Larkins and Hurkman, 1978). Thus, this class of protein bodies in monocotyledonous plants appears to represent specialised regions of dilated rough ER.

Protein bodies arising from vacuoles are best characterised in seeds of dicotyledonous plants. However, similar protein bodies (i.e. of vacuolar origin) have been characterised in seeds of several monocotyledonous plants, most notably rice (Krishnan *et al.*, 1986) and wheat (Kim *et al.*, 1988). In rice seeds both classes of protein bodies, those derived directly from rough ER and those arising from vacuolar specialisation, are present and are specialised for the storage of prolamines and glutelins respectively (Krishnan *et al.*, 1986). In developing legume seed cotyledons, the large central vacuole becomes subdivided and storage proteins are deposited within the resulting small vacuoles (Craig *et al.*, 1979). It is now well established that most, if not all, proteins destined for deposition in the legume protein body are synthesised on rough ER and transit the Golgi apparatus *en route* to their site of deposition (Chrispeels, 1985b). The close relationship of legume protein bodies to vacuoles has led to the use of the term 'protein storage vacuoles' to describe this class of protein bodies (Dorel *et al.*, 1989).

7.2.7 *Lipid bodies*

Lipid bodies can properly be considered to be endomembrane compartments because they arise by budding from the ER. The lipid body is unique in that it is bounded by a phospholipid monolayer, or half-membrane, and appears to arise by deposition of triacylglycerides between leaflets of the ER membrane lipid bilayer (Wanner *et al.*, 1982). This leads to swelling and eventual budding of unilamellar lipid-containing vesicles. Although this model appears to account for biogenesis of the lipid body half-membrane, recent evidence suggests that lipid body proteins are not synthesised on or present in the ER membrane (Herman, 1987). This suggests that lipid body proteins may be acquired from the cytosol.

7.2.8 *Peroxisomes*

Peroxisomes are unique organelles that cannot properly be considered either endomembrane or semiautonomous organelles because they do not arise from the ER or possess a semiautonomous genome. However, peroxisomes were historically thought to arise by direct vesiculation of the ER, and have therefore often been included as a component of the endomembrane system (Wanner *et al.*, 1982). Recent evidence, however, argues for peroxisome biogenesis occurring by growth and division of pre-existing peroxisomes (Trelease, 1984; Lazarow and Fujiki, 1985). It is also well documented that soluble peroxisomal proteins are acquired by direct import from the cytosol rather than by vesicle flow from the ER. Nevertheless, there is evidence that peroxisomal membrane lipid is derived from the ER and that some peroxisomal membrane proteins are glycosylated, suggesting that they may be acquired from the ER (Trelease, 1984). These considerations suggest that some aspects of peroxisome biogenesis may include transfer of ER-derived membrane and membrane proteins.

7.3 Mechanisms of protein transport

The mechanisms and pathways of protein transport within the endomembrane system have been described with increasing detail since the basic pathway was outlined by Palade (1975). As compared with yeast and mammalian cells, relatively little research has specifically addressed the mechanisms of protein transport in plant cells. Gene transfer experiments have indicated that plant proteins expressed in mammalian (Voelker *et al.*, 1986), yeast (Edens *et al.*, 1984; Rothstein *et al.*, 1984; Tague and Chrispeels, 1987) or amphibian (Hurkman *et al.*, 1981; Bassüner *et al.*, 1983; Vitale *et al.*, 1986) cells are processed and translocated within the endomembrane system, suggesting that the general mechanisms of protein translocation are conserved in all eukaryotic cells. Nevertheless, heterologous expression of plant proteins often results in mislocalisation (Bassüner *et al.*, 1983; Vitale *et al.*, 1986; Voelker *et al.*, 1986), indicating that the mechanism of protein transport, while generally similar, differs in specific details. Thus, at the present time we must refer to the wealth of literature on protein transport in non-plant cells to understand the mechanisms that probably operate in plants, while acknowledging the gaps in our understanding of details of the transport process and machinery that almost certainly distinguish protein trafficking in plants.

7.3.1 *Endoplasmic reticulum entry*

It is well established in both plant and animal cells that proteins destined for endomembrane compartments enter the system by co-translational protein translocation across the ER membrane. At this step the protein may be glycosylated by the transfer of high-mannose oligosaccharides from a lipid-linked donor to specific asparagine residues. There is also evidence that many proteins fold and oligomerise in the ER, and in at least one case correct folding and oligomerisation have been shown to be required for export from the ER (Gething *et al.*, 1986). An ER luminal protein referred to as either BiP (binding protein) or GRP78 (glucose-repressible protein–78 kDa) has been shown to bind misfolded protein and monomers of oligomeric proteins (Haas and Wabl, 1983; Bole *et al.*, 1986; Munro and Pelham, 1986). It has been proposed that BiP acts as a chaperone in facilitating folding and assembly of proteins in the ER and that it may furthermore prevent export of the protein until it has achieved its mature conformation (Pelham, 1986). Although BiP has been characterised primarily in mammalian cells, its presence in yeast has been recently confirmed (Rose *et al.*, 1989), and there is increasing evidence for its presence in plant cells (Miernyk, 1988; N. Ewing and A. Bennett, unpublished data). There is also evidence that proteins other than BiP may carry out functions related to assembly and membrane anchoring of proteins within the ER (Lusis *et al.*, 1976).

7.3.2 *Endoplasmic reticulum-to-Golgi transport*

Proteins are exported from the ER in transition vesicles that have been frequently, but not always, observed between the ER and the *cis* face of the Golgi apparatus of plant cells (Robinson, 1985). Other than the ultrastructural identification of putative ER-to-Golgi transition vesicles, little is known regarding this step of protein transport in plant cells. In mammalian cells the rate of protein transport between the ER and Golgi suggests that 50% of the ER membrane is transported to the *cis* face of the Golgi apparatus every 10 minutes (Wieland *et al.*, 1987). Because this greatly exceeds the rate of membrane biosynthesis in the ER, it has been proposed that recycling of lipid from the Golgi apparatus to the ER must occur. This suggestion was confirmed by the observation that brefeldin A, a fungal metabolite that blocks transport between the ER and Golgi, resulted in the redistribution of *cis* Golgi enzymes to the ER (Doms *et al.*, 1989). This observation was interpreted to indicate that retrograde transport from the Golgi to the ER occurs constitutively and that this recycling pathway can be revealed by selectively blocking forward (i.e. ER to Golgi) transport with brefeldin A. Thus, membrane and protein traffic between the ER and Golgi appears to be bidirectional in mammalian cells. In plant cells, however, we have only the most rudimentary observations of vesicles that may participate in similar traffic patterns between these organelles.

7.3.3 *Intra-Golgi transport*

The mechanisms of protein transport within the Golgi apparatus are completely unstudied in plant cells. This is largely because few laboratories are engaged in studying the plant Golgi apparatus but may also reflect a technical limitation in the routine biochemical separation of *cis*, medial and *trans* Golgi (Sturm *et al.*, 1987a). Consequently, biochemical markers and/or immunochemical probes for sub-Golgi compartments are not available in plants. In mammalian cells transport of cell-surface viral proteins within the Golgi apparatus has been shown to occur by sequential flow from the *cis* to the *trans* face of the Golgi apparatus (Bergman and Singer, 1983). This sequential movement could occur within a contiguous lumen of the Golgi apparatus or by vesicle flow between cisternae. The strongest support for the model of vesicle flow between Golgi cisternae comes from experiments in which Chinese hamster ovary cells having defects in specific galactosyl-, *N*-acetylglucosaminyl- or sialyltransferase activities were fused and the transfer of viral glycoprotein between distinct Golgi stacks was inferred from processing of the glycoprotein (Rothman *et al.*, 1984). These experiments demonstrated that proteins can move from one Golgi stack to another in a directional manner, and perhaps most surprisingly that the rate of inter-Golgi transfer was the same as intra-Golgi transfer. This experimental result provided strong support for a model of targeted vesicle flow between Golgi cisternae rather than by movement through a contiguous Golgi lumen (Pfeffer and Rothman, 1987). The demonstration that proteins could move between distinct Golgi complexes in a directional manner (i.e. medial to *trans* cisternae) indicates that vesicle flow is not confined to movement between adjacent cisternae of the same Golgi apparatus and therefore each vesicle must possess some form of targeting signal to the appropriate sub-Golgi cisternae. As emphasised above, there is no direct evidence that the Golgi apparatus of plants is organised or functions in a manner analogous to that which has been described in animal cells.

7.3.4 *Post-Golgi transport*

Protein transport to the cell surface, vacuole, protein bodies and perhaps other endomembrane compartments occurs by vesicle transport departing from the *trans* face of the Golgi apparatus. Griffiths and Simons (1986) have proposed that a compartment previously referred to as GERL (see Section 7.2.5) represents the region of protein sorting. These authors have suggested that this compartment be renamed the trans Golgi network (TGN) to emphasize its functional association with the Golgi apparatus . As discussed above in relation to vacuole biogenesis, a similar structure has been characterised in plant cells (Marty, 1978), although evidence for direct participation of this structure in sorting of plant endomembrane proteins is lacking.

Regardless of the exact site of protein sorting and secretory vesicle formation, secretory vesicles form near the *trans* face of the Golgi apparatus. Such vesicles have been ultrastructurally identified in plant cells, and treatments with ionophores such as monensin or nigericin that inhibit exocytosis result in an accumulation of these vesicles (Mollenhauer *et al.*, 1982; Griffing and Ray, 1985). These results have been interpreted to suggest that, as in animal cells (Mellman *et al.*, 1984), post-Golgi transport in plant cells requires acidification of the Golgi or of Golgi-derived vesicles. Many intracellular compartments, including the Golgi apparatus, are acidified by vacuolar-type H^+-ATPases (Chanson and Taiz, 1985; Moriyama and Nelson, 1989), and it is therefore interesting to note that a class of yeast vacuolar missorting mutants is deficient in vacuolar acidification and vacuolar-type H^+-ATPase subunits (Rothman *et al.*, 1989a).

Secretory vesicles have been isolated from both plant (Taiz *et al.*, 1983) and animal (deCurtis and Simons, 1989) cells. Because these vesicles are presumed to possess signals targeting themselves for fusion with specific subcellular membranes (Palade, 1983), characterisation of the vesicle surface may provide unique insights into the mechanisms of vesicle recognition and sorting.

7.4 Mechanisms of protein sorting

Although it is clear that the complex mixture of proteins entering the endomembrane system via the ER must be sorted, even the general nature of sorting signals has been controversial. Given the ultrastructural view of endomembrane protein transport, with a mixture of proteins sharing common compartments until they become progressively sorted as they move through the endomembrane pathway, one can imagine two distinct types of protein-sorting signals. The first would be signals directing the protein to 'go forward' to the next endomembrane compartment. In this model proteins lacking such 'go forward' or 'transport' signals would be excluded from transport vesicles, and thus retained in a specific compartment. Alternatively, sorting could occur using 'retention' signals that serve to retain proteins within a specific compartment, while all proteins lacking such signals would proceed to the next compartment by bulk flow. Wieland *et al.* (1987) compared the rate of bulk flow within the endomembrane system with the rate of secretion of the most rapidly secreted proteins, reasoning that this distinction between 'transport' and 'retention' signals could be resolved by demonstrating whether the rate of protein secretion was equivalent to or greater than the rate of bulk flow of a marker within the endomembrane system. Using membrane-permeant tripeptides that became glycosylated and trapped in the ER, the rate of bulk flow within the endomembrane system was determined to be slightly faster than, or equal to, secretion of the most rapidly secreted proteins. This result was interpreted to indicate that other proteins moved more slowly as a result of specific retention within compartments along the secretory route. This model of the role of specific 'retention' signals in endomembrane protein targeting has been supported by the identification of molecular signals that retain proteins within endomembrane compartments (see below). However, as Klausner (1989) has pointed out, such 'retention' signals alone cannot explain sorting at branch points in a pathway where 'go forward' or 'transport' signals must be invoked. It is likely that protein sorting cannot be described using a single type of sorting signal and that the reality of protein sorting within the cell involves complex combinations of 'retention' and 'transport' signals. Nevertheless the distinction between 'retention' and 'transport' signals is useful as a concept to describe the signals directing protein transport through common intermediate compartments.

7.4.1 *Endoplasmic reticulum entry—the signal sequence*

7.4.1.1 *Function of the signal sequence.* All proteins targeted to endomembrane compartments enter by co-translational insertion of the polypeptide into the lumen of the ER. Thus, entry into the ER is the primary targeting event in endomembrane protein transport and is directed by an amino acid sequence, commonly referred to as the signal sequence. The signal sequence is usually located at the extreme amino terminus of the protein (for an exception see Lingappa *et al.*, 1979) and therefore is the first portion of the protein synthesised. Upon emergence of the nascent amino terminus (i.e. the signal sequence) of an endomembrane-localised protein, a cytosolic ribonucleoprotein, the signal recognition particle (SRP), binds to both the nascent polypeptide signal sequence and ribosome (Walter and Blobel, 1986). Binding of SRP may arrest or slow translation but functions primarily to target the ribosomes to binding and protein-translocation sites on the ER membrane. Once bound, SRP dissociates and translation continues, with the signal sequence now associating with a specific ER membrane-bound signal sequence receptor that has been identified by photoaffinity cross-linking as a 35 kDa glycoprotein (Weidmann *et al.*, 1987; Krieg *et al.*, 1989). Further polypeptide chain elongation is accompanied by co-translational translocation of the nascent polypeptide across the ER membrane, co-translational cleavage of the signal sequence and co-translational glycosylation of glycoproteins. SRP is present in limiting amounts in many cells, suggesting that it acts catalytically by cycling between the cytosol and ER membrane.

A schematic outline of protein import to the ER is shown in Figure 7.1. Although this model has been developed from studies of non-plant cells, there are several lines of evidence that similar mechanisms operate in plants. The most direct evidence is that components of the ER import machinery, including functional signal sequences and SRP (Prehn *et al.*, 1987; Campos *et al.*, 1988), have been identified in plants. In contrast to previous observations however, SRP from maize endosperm (Campos *et al.*, 1988) or wheat-germ (Prehn *et al.*, 1987) did not arrest translation in the wheat-germ cell-free translation system. This suggests that plant SRP may not halt translation of endomembrane-targeted proteins in the cytosol and raises the possibility that proteins may be synthesised as precursors in the cytosol of plant cells and translocated across the ER post-translationally. In this regard, post-translational import to the ER has been demonstrated *in vitro* (Hansen *et al.*, 1986; Perara *et al.*, 1986). However, post-translational import is inefficient and it is likely that most import *in vivo* occurs co-translationally.

Other evidence that ER protein import in plants is similar to other eukaryotic cells comes from numerous studies demonstrating protein import using heterologous mixtures of mRNA, cell-free translation lysates and microsomal membranes. These experiments have indicated that plant mRNAs can direct co-translational import and processing in either wheat-germ or rabbit reticulocyte cell-free translation lysates coupled with either mammalian or plant microsomal membranes (Burr and Burr, 1981; Torrent *et al.*, 1986; Hattori *et al.*, 1987; DellaPenna and Bennett, 1988). Additionally, expression of plant genes in yeast (Edens *et al.*, 1984; Rothstein *et al.*, 1984; Tague and Chrispeels, 1987), frog oocytes (Hurkman *et al.*, 1981; Bassüner *et al.*, 1983; Vitale *et al.*, 1986) and mammalian cells (Voelker *et al.*, 1986) further suggests

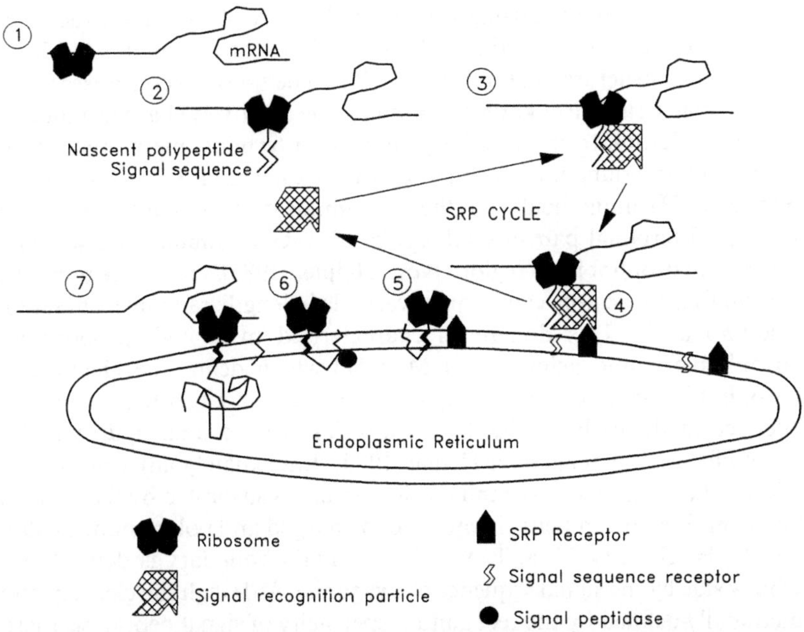

Figure 7.1. Schematic diagram of co-translational protein import to the ER. mRNA encoding proteins destined for ER import initiate translation on cytoplasmic ribosomes (1). Following emergence of the N-terminal signal sequence from the ribosome, signal recognition particle (SRP) binds to the signal sequence/ribosome complex (2, 3). Binding of SRP may halt or slow translation, although this function of SRP in plants has been disputed. The SRP–signal sequence–ribosome complex associates with the ER membrane by specific interaction between SRP and an SRP receptor (4). At this point the signal sequence also associates with an integral membrane signal sequence receptor. Subsequently, SRP dissociates from the ER, leaving ribosomes attached and the nascent polypeptide associated with the signal sequence receptor (5), which may also function as a component of the protein translocation apparatus (Krieg *et al.*, 1989). Translation proceeds with protein translocation across the ER membrane occurring co-translationally. Signal peptidase, associated with the inner face of the ER lumen, cleaves the signal sequence which is thought to remain membrane-associated with the N-terminus exposed to the cytoplasm (Kuhn, 1987) (6). Translation continues until the entire protein is synthesised and translocated into the ER (7).

that similar mechanisms function to import proteins to the ER in diverse organisms.

7.4.1.2 *Structure of the signal sequence.* The structure of signal sequences has been characterised and reviewed extensively by von Heijne (1983, 1984, 1985, 1988). The striking feature of the hundreds of known eukaryotic signal sequences is their lack of strict sequence homology. Nevertheless, comparison of a large number of signal sequences has revealed a common structural motif shared by all signal sequences.

The signal sequence comprises an N-terminal domain of 16–26 amino acids that is characterised by a strongly hydrophobic core of 7–15 amino acids. The signal sequence can be further subdivided into three domains, each with characteristic structural features (Figure 7.2). The n-region comprises the first 1–5 amino acids and in eukaryotes contains one positively charged amino acid residue in addition to the positively charged N-terminal methionine. The n-region domain boundary is defined by the last charged (D, E, R, K) or large polar (N, Q, H) amino acid from the N-terminus of the signal sequence or by the most N-terminal pair of small apolar (S, T, G, P) amino acids within the adjacent hydrophobic h-region (von Heijne, 1985). The h-region, lying adjacent to the n-region, can be quite variable in length, ranging from seven to 15 amino acids. This domain is characterised by a high proportion of hydrophobic amino acids (F, L, I, M, V, W) which define the characteristic hydrophobic core of the signal sequence (Figure 7.2). The h-region boundary is defined at the carboxy side by a transition from overall hydrophobic to hydrophilic amino acids (von Heijne, 1985). Immediately adjacent to the h-region is the c-region of 5–6 amino acids which, as suggested by the transition to an overall hydrophilicity, is enriched in charged and polar amino acids (D, E, R, K, H, G, P, Q, N, S, T, Y). The c-domain boundary is defined at the carboxy side by the signal sequence cleavage site. Although the cleavage site is functionally defined by the recognition specificity of signal peptidase, there is no strict amino acid sequence homology around this site. However, the cleavage site can be identified by the so-called (− 3, − 1) rule which recognises that small uncharged amino acids (A, C, S, T, G, P) are usually found at positions − 1 and − 3 relative to the cleavage site (von Heijne, 1983, 1984). It is

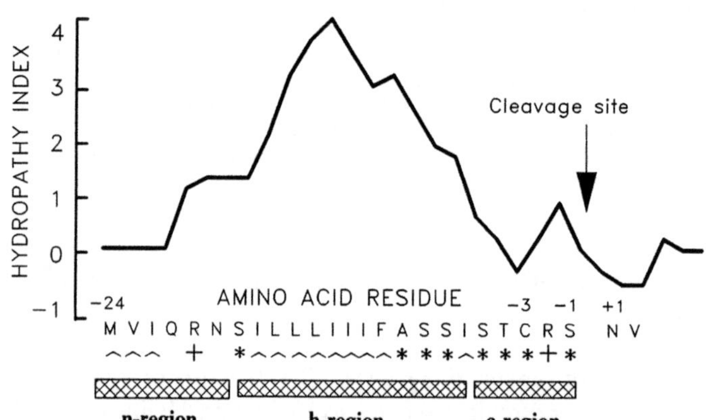

Figure 7.2. Structure of the signal sequence of tomato fruit polygalacturonase. The sequence was deduced from the nucleotide sequence of the cDNA and the signal sequence cleavage site by radiosequencing of the product of *in vitro* translation and *in vitro* processing (DellaPenna and Bennett, 1988). The hydropathy index was calculated by the method of Kyte and Doolittle (1982). The n-region is characterised by a single positively charged amino acid (+), the h-region by a core of hydrophobic amino acids (∧), and the c-region by small neutral amino acids (∗) at positions − 1 and − 3.

likely that the c-region, and specifically the amino acids at positions −1 and −3, plays an important role in signal peptidase recognition and cleavage. The n-region and h-region of the signal sequence have been proposed to function together as the primary ER targeting signal (von Heijne, 1988).

As illustrated in Figure 7.2, the signal sequence of tomato fruit polygalacturonase conforms to the general structural motif of eukaryotic signal sequences. Compilation and review of approximately 70 plant endomembrane-localised proteins have confirmed the similarity of plant signal sequences with this consensus motif (Miernyk and Nelson, 1989). While this would imply that signal sequences should be fully functional in heterologous expression systems, there are examples of signal sequences which are only partly functional when used to direct ER protein import in heterologous systems. The most extreme example is the expression of yeast carboxypeptidase Y in mammalian cells where the signal sequence failed to direct ER import (Bird *et al.*, 1987). This result suggests that, despite the overall structural similarity of signal sequences, subtle differences may be functionally significant.

7.4.2 *Endoplasmic reticulum retention*

Following entry into the ER most proteins move by bulk flow to the Golgi apparatus and destinations beyond. However, some proteins are resident in the ER. Among the best-characterised ER residents are BiP (Munro and Pelham, 1986; see Section 7.3.1), and protein disulphide isomerase (Edman *et al.*, 1985). These two proteins terminate in the same amino acid sequence and provided the first suggestion that this C-terminal sequence (KDEL) may act as an ER retention signal (Pelham, 1989a). Deletion of the C-terminal KDEL sequence or its addition to secretary proteins confirmed that KDEL is both necessary and sufficient for retention in the ER (Munro and Pelham, 1987). ER-localised proteins in yeast possess a C-terminal HDEL rather than KDEL sequence that functions similarly in ER retention (Pelham *et al.*, 1988; Rose *et al.*, 1989). Examination of the C-terminal sequence of 28 luminal ER proteins from a number of organisms identified some variation in the retention signal, including RDEL, KEEL, HIEL and HTEL (Pelham, 1989b). In maize an ER-localised auxin receptor has a C-terminal KDEL sequence (Hesse *et al.*, 1989; Inohara *et al.*, 1989), and a tomato protein related to mammalian and yeast BiP has a C-terminal HDEL sequence (N. Ewing and A.B. Bennett, unpublished data). Although few in number, the sequences of plant ER-localised proteins suggest that both C-terminal KDEL and HDEL may function to retain proteins in the ER of plant cells.

It is interesting to note that immunolocalisation of an ER-resident protein in yeast revealed the presence of the protein in the reticulate network of the ER and within the lumen of the nuclear envelope (Rose *et al.*, 1989). Similar results have been observed with a phytohaemagglutinin–KDEL chimeric protein expressed in transgenic plant cells that was localised by immunocytochemistry to the ER and the nuclear envelope (E. Herman and M. Chrispeels, personal communication). These results are consistent with the observed continuities between the ER and nuclear envelope (see above) and further suggest that the C-terminal KDEL/HDEL sequence is sufficient to target proteins to both the ER and the nuclear envelope.

Although the identification of a specific retention signal on ER-localised proteins provides a conceptually simple model for ER targeting, it is mechanistically difficult to reconcile the physical retention of a protein with its ability to diffuse freely within the ER lumen. This difficulty was resolved by experiments demonstrating that a lysosomal protein, cathepsin D, tagged with KDEL was retained in the ER but in addition was modified by the addition of mannose-6-phosphate (Pelham, 1988). Because *N*-acetylglucosaminyl-1-phosphotransferase, the enzyme responsible for donating the mannose-6-phosphate residue, is localised in a compartment distinct from the ER and presumed to be in the *cis* Golgi, this result indicated that proteins retained in the ER nevertheless enter a post-ER compartment. Based on this evidence it was proposed that ER residents are retained by continuous recycling between the ER and a post-ER compartment, probably the *cis* Golgi. Although likely to occur, similar mechanisms of ER-to-Golgi recycling have not been investigated in plants.

7.4.3 *Secretion*

Protein secretion in plant cells results from fusion of Golgi-derived and possibly ER-derived vesicles with the plasma membrane (Akazawa and Hara-Nishimura, 1985). Although the question of whether secretion bypasses the Golgi apparatus in plants has been controversial, there is clear evidence of transit through the Golgi apparatus for a number of secreted proteins (Gardiner and Chrispeels, 1975; Bennett and Christofferson, 1986; Gubler *et al.*, 1986; DellaPenna and Bennett, 1988). A parallel pathway of direct secretion from the ER to the cell surface may occur in some plant tissues but has not been fully characterized.

In animal and yeast cells secretion represents a default pathway resulting from bulk flow of proteins through the endomembrane system (Pfeffer and Rothman, 1987). The evidence for this is that the rate of secretion of several secretory proteins is similar to the rate of bulk flow of glycotripeptides (Wieland *et al.*, 1987), that saturation of vacuolar targeting mechanisms in yeast by overexpression of vacuolar proteins results in overflow being diverted to secretion (Stevens *et al.*, 1986), and that secreted bacterial proteins are also secreted when expressed in eukaryotic cells (Lund *et al.*, 1989). The situation in plant cells is less clear, although the evidence presently available favours a similar model of secretion by a default pathway. The strongest evidence for this pathway is the report of secretion of bacterial chitinase when expressed in transgenic tobacco (Lund *et al.*, 1989). Because a bacterial protein would be unlikely to possess signals for sorting or targeting in the eukaryotic endomembrane system, secretion of this protein suggests that the presence of a signal sequence alone is sufficient to direct secretion in plant cells. However, the localisation of a chimeric protein possessing a signal sequence (from phytohaemagglutinin) fused to a cytosolic protein (albumin) could not be clearly demonstrated when expressed in transgenic tobacco (Dorel *et al.*, 1989). However, it may be that albumin was not structurally competent for transport to the extracellular space. This interpretation emphasises that not all proteins, even when conferred with correct targeting and sorting signals, will be competent for transport and localisation in endomembrane compartments (Emr *et al.*, 1984).

Consistent with a default mechanism of protein sorting to the secretory pathway are experiments indicating that carbohydrate moieties of glycoproteins are not required for localisation of extracellular proteins. Evidence for this includes the observation that several secreted proteins are not glycosylated (Boston *et al.*, 1982) and that treatment with tunicamycin, an inhibitor of glycosylation, does not prevent secretion of some glycoproteins (Miyata and Akazawa, 1982; Faye *et al.*, 1986). It should be noted, however, that secretion of some plant glycoproteins is sensitive to tunicamycin (Faye and Chrispeels, 1989). The inhibition of glycosylation may render the protein incompetent to traverse the endomembrane system, perhaps by reducing its stability or solubility in the ER lumen. In this regard, failure of carrot cells to secrete invertase when treated with tunicamycin was attributed to degradation of the unglycosylated protein within the endomembrane system (Faye and Chrispeels, 1989). From studies with tunicamycin we can conclude that, while glycoprotein glycan moieties may be important in maintaining the stability, solubility, and perhaps activity of some extracellular proteins, it is unlikely that these structures participate directly in sorting events that direct proteins to the cell surface.

Beyond delivery to the extracellular matrix, little is known about targeting of proteins to specific sites within the cell wall or beyond. Many extracellular proteins with alkaline isoelectric points are ionically bound to acidic cell wall polymers and are presumably immobile in the cell wall. This may serve as a mechanism to target pectin hydrolases such as polygalacturonase (pI 9.4; Ali and Brady, 1982) to their acidic polymer substrates. Because of the structural complexity of the cell wall, it is likely that targeting of proteins to specific microdomains is required for appropriate substrate recognition and catalytic activity. One might speculate that such targeting could occur through lectin-like interactions between extracellular proteins and specific cell wall components. However, protein targeting at this level of resolution in the cell wall has not been investigated.

7.4.4 *Protein body/vacuole localisation*

7.4.4.1 *ER-derived protein bodies.* In cereal grains, especially corn, the predominant form of protein body is that derived by direct dilation of the rough ER as storage proteins are deposited (Robinson, 1985). In this sense the storage proteins are retained within the ER following entry and assemble to form macromolecular storage protein aggregates (Khoo and Wolf, 1970; Larkins and Hurkman, 1978). Examination of zein amino acid sequences (Larkins *et al.*, 1989), however, does not reveal the presence of KDEL or HDEL ER retention signals, suggesting that other mechanisms are responsible for storage protein retention and accumulation in the ER. It has been demonstrated that microinjection of total zein mRNA, or simply α-zein mRNA, into *Xenopus* oocytes results in the accumulation of zein storage protein within membrane vesicles that are similar to maize protein bodies (Hurkman *et al.*, 1981; Wallace *et al.*, 1988). This result suggests that the aggregation of zein proteins into a protein body results from primary interactions between zein polypeptides themselves, rather than with a structural component specific to the maize endosperm ER. Although α-zein mRNA was sufficient to promote the formation of 'protein bodies' in *Xenopus* oocytes, detailed studies localising specific zein subunits within maize protein bodies suggest that different zein subunits interact specifically with one another to assemble zeins into the protein body in an ordered manner (Lending and Larkins, 1989). One possible mechanism of ER retention of the maize storage proteins is that specific inter- and intra-zein interactions give rise to macromolecular structures that are functionally insoluble in the ER lumen and therefore incompetent to move forward within the endomembrane system. In this case the specificity of ER retention would reside in inter- or intra-polypeptide interactions within the ER lumen.

7.4.4.2 *Golgi-derived protein bodies (vacuoles).* A second class of cereal grain protein bodies and all legume protein bodies are derived from the subdivision of vacuoles and enlargement by fusion of Golgi-derived membrane vesicles (Craig *et al.*, 1979; Robinson, 1985). Because of their vacuolar origin, this type of protein body has been referred to as a protein storage vacuole (Dorel *et al.*, 1989) and mechanisms of protein targeting to these protein bodies are presumed to reflect general mechanisms of vacuolar protein targeting.

A eukaryotic organelle that has been likened to the plant vacuole is the lysosome of animal cells (Matile and Moor, 1968; Marty, 1978). The signal responsible for targeting of proteins to this endomembrane compartment has been clearly identified as a mannose-6-phosphate added post-translationally to glycan moieties of lysosomal proteins (Pfeffer and Rothman, 1987). Several lines of evidence, however, indicate that the mannose-6-phosphate does not function as a vacuolar sorting signal in plants. First, inhibition of glycosylation with tunicamycin or by site-directed mutagenesis of glycosylation sites does not interfere with vacuolar localisation of phytohaemagglutinin, indicating that glycan moieties are not required for vacuolar targeting (Bollini *et al.*, 1985; Voelker *et al.*, 1989). Secondly, efforts to detect mannose-6-phosphate residues on plant glycoproteins or to detect binding of vacuolar glycosidases to bovine liver phosphomannosyl receptors were unsuccessful (Gaudreault and Beevers, 1984).

As in plants, vacuolar protein localisation in yeast has been shown not to require glycans (Schwaiger *et al.*, 1982). In the case of yeast carboxypeptidase Y the vacuolar sorting signals have been localised by creating a series of gene fusions between the carboxypeptidase Y gene progressively deleted from the 3′ end and a reporter gene encoding yeast invertase. Deletions of carboxypeptidase Y to within 50 amino acids of the N-terminus resulted in vacuolar localization of invertase, but an additional 30 amino acid deletion resulted in diversion of invertase from the vacuole to the extracellular space (Johnson *et al.*, 1987). Thus the vacuolar sorting signal resides within the first 50 N-terminal amino acids of carboxypeptidase Y. Related experiments identified a point mutation altering Q(24) to K(24) that resulted in loss of vacuolar sorting. In addition, deletions of 3–12 amino acids between A(20) and K(31) resulted in varying degrees of carboxypeptidase Y mislocalization (Valls *et al.*, 1987). A number of other point mutations that affected vacuolar sorting of carboxypeptidase Y mapped to a sequence of four amino acids, QRPL(27). Together, these results indicated that a domain comprising approximately 30 amino acids near the N-terminus of carboxypeptidase Y is both necessary and sufficient for protein targeting to the yeast vacuole, and within this domain a tetrapeptide (QRPL) is essential for vacuolar localisation (Rothman *et al.*, 1989b). This vacuolar targeting domain resides within a cleavable prosequence that is lost from the protein following delivery of the protein to the vacuole.

There is considerable evidence suggesting that vacuolar sorting in plants occurs by mechanisms similar to those in yeast. Aside from general similarities in requirements for proper vacuolar localisation in yeast and plants, common targeting mechanisms have been suggested by the direct demonstration that expression in yeast of a gene encoding a plant vacuolar protein, phytohaemagglutinin, resulted in vacuolar localisation of the protein (Tague and Chrispeels, 1987). Furthermore, expression of gene fusions consisting of progressive 3′ deletions of the phytohaemagglutinin gene fused to yeast invertase have demonstrated that a targeting signal conferring localisation to the yeast vacuole resides in a limited polypeptide domain near the N-terminus of phytohaemagglutinin (Tague and Chrispeels, 1988). Although these experiments strongly suggest that the mechanism of vacuolar protein targeting in yeast is similar to that in plants, at the present time no experiments have directly identified targeting domains in plant vacuolar proteins that function in plant cells. Comparisons of the yeast carboxypeptidase Y vacuolar sorting signal with sequences of other yeast or plant vacuolar proteins indicate that the signal is not conserved at the primary sequence level, and suggest that a higher order protein structure may interact with vacuolar sorting signal receptors. It has been noted that the vacuolar targeting domain of phytohaemagglutinin resides in a domain that loops out and is exposed on the surface of the protein (Tague and Chrispeels, 1988). Similarly exposed structures may reside in regions other than the N-terminus of vacuolar proteins, and in this regard a glycosylated C-terminal domain is being tested as a potential vacuolar targeting domain of wheat-germ agglutinin (N. Raikhel, personal communication).

7.4.5 *Peroxisome localisation*

Peroxisomes are organelles that exist outside the endomembrane system, appearing to arise by fission of pre-existing organelles rather than by budding of the ER, as was once suggested (Trelease, 1984). Peroxisomal proteins are synthesised in the cytoplasm but, unlike proteins imported to the ER, are imported post-translationally. Peroxisomal import also usually occurs without subsequent cleavage of a leader sequence such as occurs following post-translational import of chloroplast or mitochondrial proteins. In these respects protein import to peroxisomes differs from both ER import and from mitochondrial and chloroplast import.

Recent experiments have localised the peroxisomal import signals of firefly luciferase to a short C-terminal domain (Gould *et al.*, 1987, 1988, 1989). A 12 amino acid sequence was initially shown to be necessary for peroxisomal localisation of firefly luciferase by progressive 3′ deletions of the luciferase gene and transgenic expression in CV-1 monkey cells. Sufficiency of this 12 amino acid sequence for peroxisomal localisation was also demonstrated by constructing gene fusions with chloramphenicol acetyltransferase and dihydrofolate reductase and demonstrating peroxisomal localisation of the reporter proteins (Gould *et al.*, 1987). Further experiments resolved the boundaries of the peroxisomal targeting signal to a tripeptide, SKL, at the C-terminus of luciferase (Gould *et al.*, 1989; Table 7.1). Comparison of the luciferase peroxisomal targeting signal with the sequence of other peroxisomal proteins revealed that many possessed an identical or similar C-terminal tripeptide also at, or in some cases near, the C-terminus (Gould *et al.*, 1988). Site-directed mutagenesis of the luciferase gene defined some of the changes in the C-terminal tripeptide that still directed peroxisomal import (Gould *et al.*, 1989; Table 7.1). Comparison of C-termini of several peroxisomal proteins and of site-directed mutagenesis of the luciferase C-terminal tripeptide indicates that there is considerable degeneracy in the peroxisomal targeting signal. The degenerate C-terminal sequence that has been demonstrated to function in peroxisomal targeting is [S, A, C]-[K, H, R]-L. Of four plant peroxisomal proteins whose sequence has been determined, two have C-terminal tripeptides corresponding to a tripeptide shown to be a functional peroxisomal

Table 7.1. Peroxisomal targeting sequences. The C-terminal tripeptide (SKL) of firefly luciferase was demonstrated to be necessary and sufficient for peroxisomal localisation (Gould *et al.*, 1987). Site-directed mutations in the SKL sequence were generated and the listed C-terminal tripeptides shown to function as peroxisomal targeting sequences. The C-terminal sequences of four plant peroxisomal proteins are listed for comparison with the tripeptides shown to function as peroxisomal targeting sequences in mammalian cells. In the two sequences of isocitrate lyase a C-terminal methionine (M*) is present, a permutation of the SKL tripeptide not previously tested for peroxisomal targeting.

Proteins	C-terminal amino acid sequence	
Firefly luciferase[1]		
Luciferase	KKGGK	SKL
Luciferase mutants	KKGGK	AKL
	KKGGK	CKL
	KKGGK	SHL
	KKGGK	SRL
Plant peroxisomal proteins		
Soybean uricase[2]	LSRLW	SKL
Brassica malate synthase[3]	YPINA	SRL
Brassica isocitrate lyase[4]	LVVAK	SRM*
Castor bean isocitrate lyase[5]	EVVAK	ARM*

Sequences shown are from [1]Gould *et al.* (1989); [2]Nguyen *et al.* (1985); [3]Comai *et al.* (1989a); [4] Comai *et al.* (1989b); [5]Beeching and Northcote (1987).

targeting signal in mammalian cells (Table 7.1). However, the sequences of isocitrate lyase from castor bean and rape seed indicate the presence of a C-terminal methionine, a permutation of the C-terminal tripeptide that has not been tested as a functional peroxisomal targeting signal. The present results of sequence analysis of plant peroxisomal proteins are consistent with the role of a C-terminal tripeptide in peroxisomal import, but it should be emphasised that this sequence has not been demonstrated to function in peroxisomal import in plant cells. In this regard it is important to note that several peroxisome-localised proteins do not possess C-terminal sequences comparable to those defined as peroxisome targeting signals (Borst, 1989). This suggests either that the [S, A, C]-[K, H, R]-L tripeptide is not a universal peroxisome targeting signal or that other as yet uncharacterised permutations of this sequence are functional in peroxisome targeting.

7.5 Glycosylation and post-translational modification

Many endomembrane proteins undergo structural modifications associated with transport through the endomembrane system in addition to co-translational removal of the signal sequence. In the case of animal lysosomal proteins, post-translational modification of a glycan residue (addition of mannose-6-phosphate) is a key component of the lysosomal targeting process. In plant cells there is no evidence that post-translational modification is as intimately related to the targeting process (see above). However, it is likely that post-translational modifications are critical steps in conferring stability and in activating some endomembrane-localised enzymes. Thus, post-translational modifications represent an important component of the system delivering stable and active proteins to endomembrane compartments. The two most common types of post-translational modification are glycosylation and proteolytic processing, although other modifications such as acylation, phosphorylation and sulphation have been shown to contribute to maturation of endomembrane proteins. Each of these forms of post-translational modification is summarised below.

7.5.1 *N-linked glycosylation*

N-linked glycosylation occurs by transfer of glycosyl residues to the amide nitrogen of peptide asparagines and is perhaps the most common form of post-translational modification of endomembrane-localised proteins. *N*-linked glycoproteins are present in all compartments and membranes associated with the endomembrane system. The mechanisms involved in biosynthesis of *N*-linked glycoproteins in plants are quite similar to those of other eukaryotes, and the assembly and processing of *N*-linked glycoproteins has been thoroughly studied in plants, animals and yeast (for extensive reviews see Elbein, 1979; Hubbard and Ivatt, 1981; Chrispeels, 1984; Kornfeld and Kornfeld, 1985; Hirschberg and Snider, 1987; Faye *et al.* 1989).

The initial addition of carbohydrates to *N*-linked glycoproteins takes place in the ER, with subsequent modifications occurring in the Golgi apparatus. This section describes the events associated with processing of *N*-linked glycans and some of the enzymes known to be involved.

7.5.1.1 *Co-translational core glycosylation.* As for signal sequence cleavage, the initial glycosylation reaction is coincident with translation of the nascent polypeptide and translocation into the lumen of the ER. The mechanism involves bulk transfer of a precursor oligosaccharide with the structure shown in Figure 7.3 from a membrane-anchored lipid carrier, dolichol pyrophosphate, to specific asparagine residues in the growing polypeptide chain. Details of the assembly of the dolichol-linked oligosaccharide precursor can be found in reviews by Elbein (1979) and Hirschberg and Snider (1987). The precursor oligosaccharide (Figures 7.3 and 7.5) is a branched structure consisting of a 'core' of nine mannose residues and two *N*-acetylglucosamine (GlcNAc) residues with an extension of three glucose (Glc) units. This entire oligosaccharide is transferred *en bloc* from the dolichol carrier to an asparagine in the acceptor protein on the luminal side of the ER membrane (Kornfeld and Kornfeld, 1985; Hirschberg and Snider, 1987).

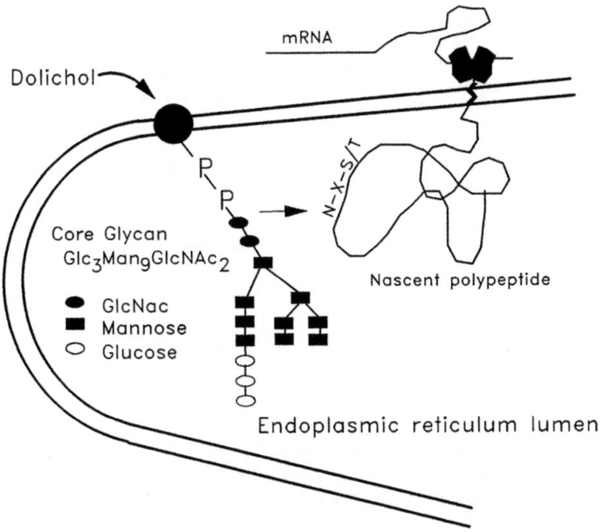

Figure 7.3. Co-translational *N*-linked core glycosylation. The precursor oligosaccharide ($Glc_3Man_9GlcNAc_2$) is transferred *en bloc* from the lipid-linked donor, dolichol pyrophosphate, to the amide nitrogen of asparagine in the consensus sequence N-*X*-S/T in the nascent polypeptide.

Not all asparagine residues in glycoproteins become glycosylated. The oligosaccharide precursor is transferred only to asparagine residues contained within the consensus sequence N-*X*-S/T, where *X* is any amino acid except possibly proline or aspartic acid. This consensus site is common to all eukaryotes. However, not all glycosylation consensus sites are utilised. Glycosylation may be limited to certain asparagine residues by accessibility of the asparagine within the consensus site to the enzyme catalysing oligosaccharide transfer from the dolichol donor to the protein, or by the availability of the dolichol donor itself (Kornfeld and Kornfeld, 1985).

7.5.1.2 *Post-translational glycan processing.* Although all glycoproteins receive identical oligosaccharide side chains in the ER, the composition of the side chains in mature glycoproteins varies considerably, and the mature forms of different *N*-linked glycans in the same protein may have different structures. These variations reflect differential post-translational modification of the glycan side chains. The *N*-linked glycans present in mature glycoproteins are generally of two types: either the 'high-mannose' type, so called because it retains at least five of the original nine mannose residues present in the precursor glycan as well as the two GlcNAc residues ($Man_5GlcNAc_2$), or the 'complex' type in which the precursor glycan has been trimmed to three mannose and two GlcNAc residues ($Man_3GlcNAc_2$; see Figure 7.5), and then further modified by addition of other sugars. The mature forms of phyto-

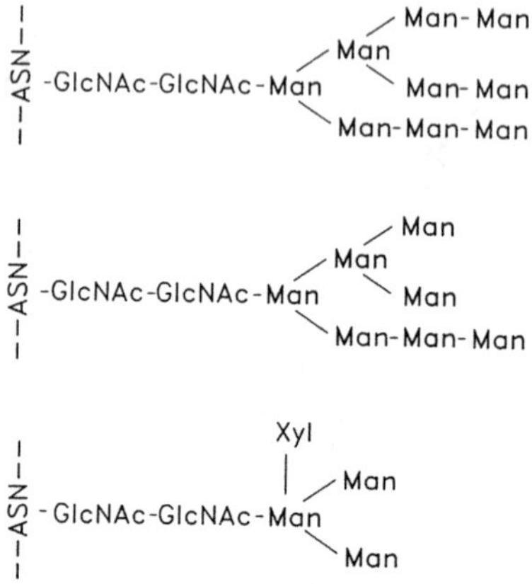

Figure 7.4. Glycan side chains of phaseolin. Mature phaseolin has one complex (lower) and two high-mannose (upper, middle) glycan side chains that arise by differential post-translational processing of the precursor oligosaccharide. (Redrawn from Sturm *et al.*, 1987).

haemagglutinin (Vitale *et al.*, 1984; Sturm and Chrispeels, 1986) and phaseolin (Sturm *et al.*, 1987b), seed storage proteins of *Phaseolus vulgaris*, both have high-mannose and complex oligosaccharide side chains. The side chains of phaseolin are shown in Figure 7.4. There are many variations of these structures in different glycoproteins. Additionally, differential processing of glycan side chains on the same polypeptide can yield different forms of the same glycoprotein (DellaPenna and Bennett, 1988).

Structural analysis of *N*-linked oligosaccharides present in glycoproteins has been facilitated by the use of hydrolytic enzymes with differing specificities for the glycan side chains. The most commonly used enzyme is endo-β-*N*-acetylglucosaminidase H (endo H). This enzyme cleaves high-mannose oligosaccharides but not complex oligosaccharides from the peptide backbone (Kornfeld and Kornfeld, 1985), and susceptibility to endo H can be conveniently assayed by increased electrophoretic mobility of the glycoprotein on sodium dodecyl sulphate (SDS) gels. Loss of endo H sensitivity during glycoprotein maturation is usually interpreted as an indication that the protein has passed through the Golgi apparatus, since the enzymes responsible for trimming of the high-mannose core, and hence loss of sensitivity to the endo H (see Figure 7.5), are localised in this compartment (Sturm *et al.*, 1987a). Lectins, which are plant-derived proteins with specific carbohydrate-binding properties, have also been useful in elucidation of glycan structure (Osawa and Tsuji, 1987).

The processing of the *N*-linked glycan moiety added in the ER follows a sequential series of steps as the glycosylated proteins move through the endomembrane pathway. The precursor oligosaccharide shown in Figures 7.3 and 7.5 is usually a short-lived species due to the rapid removal of the three glucose residues in the ER. Two glucosidases, glucosidase I (Szumillo *et al.*,

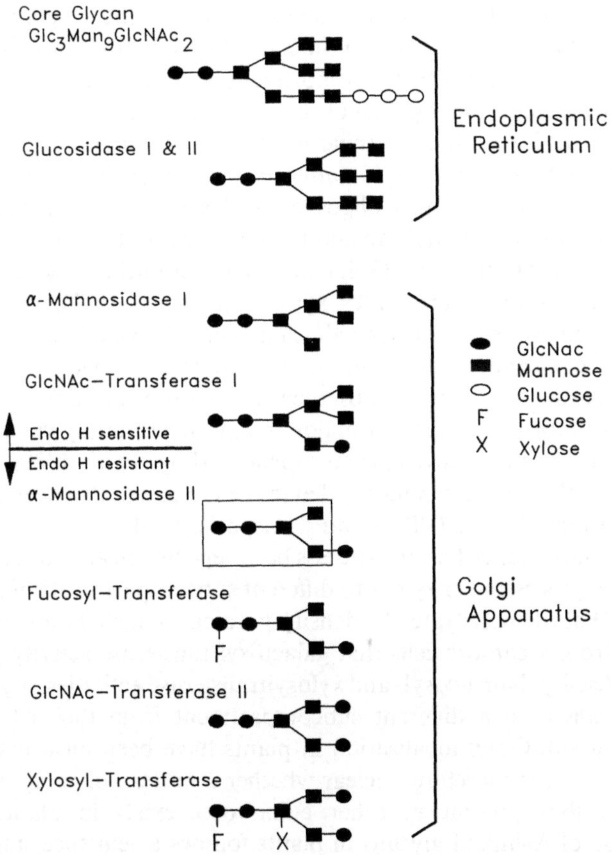

Figure 7.5. Proposed steps in sequential processing of *N*-linked glycans in plants. Following the action of α-mannosidase II, the glycan moiety loses its sensitivity to endoglycosidase H and acquires the characteristics of a complex glycan by addition of other sugars such as fucose and xylose. The core structure common to all complex *N*-linked glycans is boxed. (Adapted from Faye *et al.*, 1989).

1986a) and glucosidase II (Szumillo *et al.*, 1986b), have been characterised in microsomal fractions of mung bean seedlings and shown to be responsible for this activity in plant cells. The removal of a single mannosyl residue also occurs in the ER of animals and yeast, but this does not appear to occur in plants (Sturm *et al.*, 1987a).

Further processing of the *N*-linked oligosaccharide occurs following vesicular transport of the glycoprotein to the Golgi apparatus. In animal and yeast cells, the glycosidases and glycosyltransferases involved in further trimming of the precursor glycan moiety and addition of other sugar residues have been shown by density-gradient fractionation and immunocytochemical methods to be compartmentalised in different cisternae of the Golgi apparatus

(Farquhar, 1985; Pfeffer and Rothman, 1987). As the glycoproteins move through the Golgi from the *cis* to the medial and finally *trans* cisternae, they are modified sequentially by the complement of processing enzymes present in each compartment. The removal of the outer mannose residues occurs in the *cis* Golgi to yield the side chains characteristic of high-mannose glycoproteins. Further trimming of the high-mannose core to yield the $Man_3GlcNAc_2$ structure observed in complex oligosaccharides occurs in the medial cisternae, as does the addition of GlcNAc and fucose residues. The late steps in glycan processing occur in the *trans* Golgi and include addition of galactose and, in plant cells and some invertebrates, xylose (Faye *et al.*, 1987). As they traverse the Golgi apparatus, the glycan side chains may be acted on by all, some or none of the enzymes catalysing these reactions to yield the variety of oligosaccharide side chains observed in mature glycoproteins. Protein conformation probably plays a major role in determining which processing enzymes can act on a particular glycoprotein (Faye *et al.*, 1989). Removal of some terminal GlcNAc residues added in the Golgi has also been proposed to occur in protein bodies (Vitale and Chrispeels, 1984).

In only one instance has any success been reported in attempting to localise the various processing enzymes to different subcompartments of plant Golgi. Ali *et al.* (1986) demonstrated by density-gradient centrifugation of membrane fractions from sycamore cells that galactosyltransferase activity peaked at a different density than fucosyl- and xylosyltransferase activities, suggesting that it was localised in a different subcompartment from the last two. Other attempts at sub-Golgi localisation in plants have been unsuccessful (Sturm *et al.*, 1987a). It is therefore unclear whether a pattern of compartmentation similar to that present in other eukaryotes exists in plants; however, maturation of *N*-linked glycans in plants follows a sequence similar to that observed in other eukaryotes. Several enzymes involved in glycoprotein processing in plant Golgi have been characterised and found to have activities which parallel those of other organisms. α-mannosidase I, which removes up to four α-1,2-linked mannose units from the deglucosylated precursor, has been purified from mung bean seedlings (Szumillo *et al.*, 1986b). α-mannosidase II, which cleaves the next two mannose residues to yield the core structure found in complex glycans (see Figure 7.5), has also been identified (Johnson and Chrispeels, 1987). These two enzymes are thus responsible for trimming of the high-mannose core. *N*-acetylglucosaminyltransferase (Chrispeels, 1985b), fucosyltransferase (Sturm and Kindl, 1983) and xylosyltransferase (Johnson and Chrispeels, 1987), which catalyse the transfer of GlcNAc, fucose and xylose, respectively, to the *N*-linked oligosaccharides to form complex glycans, have also been characterised in plants. All of these enzymes have been shown by density-gradient centrifugation to be localised in the Golgi (Chrispeels, 1985b; Johnson and Chrispeels, 1987; Sturm *et al.*, 1987a). Based on the substrate specificities of the glycosidases and glycosyltransferases present in plant cells, a model (outlined in Figure 7.5) has been proposed detailing the steps involved in sequential conversion in plants of the high-mannose precursor to a complex glycan (Johnson and Chrispeels, 1987; Faye *et al.*, 1989). The major difference in this conversion between plants and animals is that xylose, a common constituent of *N*-linked glycans in plants, is rarely found in the complex glycoproteins of animal cells, whereas sialic acid, a common constituent of *N*-linked glycans in animals, is not found in plants (Johnson and Chrispeels, 1987).

7.5.1.3 *Function of N-linked glycosylation.* Despite the widespread occurrence of *N*-linked glycans in endomembrane proteins, their functions in plant cells remain elusive, although some insight has been gained from other systems. The available information has been obtained largely by the use of inhibitors of glycosylation, most notably tunicamycin, which prevents assembly of the dolichol-linked glycan precursor (Elbein, 1987). More recent studies have made use of recombinant DNA techniques to modify consensus glycosylation sites and study the effects on the encoded proteins. Such studies have indicated that oligosaccharides do not appear to play a direct role in protein sorting either to the vacuole or to the plasma membrane in plants, as discussed earlier. However, in some cases *N*-linked glycans may confer competency for protein movement through the endomembrane pathway, for instance by promoting proper folding or oligomer assembly (Olden *et al.*, 1985). While this appears to be true for some eukaryotic glycoproteins, such as vesicular stomatitis virus G-protein (Machamer *et al.*, 1985), it is not so for the bean seed storage protein phytohaemagglutinin, which is both properly transported to the vacuole and assembled in the absence of glycosylation (Voelker *et al.*, 1989). Glycosylation has also been shown to affect stability and activity of some endomembrane proteins (Olden *et al.*, 1985). It is likely that *N*-linked glycans serve multiple functions, but until many more studies are conducted their significance will remain largely a matter of speculation.

7.5.2 *O-linked glycosylation*

O-linked glycosylation differs from *N*-linked glycosylation in that sugar residues are linked to the hydroxylated amino acids serine, threonine, hydroxyproline and hydroxylysine. This form of endomembrane protein modification is less common than *N*-linked glycosylation and has been less well characterised. In plants, *O*-glycosidic linkages to peptidyl serines, threonines and hydroxyprolines have been identified, and glycosylation of hydroxyproline residues appears to be unique to plants (Selvendran and O'Neill, 1982). Addition and elongation of *O*-linked glycans occurs post-translationally by transfer of arabinose, galactose, *N*-acetylgalactosamine and other monosaccharides from their UDP conjugates to the peptide backbone. This is in contrast to *N*-linked glycosylation in which a preformed oligosaccharide is transferred *en bloc* from a membrane-anchored precursor. Unlike

N-linked glycosylation, *O*-glycosylation does not appear to involve a consensus sequence (Hanover and Lennarz, 1981); however, since not all potential hydroxylated amino acids are *O*-glycosylated it is likely that protein conformation plays a role in determining *O*-linked glycosylation patterns.

O-linked glycosylation is particularly prominent among extracellular matrix proteins in both plants and animals. In plants, the best-characterised *O*-linked glycoprotein is extensin, a structural protein of the cell wall. Extensin undergoes addition of three or four arabinose units to hydroxyproline residues and a single galactose unit to serine residues (Figure 7.6).

Most studies suggest that *O*-linked glycosylation of cell wall glycoproteins in plants occurs primarily in the Golgi. For example, based on membrane fractionation data and immunocytochemical evidence, Gardiner and Chrispeels (1975) concluded that hydroxyproline arabinosyltransferase was localised in the Golgi of carrot root tissue. Similar results were obtained with suspension-cultured cells of potato (Owens and Northcote, 1981) and tobacco (Kawasaki, 1981). Recently, however, characterisation of two *O*-glycosyltransferases in the unicellular green alga *Chlamydomonas reinhardii* indicated that hydroxyproline arabinosyltransferase activity co-fractionated with the endoplasmic reticulum, whereas hydroxyproline galactosyltransferase activity was associated with Golgi membranes (Zhang *et al.*, 1989). Hydroxyproline arabinosylation has also been reported to occur in the ER of maize roots and sugar cane (Andreae *et al.*, 1988). Similar discrepancies in the localisation of *O*-glycosylation activity have been reported in animal systems (Carraway and Hull, 1989). Such disparate results may reflect differences in enzyme localisation or membrane organisation among different cell types or taxa.

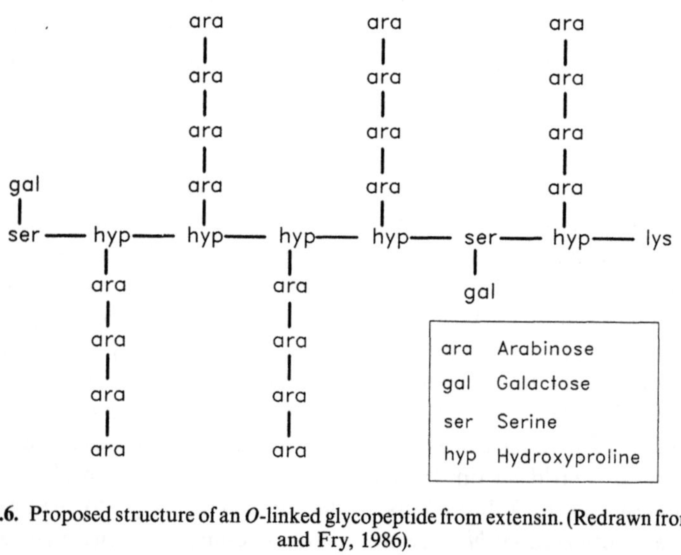

Figure 7.6. Proposed structure of an *O*-linked glycopeptide from extensin. (Redrawn from Wilson and Fry, 1986).

The functions of *O*-linked glycans have been more difficult to study than those of *N*-linked glycans because there are no known specific inhibitors of *O*-glycosylation and no apparent consensus signal that is easily modified by molecular genetic manipulation (Hanover and Lennarz, 1981). Several functions have been proposed for the *O*-linked side chains of extensin (Wilson and Fry, 1986). The glycans may be necessary to confer a rigid rod-like conformation to the molecule. They may also provide sites of attachment to other cell wall polysaccharides and thus be an important influence on physical properties of the cell wall. A specific role for arabinose residues in protecting against degradation of cell wall glycoproteins during disease infection was proposed based on the observations that arabinosylation could prevent proteolysis of cell wall glycoproteins and that the extent of hydroxyproline arabinosylation increased following fungal infection (Wilson and Fry, 1986).

7.5.3 *Proteolytic processing*

A large number of plant endomembrane proteins are synthesised as high molecular weight precursors which undergo limited post-translational proteolysis during maturation. The biological significance of such cleavages in plants is unclear. It has been shown in animal systems that proteolytic processing is important in the activation of neuropeptides and peptide hormones (Loh *et al.*, 1984). In yeast, the activity of a number of vacuolar hydrolases is dependent on the presence of a functional protease which processes the proenzymes to their mature forms (Rothman *et al.*, 1989b). The functional role of post-translational proteolysis in maturation of plant endomembrane proteins has been little studied, probably in part because the most abundant and therefore best-characterised endomembrane proteins are storage proteins for which a specific biological activity has not been defined. It has been observed, however, that the jackbean lectin concanavalin A acquires its carbohydrate-binding properties following cleavage of the precursor molecule (Chrispeels *et al.*, 1988). The fact that many non-plant endomembrane proteins are synthesised as inactive precursors suggests that the presence of extra domains in the precursors may in some cases constitute a mechanism for preventing biological activity during transit through the endomembrane pathway. As indicated earlier in this review, such extra domains may also incorporate targeting signals which specify subcellular localisation or may confer competency to traverse the endomembrane pathway, for example by incorporation of glycosylation signals or by endowing the precursor with charge characteristics favourable for transport. It is interesting to note in this context that the propeptides of several plant proproteins are highly charged and may considerably alter the charge characteristics of the proteins. For example, the tomato polygalacturonase proenzyme has a calculated pI of about 7, while the mature protein, which is localised in the cell wall, has a pI of about 10 (D. DellaPenna and A.B. Bennett,

Table 7.2. Post-translational proteolytic processing sites of plant endomembrane protein precursors. The amino acid on the N-terminal and C-terminal sides of a proteolytic cleavage site is shown in the first and second columns respectively. Most of the cleavage sites have been identified by comparison of the amino acid sequence of the mature protein with the sequence deduced from a genomic or cDNA clone. Proteins are grouped by processing site to emphasise the frequency with which a given cleavage site occurs among plant endomembrane proteins.

Processing site				
N-terminal amino acid	C-terminal amino acid	Protein	Type of Processing*	Reference
N	G	Pumpkin 11S globulin	S	Hayashi *et al.* (1988)
		Tomato polygalacturonase	N	DellaPenna and Bennett (1988)
		Pea legumin	S	Lycett *et al.* (1984)
		Rice glutelin	S	Takaiwa *et al.* (1986, 1989)
		Cotton β-globulin	S	Chlan *et al.* (1986)
		Arabidopsis 12S globulin	S	Pang *et al.* (1988)
		Soybean glycinin	S	Marco *et al.* (1984)
		Helianthinin	S	Vonder Haar *et al.* (1988)
		Cruciferin	S	Simon *et al.* (1985)
		Oat 12S globulin	S	Shotwell *et al.* (1988)
	P	Napin	S	Ericson *et al.* (1986)
		Brazil nut 2S albumin	S	Krebbers *et al.* (1988)
		Arabidopsis 2S albumin	N	Krebbers *et al.* (1988)
	A	Ricin	S	Lamb *et al.* (1985)
		Concanavalin A	S	Chrispeels *et al.* (1986)
	V	Pea seed lectin	S	Higgins *et al.* (1983b)
		Concanavalin A	S	Chrispeels *et al.* (1986)
	I	Sunflower storage albumin	N	Allen *et al.* (1987)
	L	Tomato proteinase inhibitor I	N	Graham *et al.* (1985a, b)
	D	Pea vicilin	S	Lycett *et al.* (1983)
	E	Concanavalin A	S	Chrispeels *et al.* (1986)
	S	Napin	N	Krebbers *et al.* (1988)
	Q	Brazil nut 2S albumin	N, S	Krebbers *et al.* (1988)

Q	K	Potato inhibitor I	N	Cleveland *et al.* (1987)
	R	Soybean glycinin	C	Marco *et al.* (1984)
	W	Napin	S	Krebbers *et al.* (1988)
D	D	Pea vicilin	C	Higgins *et al.* (1988)
	L	*Dolichos* seed lectin	C	Schnell and Etler (1987)
	I	Tomato pectinesterase	N	Ray *et al.* (1988)
	Y	Tomato pectinesterase	C	Ray *et al.* (1988)
E	I	Pea vicilin	C	Lycett *et al.* (1983)
R	G	Brazil nut 2S albumin	S	Altenbach *et al.* (1987)
	I	Napin	N	Ericson *et al.* (1986)
K	D	Pea vicilin	S	Higgins *et al.* (1988)
	L	Thionin	C	Gausing *et al.* (1987)
	R	Soybean glycinin	S	Marco *et al.* (1984)
A	L	Thaumatin	C	Cornelissen (1986)
		Aleurain	N	Rogers *et al.* (1988)
	S	Sporamin	N	Murakami *et al.* (1986)
S	L	Polygalacturonase	C	Sheehy *et al.* (1987)
		Horseradish peroxidase C	C	Fujiyama *et al.* (1988)
	Y	Napin	C	Ericson *et al.* (1986)
	P	Brazil nut 2S albumin	S	Krebbers *et al.* (1988)
	I	Brazil nut 2S albumin	C	Krebbers *et al.* (1988)
	F	*Arabidopsis* 2S albumin	C	Krebbers *et al.* (1988)

*N and C indicate N-terminal or C-terminal fragment lost as a result of processing. S indicates that proteolytic cleavage results in the formation of subunits, which are often joined by disulphide bridges. In some cases, subunits are joined by a linker which is lost upon cleavage of the precursor into subunits.

unpublished data). Thus, while extra domains may be required for protein delivery to the proper subcellular compartment, their removal may in some cases be necessary for protein function.

Identification of proteolytic processing domains in endomembrane proteins has been vastly accelerated in recent years by molecular cloning of DNA sequences encoding many of these proteins. Comparison of amino acid sequences deduced from genomic or cDNA clones with those obtained by sequencing of mature proteins has shown that post-translational proteolysis is a common feature of endomembrane protein maturation in plants. Table 7.2 lists most of the plant endomembrane proteins for which cleavage sites have been defined. (It should be mentioned, however, that proteolysis during purification of proteins for sequencing may yield artifacts leading to erroneous identification of a processing site.) Many of these cleavages result in loss of a peptide from the precursor molecule, as in the case of tomato polygalacturonase, in which both amino and carboxy stretches of amino acids are post-translationally cleaved from the precursor (Sheehy *et al.*, 1987; DellaPenna and Bennett, 1988). In Table 7.2, such processing events are indicated by 'N' (amino-terminal fragment lost) or 'C' (carboxy-terminal fragment lost). In contrast, many of the seed storage protein precursors, such as provicillin, are processed to yield subunits which are often joined by disulphide bridges (Gatehouse *et al.*, 1984). Such cleavages are indicated by 'S' in the table.

Proteolytic processing may be due to the activity of endoproteases, which cleave specific sites within the protein, or to exoproteases, which remove one or a few amino acids at a time from the carboxy or amino terminus. Where the proteolytic event results in loss of a peptide from the protein rather than cleavage into subunits without loss of intervening sequences, it is difficult to distinguish endo-from exoproteolytic activity, particularly in plants where the processing proteases are almost completely uncharacterised. The most common endoproteolytic processing site in plant endomembrane proteins is N-G (Table 7.2). Vonder Haar *et al.* (1988) have shown that the sequence N-G-L is highly conserved at the α/β cleavage site of 11S and 12S globulin-type seed storage proteins, suggesting that a specific endoprotease may be responsible for the processing activity. It has been proposed that a protease with specificity for cleavage on the carboxy side of asparagine residues may exist in dicotyledonous plants, since such cleavages are common among endomembrane proteins of dicots., as indicated in Table 7.2. In one study, a vacuolar extract from developing pumpkin seeds was shown to contain proteolytic activity which cleaved the pumpkin 11S globulin precursor, which is processed at an N-G site, into its constituent subunits. This same extract also processed the pumpkin trypsin inhibitor precursor to its mature size. Both activities exhibited properties characteristic of thiol proteases, leading to the hypothesis that the same enzyme was responsible for cleavage of both proteins (Hara-Nishimura and Nishimura 1987). The processing protease has yet to be isolated. Whether the pumpkin extract could cleave non-pumpkin precursor proteins with N-*X* processing sites was not investigated, although a castor bean protein body extract was also able to process correctly the pumpkin proglobulin (Hara-Nishimura and Nishimura 1987), lending support to the speculation that a protease which cleaves at N-*X* sites is present in many plants.

In another study, a thiol endoprotease purified from lupin seeds and thought to be important in maturation of lupin seed storage proteins was shown to have activity, suggesting a specificity for cleavage in the vicinity of paired arginine residues (Duranti *et al.*, 1987). Post-translational processing near paired arginines also occurs in proglycinin (Marco *et al.*, 1984) and pea legumin (Gatehouse *et al.*, 1984) and has been proposed in cotton α-globulin (Chlan *et al.*, 1986). Proteolytic processing on the carboxy side of basic residues is common in mammalian (Loh et al., 1984) and yeast (Mizuno *et al.*, 1984) endomembrane proteins, and may also be widespread in plants. Little else is known about the proteases responsible for post-translational cleavage of plant endomembrane proteins, although structural motifs around processing sites such as the N-*X* motif discussed above should provide some clues as to the possible specificities of these yet-to-be-identified enzymes.

Proteolytic processing often occurs late in the maturation process. Pulse-chase experiments on several legume storage proteins indicate that proteolytic processing of these proteins most likely occurs in the vacuoles or protein bodies (Chrispeels *et al.*, 1982; Higgins *et al.*, 1983a) and that an acidic pH is required for maximum proteolytic activity (Stinissen *et al.*, 1985). In addition, proteolytic acitivity which converts pumpkin proglobulin to its mature form has been identified in vacuole extracts of pumpkin and castor bean, as indicated above (Hara-Nishimura and Nishimura, 1987). However, these studies have not unequivocally ruled out the possibility that processing occurs in an acidic post-Golgi compartment just prior to deposition of the storage proteins in the protein bodies (Hara-Nishimura *et al.*, 1985). The site of proteolytic processing of non-storage endomembrane proteins is completely unexplored in plants.

7.5.4 *Other post-translational modifications*

A wide variety of other types of post-translational processing occurs in eukaryotic cells (Uy and Wold, 1977). Acylation, phosphorylation, sulphation and methylation all occur post-translationally in some non-plant proteins, but have been studied little in plant endomembrane proteins. Hydroxylation of proline residues is prominent in structural proteins of the cell wall. The enzyme responsible, prolyl hydroxylase, has been localised in the ER membranes of maize roots (Sauer and Robinson, 1985). Disulphide bond formation is widespread, particularly among seed storage proteins. In the case of the pumpkin 11S globulin, disulphide bond formation occurs in the ER and serves to maintain attachment of the two subunits upon proteolytic cleavage in the vacuole (Hara-Nishimura *et al.*, 1985). In barley, a post-translational modification resulting in a lowering of the pI of secreted α-amylase without a concomitant shift in molecular weight has been identified in aleurone tissue (Jacobsen *et al.*, 1989). This change in pI may be due to acetylation or deamidation of a higher pI precursor (L. Sticher and R.L. Jones, personal communication). A unique modification which has been documented in concanavalin A of jack bean is the post-translational formation of a peptide bond between two polypeptides derived from the same precursor by proteolytic processing. This is the only reported example of polypeptide ligation in eukaryotic cells (Bowles *et al.*, 1986; Chrispeels *et al.*, 1986). Finally, oligomerisation is a common form of post-translational processing in plants. Oligomer formation occurs in both the ER and protein bodies (Chrispeels, 1984). It will be interesting to learn from future research whether oligomerisation in the endomembrane system of plants is controlled by 'chaperones' such as the mammalian BiP (described in Section 7.3.1) which takes part in the assembly of proteins in the ER (Pelham, 1986). A cDNA clone with homology to BiP has been isolated from tomato and is thought to be localised in the ER (N. Ewing, J.D. Meyer and A.B. Bennett, unpublished). Similar proteins have been implicated in the association of the large and small subunits of RuBP carboxylase to form the functional holoenzyme (Ellis and Hemmingsen, 1989) and the role of molecular chaperones in the delivery of properly assembled proteins to endomembrane compartments is likely to become an increasingly important research topic.

References

Akazawa, T. and Hara-Nishimura, I. (1985) Topographic aspects of biosynthesis, extracellular secretion, and intracellular storage of proteins in plant cells. *Ann. Rev. Plant Physiol.* **36**: 441.

Ali, S., Mitsui, T. and Akazawa, T. (1986) Golgi-specific localization of transglycosylases engaged in glycoprotein biosynthesis in suspension-cultured cells of Sycamore (*Acer pseudoplatanus* L.). *Arch. Biochem. Biophys.* **251**: 421.

Ali, Z.M. and Brady, C.J. (1982) Purification and characterization of the polygalacturonases of tomato fruits. *Aust. J. Plant Physiol.* **9**: 155.

Allen R.D., Cohen, E.A., Vonder Haar, R.A., Adams, C.A., Ma, D.P., Nessler, C.L. and Thomas, T.L. (1987) Sequence and expression of a gene encoding an albumin storage protein in sunflower. *Mol. Gen. Genet.* **210**: 211.

Altenbach, S.B., Pearson, K.W., Leung, F.W. and Sun, S.S.M. (1987) Cloning and sequence analysis of a cDNA encoding a Brazil nut protein exceptionally rich in methionine. *Plant Mol. Biol.* **8**: 239.

Andreae, M., Lang, W.C. Barg, C. and Robinson, D.G. (1988) Hydroxyproline-arabinosylation in the endoplasmic reticulum of maize roots. *Plant Sci.* **56**: 205.

Bassüner, R., Huth, A., Manteuffel, R. and Rapoport, T.R. (1983) Secretion of plant storage globulin polypeptides by *Xenopus laevis* oocytes. *Eur. J. Biochem.* **133**: 321.

Beeching, J.R. and Northcote, D.H. (1987) Nucleic acid (cDNA) and amino acid sequences of isocitrate lyase from castor bean. *Plant Mol. Biol.* **8**: 471.

Bennett, A.B. and Christofferson, R.E. (1986) Synthesis and processing of cellulase from ripening avocado. *Plant Physiol.* **81**: 830.

Bergmann, J.E. and Singer, S.J. (1983) Immunoelectron microscopic studies of the intracellular transport of the membrane glycoprotein (G) of vesicular stomatitis virus in infected Chinese hamster ovary cells. *J. Cell. Biol.* **97**: 1777.

Bird, P., Gething, M.-J. and Sambrook, J. (1987) Translocation in yeast and mammalian cells: Not all signal sequences are functionally equivalent. *J. Cell Biol.* **105**: 2905.

Bole, D.G., Hendershot, L.M. and Kearney, J.F. (1986) Posttranslational association of immunoglobulin heavy chain binding protein with nascent heavy chains in nonsecreting and secreting hybridomas. *J. Cell Biol.* **102**: 1558.

Bollini, R., Ceriotti, A., Daminati, M.G. and Vitale, A. (1985) Glycosylation is not needed for the intracellular transport of phytohemagglutinin in developing *Phaseolus vulgaris* cotyledons and for the maintenance of its biological activities. *Physiol. Plant.* **65**: 15.

Borst, P. (1989) Peroxisome biogenesis revisited. *Biochim. Biophys. Acta* **1008**: 1.

Boston, R.S., Miller, T.J., Mertz, J.E. and Burgess, R.R. (1982) In vitro synthesis and processing of wheat α-amylase. *Plant Physiol.* **69**: 150.

Bowles, D.J., Marcus, J.E., Pappin, D.J.C., Finlay, J.B.C. Eliopoulos, E., Maycox, P.R. and Burgess, J. (1986) Post-translational processing of concanavalin A precursors in jackbean cotyledons. *J. Cell Biol.* **102**: 1284.

Burr, F.A. and Burr, B. (1981) *In vitro* uptake and processing of prezein and other maize preproteins by maize membranes. *J. Cell Biol.* **90**: 427.

Campos, N., Palau, J., Torrent, M. and Ludevid, D. (1988) Signal recognition-like particles are present in maize. *J. Biol. Chem.* **263**: 9646.

Carraway, K.L. and Hull, S.R. (1989) O-glycosylation pathway for mucin-type glycoproteins. *BioEssays* **10**: 117.

Chanson, A. and Taiz, L. (1985) Evidence for an ATP-dependent proton pump localized on the Golgi of corn coleoptiles. *Plant Physiol.* **78**: 232.

Chirico, W.J. Waters, M.G. and Blobel, G. (1988) 70K heat shock related proteins stimulate protein translocation into microsomes. *Nature* **332**: 805.

Chlan, C.A., Pyle, J.B., Legocki, A.B. and Dure III, L. (1986) Developmental biochemistry of cottonseed embryogenesis and germination. XVIII. cDNA and amino acid sequences of members of the storage protein families. *Plant Mol. Biol.* **7**: 475.

Chrispeels, M.J. (1983) The Golgi apparatus mediates the transport of phytohemagglutinin to the protein bodies in bean cotyledons. *Planta* **158**: 140.

Chrispeels, M.J. (1984) Biosynthesis, processing and transport of storage proteins and lectins in cotyledons of developing legume seeds. *Phil. Trans. R. Soc. Lond.* **304**: 309.

Chrispeels, M.J. (1985*a*) The role of the Golgi apparatus in the transport and post-translational modifications of vacuolar (protein body) proteins. *Oxford Survey Plant Mol. Cell Biol.* **2**: 43.

Chrispeels, M.J. (1985*b*) UDP-GlcNAc: glycoprotein GlcNAc-transferase is located in the Golgi apparatus of developing bean cotyledons. *Plant Physiol.* **78**: 835.

Chrispeels, M.J., Hartl, P.M, Sturm, A. and Faye, L. (1986) Characterization of the endoplasmic reticulum-associated precursor of concanavalin A. *J. Biol. Chem.* **261**: 10021.

Chrispeels, M.J., Higgins, T.J.V. and Spencer, D. (1982) Assembly of storage protein oligomers in the endoplasmic reticulum and processing of the polypeptides in the protein bodies of developing pea cotyledons. *J. Cell Biol.* **93**: 306.

Cleveland, T.E., Thornburg, R.W. and Ryan, C.A. (1987) Molecular characterization of a wound-inducible inhibitor I gene from potato and the processing of its mRNA and protein. *Plant Mol. Biol.* **8**: 199.

Comai, L., Baden, C.S. and Harada, J.J. (1989*a*) Deduced sequence of a malate synthase polypeptide encoded by a subclass of the gene family. *J. Biol. Chem.***264**: 2778.

Comai, L., Dietrich, R.A., Maslyar, D.J., Baden, C. and Harada, J.J. (1989*b*) Coordinate expression of transcriptionally regulated isocitrate lyase and malate synthase genes in *Brassica napus* L. *Plant Cell* **1**: 293.

Cornelissen, B.J.C. (1986) A tobacco mosaic virus-induced tobacco protein is homologous to the sweet-tasting protein thaumatin. *Nature* **321**: 531.

Craig, S., Goodchild, D.J. and Hardham, A.R. (1979) Structural aspects of protein accumulation in developing pea cotyledons. I. Qualitative and quantitative changes in parenchyma cell vacuoles. *Aust. J. Plant Physiol.* **6**: 81.

deCurtis, I. and Simons, K. (1989) Isolation of exocytic carrier vesicles from BHK cells. *Cells* **58**: 719.

DellaPenna, D. and Bennett, A.B. (1988) *In vitro* synthesis and processing of tomato fruit polygalacturonase. *Plant Physiol.* **86**: 1057.

Dingwall, C. and Laskey, R.A. (1986) Protein import into the cell nucleus. *Ann. Rev. Cell Biol.* **2**: 367.

Doms, R.W., Russ, G. and Yewdell, J.W. (1989) Brefeldin A redistributes resident and itinerant Golgi proteins to the endoplasmic reticulum. *J. Cell Biol.* **109**: 61.

Dorel, C., Voelker, T.A., Herman, E.M. and Chrispeels, M.J. (1989) Transport of proteins to the plant vacuole is not by bulk flow through the secretory system, and requires positive sorting information. *J. Cell Biol.* **108**: 327.

Duranti, M., Gatehouse, J.A., Boulter, D. and Cerletti, P. (1987) *In vitro* proteolytic processing of pea and jack bean storage proteins by an endopeptidase from lupin seeds. *Phytochemistry* **26**: 627.

Edens, L., Heslinga, L., Klok, R., Ledeboer, A.M., Maat, J., Toonen, M.Y., Visser, C. and Verrips, C.T. (1982) Cloning of cDNA encoding the sweet-tasting plant thaumatin and its expression in *Escherichia coli. Gene* **18**: 1.

Edens, L., Bom, I., Ledeboer, A.M., Maat, J., Toonen, M.Y., Visser, C. and Venips, C.T. (1984) Synthesis and processing of the plant protein thaumatin in yeast. *Cell* **37**: 629.

Edman, J.C., Ellis, L., Blacker, R.W., Roth, R.A. and Rutter, W.J. (1985) Sequence of protein disulphide isomerase and implications of its relationship to thioredoxin. *Nature* **317**: 267.

Elbein, A.D. (1979) The role of lipid-linked saccharides in the biosynthesis of complex carbohydrates. *Ann. Rev. Plant Physiol.* **30**: 239.

Elbein, A.D. (1987) Inhibitors of the biosynthesis and processing of N-linked oligosaccharide chains. *Ann. Rev. Biochem.* **56**: 497.

Ellis, R.J. and Hemmingsen, S.M. (1989) Molecular chaperones: proteins essential for the biogenesis of some macromolecular structures. *Trends Biochem. Sci.* **14**: 339.

Emr, S.D., Schauer, I., Hansen, W., Esmon, P. and Schekman, R. (1984) Invertase β-galactosidase hybrid proteins fail to be transported from the endoplasmic reticulum in *Saccharomyces cerevisiae. Mol. Cell. Biol.* **4**: 2347.

Ericson, M.L., Rödin, J., Lenman, M., Glimelius, K., Josefsson, L.-G. and Rask, L. (1986) Structure of the rapeseed 1.7 storage protein, napin, and its precursor. *J. Biol. Chem.* **261**: 14576.

Farquhar, M.G. (1985) Progress in unraveling pathways of Golgi traffic. *Ann. Rev. Cell Biol.* **1**: 447.

Faye, L. and Chrispeels, M.J. (1989) Apparent inhibiton of β-fructosidase secretion by tunicamycin may be explained by breakdown of the unglycosylated protein during secretion. *Plant Physiol.* **89**: 845.

Faye, L., Johnson, K.D. and Chrispeels, M.J. (1986) Oligosaccharide side chains of glycoproteins that remain in the high-mannose form are not accessible to glycosidases. *Plant Physiol.* **81**: 206.

Faye, L., Johnson, K.D., Sturm, A. and Chrispeels, M.J. (1989) Structure, biosynthesis, and function of asparagine-linked glycans on plant glycoproteins. *Physiol.* Plant. **75**: 309.

Fujiyama, K., Takemura, H., Shibayama, S., Kobayashi, K., Choi, J.-K., Shinmyo, A., Takano, M., Yamada, Y. and Okada, H. (1988) Structure of the horseradish peroxidase isozyme C genes. *Eur. J. Biochem.* **173**: 681.

Gardiner, M. and Chrispeels, M.J. (1975) Involvement of the Golgi apparatus in the synthesis and secretion of hydroxyproline-rich cell wall glycoproteins. Plant Physiol. **55**: 536.

Gatehouse, J.A., Croy, R.R.D. and Boulter, D. (1984) The synthesis and structure of pea storage proteins. *CRC Crit. Rev. Plant Sci.* **1**: 287.

Gaudreault, P.-R. and Beevers, L. (1984) Protein bodies and vacuoles as lysosomes. *Plant Physiol.* **76**: 228.

Gausing, K. (1987) Thionin genes specifically expressed in barley leaves. *Planta* **717**: 241.

Gething, M.-J., McCammon, K. and Sambrook, J. (1986) Expression of wild-type and mutant forms of influenza hemagglutinin: The role of folding in intracellular transport. *Cell* **46**: 939.

Gould, S.J., Keller, G.-A. and Subramani, S. (1987) Identification of a peroxisomal targeting signal at the carboxy terminus of firefly luciferase. *J. Cell Biol.* **105**: 2923.

Gould, S.J., Keller, G.-A. and Subramani, S. (1988) Identification of peroxisomal targeting signals located at the carboxy terminus of four peroxisomal proteins. *J. Cell Biol.* **107**: 897.

Gould, S.J., Keller, G.-A., Hosken, N., Wilkinson, J. and Subramani, S. (1989) A conserved tripeptide sorts proteins to peroxisomes. *J. Cell Biol.* **108**: 1657.

Graham, J.S., Pearce, G., Merryweather, J., Titani, K., Ericsson, L. and Ryan, C.A. (1985*a*)

Wound-induced proteinase inhibitors from tomato leaves. I. The cDNA-deduced primary structure of pre-inhibitor I and its post-translational processing. *J. Biol. Chem.* **260**: 6555.

Graham, J.S., Pearce, G., Merryweather, J., Titani, K., Ericsson, L.H. and Ryan, C.A. (1985*b*) Wound-induced proteinase inhibitors from tomato leaves. II. the cDNA-deduced primary structure of pre-inhibitor II. *J. Biol. Chem.* **260**: 6561.

Greenwood, J.S. and Chrispeels, M.J. (1985) Immunocytochemical localization of phaseolin and phytohemagglutinin in the endoplasmic reticulum and Golgi complex of developing bean cotyledons. *Planta* **164**: 295.

Griffing, L.R. and Ray, P. (1985) Involvement of monovalent cations in Golgi secretion by plant cells. *Eur. J. Cell. Biol.* **36**: 24.

Griffiths, G. and Simons, K. (1986) The trans Golgi network: sorting at the exit site of the Golgi complex. *Science* **234**: 438.

Gubler, F., Jacobsen, J.V. and Ashford, A.E. (1986) Involvement of the Golgi apparatus in the secretion of α-amylase from gibberellin-treated barley aleurone cells. *Planta* **168**: 447.

Haas, I.G. and Wabl, M. (1983) Immunoglobulin heavy chain binding proein. *Nature* **306**: 387.

Hanover, J.A. and Lennarz, W.J. (1981) Transmembrane assembly of membrane and secretory glycoproteins. *Arch. Biochem. Biophys.* **211**: 1.

Hansen, W., Garcia, P.D. and Walter, P. (1986) *In vitro* protein translocation across the yeast endoplasmic reticulum: ATP-dependent post-translational translocation of the prepro-α-factor. *Cell* **45**: 397.

Hara-Nishimura, I. and Nishimura, M. (1987) Proglobulin processing enzyme in vacuoles isolated from developing pumpkin cotyledons. *Plant Physiol.* **85**: 440.

Hara-Nishimura, I., Nishimura, M. and Akazawa, T. (1985) Biosynthesis and intracellular transport of 11S globulin in developing pumpkin cotyledons. *Plant Physiol.* **77**: 747.

Hattori, T., Ichihara, S. and Nakamura, K. (1987) Processing of a plant vacuolar protein precursor *in vitro*. *Eur. J. Biochem* **166**: 533.

Hayashi, M., Mori, H., Nishimura, M., Akazawa, T. and Hara-Nishimura, I. (1988) Nucleotide sequence of cloned cDNA coding for pumpkin 11-S globulin β subunit. *Eur. J. Biochem.* **172**: 627.

von Heijne, G. (1983) Patterns of amino acids near signal-sequence cleavage sites. *Eur. J. Biochem.* **133**: 17.

von Heijne, G. (1984) How signal sequences maintain cleavage specificity. *J. Mol. Biol.* **173**: 243.

von Heijne, G. (1985) Signal sequences: the limits of variation. *J. Mol. Biol.* **184**: 99.

von Heijne, G. (1988) Role of the signal peptide in protein export. ISI Atlas of Science: *Biochemistry* 1988: 205.

Herman, E.M. (1987) Immunogold-localization and synthesis of an oil-body membrane protein in developing soybean seeds. *Planta* **172**: 336.

Hesse, T., Feldwisch, J., Balshüseman, D., Bauw, G., Puype, M., Vandekerckhove, J., Löbler, M., Klämbt, D., Schell, J. and Palme, K. (1989) Molecular cloning and structural analysis of a gene from *Zea mays* (L.) coding for a putative receptor for the plant hormone auxin. *EMBO J.* **8**: 2453.

Higgins, T.J.V., Chandler, P.M., Zurawshi, G., Button, S.C. and Spencer, D. (1983a) The biosynthesis and primary structure of pea seed lectin. *J. Biol. Chem.* **258**: 9544.

Higgins, T.J.V., Chrispeels, M.J., Chandler, P.M. and Spencer, D. (1983*b*) Intracellular sites of synthesis and processing of lectin in developing pea cotyledons. *J. Biol. Chem.* **258**: 9550.

Higgins, T.J.V., Newbigin, E.J., Spencer, D., Llewellyn, D.J. and Craig, S. (1988) The sequence of a pea vicilin and its expression in transgenic tobacco plants. *Plant Mol. Biol.* **11**: 683.

Hirschberg, C.B. and Snider, M.D. (1987) Topography of glycosylation in the rough endoplasmic reticulum and Golgi apparatus. *Ann. Rev. Biochem.* **56**: 63.

Horn, M.A., Heinstein, P.F. and Low, P.S. (1989) Receptor-mediated endocytosis in plant cells. *Plant Cell* **1**: 1003.

Hubbard, S.C. and Ivatt, R.J. (1981) Synthesis and processing of asparagine-linked oligosaccharides. *Ann. Rev. Biochem.* **50**: 555.

Hurkman, W.J., Smith, L.D., Richter, D. and Larkins, B.A. (1981) Subcellular compartmentation of maize storage proteins in *Xenopus* oocytes. *J. Cell. Biol.* **89**: 292.

Inohara, N., Shimomura, S., Fukui, T. and Futai, M. (1989) Auxin-binding protein located in the endoplasmic reticulum of maize shoots: Molecular cloning and primary structure. *Proc. Natl Acad. Sci. USA* **86**: 3564.

Jacobsen, J.V., Bush, D.S., Sticher, L. and Jones, R.L. (1988) Evidence for precursor forms of the low isoelectric point α-amylase isozymes secreted by barley aleurone cells. *Plant Physiol.* **88**: 1168.

Johnson, K.D. and Chrispeels, M.J. (1987) Substrate specificities of N-acetylglucosaminyl-, fucosyl, and xylosyltransferases that modify flycoproteins in the Golgi apparatus of bean cotyledons. *Plant Physiol.* **84**: 1301.

Johnson, L.M., Bankaitis, V.A. and Emr, S.D. (1987) Distinct sequence determinants direct intracellular sorting and modification of a yeast vacuolar protease. *Cell* **48**: 875.

Kawasaki, S. (1981) Synthesis of arabinose-containing cell wall precursors in suspension-cultured tobacco cells. I. Intracellular site of synthesis and transport. *Plant Cell Physiol.* **22**: 431.

Khoo, V. and Wolf, J.J. (1970) Origin and development of protein granules in maize endosperm. *Am. J. Bot.* **57**: 4042.

Kim, W.T., Franceschi, V.R., Krishnan, H.B. and Okita, T.W. (1988) Formation of wheat protein bodies: involvement of the Golgi apparatus in gliadin transport. *Planta* **176**: 173.

Klausner, R.D. (1989) Sorting and traffic in the central vacuolar system. *Cell* **57**: 703.

Kornfeld, R. and Kornfeld, S. (1985) Assembly of asparagine-linked oligosaccharides. *Ann. Rev. Biochem.* **54**: 631.

Krebbers, E., Herdies, L., Clercq, A.D., Seurinck, J., Leemans, J., Van Damme, J., Segura, M., Gheysen, G., Montagu, M.V. and Vandekerckhove. (1988) Determination of the processing sites of an *Arabidopsis* 2S albumin and characterization of the complete gene family. *Plant Physiol* **87**: 859.

Krieg, U.C., Johnson, A.E. and Walter, P. (1989) Protein translocation across the endoplasmic reticulum membrane: identificaiton by photocross-linking of a 39kD integral membrane glycoprotein as part of a putative translocation tunnel. *J. Cell Biol.* **109**: 2033.

Krishnan, H.B., Franceschi, V.R. and Okita, T.W. (1986) Immunochemical studies on the role of the Golgi complex in protein-body formation in rice seeds. *Planta* **169**: 471.

Kuhn, A. (1987) Bacteriophage M13 procoat protein inserts into the plasma membrane as a loop structure. *Science* **238**: 1413.

Kyte, T. and Doolittle, R.F. (1982) A simple method for displaying the hydropathic character of a protein. *J. Mol. Biol.* **157**: 105.

Lamb, F.I., Roberts, L.M. and Lord, J.M. (1985) Nucleotide sequence of cloned cDNA coding for preproricin. *Eur. J. Biochem.* **148**: 265.

Larkins, B.A. and Hurkman, W.J. (1978) Synthesis and deposition of zein in protein bodies of maize endosperm. *Plant Physiol.* **62**: 256.

Larkins, B.A., Wallace, J.C., Lending, C.R., Galili, G. and Kawata, E.E. (1989) Characterization and modifications of maize storage proteins. In *Biotechnology and Food Quality*, eds. Kung, S.-D., Bills, D.D. and Quatrano, R. Butterworths, London. 67.

Lazarow, P.B. and Fujiki, Y. (1985) Biogenesis of peroxisomes. *Ann. Rev. Cell Biol.* **1**: 489.

Lending, C.R. and Larkins, B.A. (1989) Changes in the zein composition of protein bodies during maize endosperm development. *Plant Cell* **1**: 1011.

Lingappa, V.R., Lingappa, J.R. and Blobel, G. (1979) Chicken ovalbumin contains an internal signal sequence. *Nature* **281**: 117.

Loh, Y.P., Brownstein, M.J. and Gainer, H. (1984) Proteolysis in neuropeptide processing and other neural functions. *Ann. Rev. Neurosci.* **7**: 189.

Lund, P., Lee, R.Y. and Dunsmuir, P. (1989) Bacterial chitinase is modified and secreted in transgenic tobacco. *Plant Physiol.* **91**: 130.

Lusis, A.J., Tomino, S. and Paigen, K. (1976) Isolation, characterization and radioimmunoassay of murine egasyn, a protein stabilizing glucuronidase membrane binding. *J. Biol. Chem.* **251**: 7753.

Lycett, G.W., Delauney, A.J., Gatehouse, J.A., Gilroy, J., Croy, R.R.D. and Boulter, D. (1983) The vicilin gene family of pea (*Pisum sativum* L.) a complete cDNA coding sequence for preprovicilin. *Nucl. Acids Res.* **11**: 8.

Lycett, G.W., Croy, R.R.D., Shirsat, A.H. and Boulter, D. (1984) The complete nucleotide sequence of a legumin gene from pea (*Pisum sativum* L.). *Nucl. Acids Res.* **12**: 4493.

Machamer, C.E., Florkiewicz, R.Z. and Rose, J.K. (1985) A single N-linked oligosaccharide at either of the two normal sites is sufficient for transport of vesicular stomatitis virus G protein to the cell surface. *Mol. Cell. Biol.* **5**: 3074.

Marco, Y.A., Thanh, V.H., Tumer, N.E., Scallon, B.J., and Nielsen, N.C. (1984) Cloning and structural analysis of DNA encoding and A_2B_{1a} subunit of glycinin. *J. Biol. Chem.* **259**: 13436.

Marty, F. (1978) Cytochemical studies on GERL, provacuoles, and vacuoles in root meristematic cells of *Euphorbia*. *Proc. Natl Acad. Sci. USA* **75**: 852.

Matile, P. and Moor, H. (1968) Vacuolation: origin and development of the lysosomal apparatus in root-tip cells. *Planta* **80**: 159.

Mellman, I., Fuchs, R. and Helenius, A. (1986) Acidification of the endocytic and exocytic pathways. *Ann. Rev. Biochem.* **55**: 663.

Miernyk, J.A. (1988) Is there a BiP-like protein in the endoplasmic reticulum of plant cells? *J. Cell Biol.* **107**: 763a.

Miernyk, J.A. and Nelson, T.C. (1989) Patterns of amino acids near the signal-sequence cleavage sites of plant secretory protein precursors. *Plant Physiol.* **89(S)**: 372.

Miyata, S. and Akazawa, T. (1982) Enzymic mechanism of starch breakdown in germinating rice seeds. *Plant Physiol.* **70**: 147.

Mizuno, K. and Matsuo, H. (1984) A novel protease from yeast with specificity towards paired basic residues. *Nature* **309**: 558.

Mollenhauer, H.H., Morre, D.J. and Norman, J.O. (1982) Ultrastructural observations of maize root tips following exposure to monensin. *Protoplasma* **114**: 117.

Moore, T.S. (1982) Phospholipid biosynthesis. *Ann. Rev. Plant Physiol* **33**: 235.

Moriyama, Y. and Nelson, N. (1989) H^+-translocating ATPase in Golgi apparatus. Characterization as vacuolar H^+-ATPase and its subunit structures. *J. Biol. Chem.* **264**: 18445.

Munro, S. and Pelham, H.R.B. (1986) An HSP70-like protein in the ER: Identity with the 78kD glucose-regulated protein and immunoglobulin heavy chain binding protein. *Cell* **46**: 291.

Munro, S. and Pelham, H.R.B. (1987) A C-terminal signal prevents secretion of luminal ER proteins. *Cell* **48**: 899.

Murakami, S., Hattori, T. and Nakamura, K. (1986) Structural differences in full-length cDNAs for two classes of sporamin, the major soluble protein of sweet potato tuberous roots. *Plant Mol. Biol.* **7**: 343.

Nguyen, T., Zelechowska, M., Foster, V., Bergmann, H. and Verma, D.P.S. (1985) Primary structure of the soybean nodulin-35 gene encoding uricase II localized in the peroxisomes of uninfected cells of nodules. *Proc. Natl Acid. Sci. USA* **83**: 1232.

Novikoff, P.M., Novikoff, A.B., Quintana, N. and Hauw, J.J. (1971) Golgi apparatus, GERL, and lysosomes of neurons in rat dorsal root ganglia, studied by thick section and thin section cytochemistry. *J. Cell Biol.* **50**: 859.

Olden, K., Bernard, B.A., Humphries, M.J., Yeo, T.-K., White, S.L., Newton, S.A., Bauer, H.C. and Parent, J.B. (1985) Function of glycoprotein glycans. *Trends Biochem. Sci.* **10**: 78.

Osawa, T. and Tsuji, T. (1987) Fractionation and structural assessment of oligosaccharides and glycopeptides by use of immobilized lectins. *Ann. Rev. Biochem.* **56**: 21.

Owens, R.J. and Northcote, D.H. (1981) The location of arabinosyl:hydroxyproline transferase in the membrane system of potato tissue culture cells. *Biochem. J.* **195**: 661.

Palade, G. (1975) Intracellular aspects of the process of protein synthesis. *Science* **189**: 347.

Palade, G.E. (1983) Membrane biogenesis: an overview. *Meth. Enzymol.* **96**: XXIV.

Pang, P.P., Pruitt, R.E. and Meyerowitz, E.M. (1988) Molecular cloning, genomic organization, expression and evolution of 12S seed storage protein genes of *Arabidopsis thaliana*. *Plant Mol. Biol.* **11**: 805.

Pelham, H.R.B. (1986) Speculations on the functions of the major heat shock and glucose-regulated protein. *Cell.* **46**: 959.

Pelham, H.R.B. (1988) Evidence that luminal ER proteins are sorted from secreted proteins in a post-ER compartment. *EMBO J.* **7**: 913.

Pelham, H.R.B. (1989*a*) Heat shock and the sorting of luminal ER proteins. *EMBO J.* **8**: 3171.

Pelham, H.R.B. (1989*b*) Control of protein exit from the endoplasmic reticulum. *Ann. Rev. Cell Biol.* **5**: 1.

Pelham, H.R.B., Hardwick, K.G. and Lewis, M.J. (1988) Sorting of soluble ER proteins in yeast. *EMBO J.* **7**: 1757.

Perara, E., Rothman, R. and Lingappa, V. (1986) Uncoupling translocation from translation: Implications for transport of proteins across membranes. *Science* **232**: 348.

Pesacreta, T.C. and Lucas, W.J. (1984) Plasma membrane coat and a coated vesicle-associated reticulum of membrane: Their structure and possible interrelationship in *Chara corallina*. *J. Cell Biol.* **98**: 1537.

Pfeffer, S.R. and Rothman, J.E. (1987) Biosynthetic protein transport and sorting by the endoplasmic reticulum and Golgi. *Ann. Rev. Biochem.* **56**: 829.

Prehn, S., Wiedmann, M., Rapoport, T.A. and Zwieb, L. (1987) Protein translocation across wheat germ microsomal membranes requires an SRP-like component. *EMBO J.* **6**: 2093.

Raikhel, N.V. and Wilkins, T.A. (1987) Isolation and characterization of a cDNA clone encoding wheat germ agglutinin. *Proc. Natl Acad. Sci. USA* **84**: 6745.

Ray, J., Knapp, J., Grierson, D., Bird, C. and Schuch, W. (1988) Identification and sequence determination of a cDNA clone for tomato pectin esterase. *Eur. J. Biochem.* **174**: 119.

Robinson, A., Kaderbhai, M.A. and Austen, B.M. (1987) Identification of signal sequence binding proteins integrated into the rough endoplasmic reticulum membrane. *Biochem. J.* **242**: 767.

Robinson, D.G. (1985) *Endo- and Plasma Membranes of Plant Cells.* John Wiley & Sons, New York.

Rogers, J.C., Dean, D. and Heck, G.R. (1985) Aleurain: A barley thiol protease closely related to mammalian cathepsin H. *Proc. Natl. Acad. Sci. USA* **82**: 6512.

Rose, M.D., Misra, L.M. and Vogel, J.P. (1989) KAR2, a karyogamy gene, is the yeast homolog of the mammalian BiP/GRP78 gene. *Cell* **57**: 1211.

Rothman, J.E., Miller, R.L. and Urbani, L.J. (1984) Intercompartmental transport in the Golgi complex is a dissociative process: Facile transfer of membrane protein between two Golgi populations. *J. Cell Biol.* **99**: 260.

Rothman, J.H., Yamashiro, C.T., Raymond, C.K., Kane, P.M. and Stevens, T.H. (1989*a*) Acidification of the lysosome-like vacuole and the vacuolar H^+-ATPase are deficient in two yeast mutants that fail to sort vacuolar proteins. *J. Cell Biol.* **109**: 93.

Rothman, J.H., Yamashiro, C.T., Kane, P.M. and Stevens, T.H. (1989*b*) Protein targeting to the yeast vacuole. *Trends Biochem. Sci.* **14**: 347.

Rothstein, S.J., Lazarus, C.M., Smith, W.E., Baulcombe, D.C. and Gatenby, A.A. (1984) Secretion of a wheat α-amylase expressed in yeast. *Nature* **308**: 662.

Sauer, A. and Robinson, D.G. (1985) Intracellular localization of posttranslational modifications in the synthesis of hydroxyproline-rich glycoproteins. Peptidyl proline hydroxylation in maize roots. *Planta* **164**: 287.

Schnell, D.J. and Etzler, M.E. (1987) Primary structure of the *Dolichos biflorus* seed lectin. *J. Biol. Chem.* **262**: 7220.

Schwaiger, H.A., Hasilik, A., von Figura, K., Wiemken, A. and Tanner, W. (1982) Carbohydrate-free carboxypeptidase Y is transferred into the lysosome-like yeast vacuole. *Biochem. Biophys. Res. Commun.* **104**: 950.

Selvendran, R.R. and O'Neill, M.A. (1982) Plant glycoproteins. *Encyclo. Plant Physiol.* **13A**: 515.

Sheehy, R.E., Pearson, J., Brady, C.J. and Hiatt, W.R. (1987) Molecular characterization of tomato fruit polygalacturonase. *Mol. Gen. Genet.* **208**: 30.

Shinshi, H., Wenzler, H., Neuhaus, J.-M., Felix, G., Hofsteenge, J. and Meins, J.F. (1988) Evidence for N- and C-terminal processing of a plant defense-related enzyme: Primary structure of tobacco prepro-β-1,3-glucanase. *Proc. Natl Acad. Sci. USA* **85**: 5541.

Shotwell, M.A., Afonso, C., Davies, E., Chesnut, R.S. and Larkins, B.A. (1988) Molecular characterization of oat seed globulins. *Plant Physiol.* **87**: 698.

Simon, A.E., Tenbarge, K.M., Scofield, S.R., Finkelstein, R.R. and Crouch, M.L. (1985) *Plant Mol. Biol.* **5**: 191.

Stevens, T.H., Rothman, J.H., Payne, G.S. and Shekman, R. (1986) Gene dosage-dependent secretion of yeast vacuolar carboxypeptidase Y. *J. Cell Biol.* **102**: 1551.

Stinissen, H.M., Chrispeels, M.J. and Peumans, W.J. (1985) Biosynthesis of lectin in roots of germinating and adult cereal plants. *Planta* **164**: 278.

Sturm, A. and Chrispeels, M.J. (1986) The high mannose oligosaccharide of phytohemagglutinin is attached to asparagine 12 and the modified oligosaccharide to asparagine 60. *Plant Physiol.* **80**: 320.

Sturm, A., Johnson, K.D., Szumilo, T., Elbein, A.D. and Chrispeels, M.J. (1987*a*) Subcellular localization of glycosidases and glycosyltransferases involved in the processing of N-linked oligosaccharides. *Plant Physiol.* **85**: 741.

Sturm, A., Van Kuik, J.A., Vliegenthart, J.F.G. and Chrispeels, M.J. (1987*b*) Structure, position, and biosynthesis of the high mannose and complex oligosaccharide side chains of the bean storage protein phaseolin. *J. Biol. Chem.* **262**: 13392.

Sturm, A., Voelker, T.A., Herman, E.M. and Chrispeels, M.J. (1988) Correct glycosylation, Golgi-processing, and targeting to protein bodies of the vacuolar protein phytohemagglutinin in transgenic tobacco. *Planta* **175**: 170.

Szumilo, T., Kaushal, G.P. and Elbein, A.D. (1986*a*) Purification and properties of glucosidase I from mung bean seedlings. *Arch. Biochem. Biophys.* **247**: 261.
Szumilo, T., Kaushal, G.P., Hori, H. and Elbein, A.D. (1986*b*) Purification and properties of a glycoprotein processing α-mannosidase from mung bean seedlings. *Plant Physiol.* **81**: 383.
Tague, B.W. and Chrispeels, M.J. (1987) The plant vacuolar protein, phytohemagglutinin, is transported to the vacuole of transgenic yeast. *J. Cell. Biol.* **105**: 1971.
Tague, B.W. and Chrispeels, M.J. (1988) Identification of a plant vacuolar protein targeting signal in yeast. *Plant Physiol.* **86(S)**: 506.
Taiz, L., Murry, M. and Robinson, D.G. (1983) Identification of secretory vesicles in homogenates of pea stem segments. *Planta* **158**: 534.
Takaiwa, F., Kikuchi, S. and Oono, K. (1986) The structure of rice storage protein glutelin precursor deduced from cDNA. FEBS Lett. **206**: 33.
Takaiwa, F., Kikuchi, S. and Oono, K. (1989) The complete nucleotide sequence of new type cDNA coding for rice storage protein glutelin. *Nucl. Acids Res.* **17**: 3289.
Torrent, M., Poca, E., Campos, N., Ludevid, M.D. and Palau, J. (1986) In maize, glutelin-2 and low molecular weight zeins are synthesized by membrane-bound polyribosomes and translocated into microsomal membranes. *Plant Mol. Biol.* **7**: 393.
Trelease, R.B. (1984) Biogenesis of glyoxysomes. *Ann. Rev. Plant Physiol.* **35**: 321.
Uy, R. and Wold, F. (1977) Posttranslational covalent modifications of proteins. *Science* **198**: 890.
Valls, L., Hunter, C.P. and Rothman, J.H. (1987) Protein sorting in yeast: the localization determinant of yeast vacuolar carboxypeptidase Y resides in the propeptide. *Cell* **48**: 887.
Vitale, A. and Chrispeels, M.J. (1984) Transient N-acetylglucosamine in the biosynthesis of phytohemagglutinin: Attachment in the Golgi apparatus and removal in protein bodies. *J. Cell Biol.* **99**: 133.
Vitale, A., Warner, T.G. and Chrispeels, M.J. (1984) *Phaseolus vulgaris* phytohemagglutinin contains high-mannose and modified oligosaccharide chains. *Planta* **160**: 256.
Vitale, A., Sturm, A. and Bollini, R. (1986) Regulation of processing of a plant glycoprotein in the Golgi complex: A comparative study using *Xenopus* oocytes. *Planta* **169**: 108.
Voelker, T.A., Florkiewicz, R.Z. and Chrispeels, M.J. (1986) Secretion of phytohemagglutinin by monkey COS cells. *Eur. J. Biol.* **42**: 218.
Voelker, T.A., Herman, E.M. and Chrispeels, M.J. (1989) *In vitro* mutated phytohemagglutinin genes expressed in tobacco seeds: role of glycans in protein targeting and stability. *Plant Cell* **1**: 95.
Vonder Haar, R.A., Allen, R.D., Cohen, E.A., Nessler, C.L. and Thomas, T.L. (1988) Organization of the sunflower 11S storage protein gene family. *Gene* **74**: 433.
Wallace, J.C., Galili, G., Kawata, E.E., Cuellar, R.E., Shotwell, M.A. and Larkins, B.A. (1988) Aggregation of lysine-containing zeins in protein bodies in *Xenopus* oocytes. *Science* **240**: 662.
Walter, P. and Blobel, G. (1986) Mechanism of protein translocation across the endoplasmic reticulum. *Biochem. Soc. Symp.* **47**: 183.
Wanner, G., Vigil, E.L. and Theimer, R.R. (1982) Ontogeny of microbodies (glyoxysomes) in cotyledons of dark-grown watermelon (*Citrullus vulgaris* Schrad.) seedlings. *Planta* **156**: 314.
Weidmann, M., Kurzchalia, T.V., Hartmann, E. and Rapaport, A. (1987) A signal sequence receptor in the endoplasmic reticulum membranes. *Nature* **328**: 830.
Wieland, F.T., Gleason, M.L., Serafomo, T.A. and Rothman, J.E. (1987) The rate of bulk flow from the endoplasmic reticulum to the cell surface. *Cell* **50**: 289.
Wilkins, T.A. and Raikhel, N.V. (1989) Expression of rice lectin is governed by two temporally and spatially regulated mRNAs in developing embryos. *Plant Cell* **1**: 541.
Wilson, L.G. and Fry, J.C. (1986) Extension—a major cell wall glycoprotein. *Plant Cell and Environ.* **9**: 239.
Zhang, Y-H., Lang, W.C. and Robinson, D.G. (1989) *In vitro* localization of hydroxyproline O-glycosyl transferases in *Chlamydomonas reinhardii. Plant Cell Physiol.* **30**: 617.

8 Identification and characterisation of tissue-specific genes from flowers

A.G. SMITH, S.M. BROWN and C.S. GASSER

8.1 Introduction

The reproductive process in plants requires the production of specialised organs to foster the development of the gametophytes and ensure fertilisation. The initiation and formation of these organs is regulated through a complex mechanism that has both genetic and environmental components (Bernier and Kinet, 1985). For example, some plants require specific photoperiods in order to initiate the flowering process. The genetic component flowering is evident from the numerous mutations that identify genes affecting flower morphology or function (Rick and Butler, 1956; Kaul and Murthy, 1985; Haughn and Somerville, 1988). Although the majority of the mutations are inherited as simple recessive traits and many of the mutations have been thoroughly described morphologically and genetically, the function of the mutant gene and the mechanism through which altered development occurs is not known. Progress on the determination of the molecular basis of such floral mutants has been impeded by the lack of a simple method for the isolation of the affected gene on the basis of phenotype and mapping alone.

Our strategy for the analysis of the genetic component of flower development has been to capitalise on the unique nature of the floral organs relative to the vegetative organs to identify cDNA clones that represent genes preferentially or exclusively expressed in flowers. These cDNAs serve as molecular markers for floral development and are useful in addressing the following questions. What are the mechanisms that regulate flower-specific gene expression? What is the nature of the flower-specific gene products, what are their functions and how do they interact during flower development? Is it possible to alter either the regulation or function of these flower-specific genes to produce changes in the flowering process? The answers to these questions would be used to synthesise a more complete model of the flowering process. However, to elucidate all the complexities and levels of coordination that are essential to the initiation, morphogenesis and function of flowers, the utilisation of a variety of strategies is required.

The morphogenesis of unique floral organs from newly organised floral meristems requires the synthesis of mRNAs that are similarly specialised and unique, relative to the mRNA populations in the vegetative organs. Several methods have been used to estimate the extent of the differences in gene expression that occur between the floral and vegetative organs. One- and two-dimensional gel electrophoresis has been used to measure directly the complexity of proteins produced in floral organs and to analyse indirectly the mRNA population through the synthesis of proteins by *in vitro* translation of floral mRNAs. The comparison of the protein profiles of reproductive and vegetative organs has shown both quantitative and qualitative differences in the proteins among the floral and vegetative organs and during different periods of floral development (Appels *et al.*, 1982; Sawhney *et al.*, 1985; Gasser *et al.*, 1988). Organ- and stage-specific proteins were clearly visible among a background of proteins that did not vary between the different organs. For example, in a study to determine the level of differential gene expression between the sporophyte and the male gametophyte, isozyme profiles of tomato pollen and vegetative tissues were compared. Sixty per cent of the isozymes found in the sporophyte were also found in the pollen. Of the 19 pollen isozymes that were tested, only one was not also found in the sporophyte (Tanksley *et al.*, 1981).

The magnitude of differential gene expression between vegetative and reproductive organs has been determined in solution hybridisation experiments that compare mRNA population between organs. In tobacco, it was found that, of the 26 000 and 25 000 diverse mRNA species in anthers and ovaries respectively, there were 10 500 mRNA species in anthers and 10 000 mRNA species in ovaries that were not present in the mRNA population from leaf tissues (Kamalay and Goldberg, 1980). Using similar methods to measure differential gene expression in *Tradescantia*, it was found that 64% of the mRNA sequences in pollen hybridise to those in shoots, and 60% of the shoot mRNAs hybridised to those in pollen (Willing and Mascarenhas, 1984). An understanding of the regulatory mechanisms responsible for the floral-specific expression of genes would provide a crucial key in the formulation of a model for the regulation of flower development in general. In addition, characterisation of the nature and function of the protein products of these genes would enhance our understanding of the processes of floral organ and tissue differentiation. We will discuss success in the identification of flower-specific cDNAs by a method of differential screening that exploits the preferential or exclusive expression of mRNAs in floral organs relative to the vegetative organs.

8.2 Identification of flower-specific genes

Tomato (*Lycopersicon esculentum* L.) cultivar VF36 was selected as a plant well suited for the study of flower development for a variety of reasons. Tomato has a long history as a popular research organism which provides a large base of published data on the physiology and morphology of fruits and flowers (Picken *et al.*, 1985; Sawhney and Bhadula, 1988). Also, a large number of mutant tomato plants have been identified that show infertility or abnormal flower development. These mutants can be used to examine the relationship between altered flower development and the regulation and function of the flower-specific genes. Isolation of floral material from tomato is simple compared with other plant species. Floral tissue is readily available from plants grown year-round, since the cultivar VF36 is day-neutral in its flowering requirement. The tomato flowers are large enough to allow easy dissection of the reproductive organs, and many flowers are produced on a single plant, allowing for the isolation of sufficient quantities of floral organs for biochemical analysis. A routine *Agrobacterium tumefaciens*-based method of gene transfer is available for introduction of genes into tomato (McCormick *et al.*, 1986). The ability to introduce modified or chimeric genes back into tomato will allow the regulation and function of these genes to be analysed in the same plant from which they were isolated. These properties make the tomato an excellent choice for the molecular analysis of flower development.

Genes that are tissue-specific or developmentally regulated have been isolated using a variety of methods. A direct method of gene isolation by tagging the gene of interest with an indigenous or introduced transposable element (Baker *et al.*, 1986; Döring and Starlinger, 1986) or with T-DNA introduced by infection with *Agrobacterium* (Feldman *et al.*, 1989) has been used (see Chapter 2). Both of these methods have general requirements for them to be effective. These requirements include: a high efficiency of gene inactivation when the transposable element moves into a gene, a transposon that is very mobile and inserts into many loci, a phenotype that can be readily scored after the tagging and, in the case of using the T-DNA, transformation and regeneration methods that allow many independent transgenic plants to be analysed. These requirements limit the applicability of gene tagging for the recovery of genes affecting flowering to a few species. Other genetic approaches applicable to the isolation of flower-specific genes could take advantage of plants with altered flower development obtained by natural variation or chemical mutagenesis. Numerous examples of plants in which single gene mutations lead to homeotic transformations of floral organs, abnormal floral organs or sterility have been described (Rick and Butler, 1956; Laser and Lersten, 1972; Koornneef *et al.*, 1983; Kaul and Murthy, 1985; Haughn and Somerville, 1988; Bowman *et al.*, 1989). Although the existence of these mutants demonstrates the importance of the genetic component in flower formation, the isolation of such phenotypically identified genes for further study raises important technical problems. One approach to cloning such genes is to identify a restriction fragment length polymorphism (RFLP) or previously cloned gene that is closely linked to the mutant locus. The method of chromosome walking may then be used to isolate sequentially clones from a genomic library that could potentially contain the gene of interest. This difficult method will be facilitated by the availability of increasingly refined RFLP maps (Bernatzky *et al.*, 1986; Nam *et al.*, 1989) and the ability to clone very large DNA fragments in yeast artificial chromosomes (Burke *et al.*, 1987).

Clones for tissue-specific genes can be directly isolated if the protein products of such genes can be identified and purified. The purified protein can be used to produce a specific antibody probe. The antibodies against the tissue-specific gene product would then be used to screen a cDNA library for clones expressing this protein. This method was used to isolate a cDNA clone corresponding to a developmentally regulated gene encoding polygalacturonase in tomato fruit (Sheehy *et al.*, 1987). The purified protein can also be sequenced to determine the amino acid composition of the protein. This sequence can be used to predict the nucleotide sequence of the gene, from which an oligonucleotide probe can be synthesised and used to screen a cDNA library directly for homologous sequences. This method has been applied successfully in the isolation of a cDNA clone for a glycoprotein that is associated with the expression of self-incompatibility in tobacco (Anderson *et al.*, 1986).

For our studies we were interested in examining a broad set of flower-specific genes that were regulated according to similar developmental programmes. We selected differential screening of cDNA libraries cloned in bacteriophages (Huynh *et al.*, 1985) as a method that was both rapid and simple for isolating a set of such genes from different developmental stages of the male and female reproductive organs of tomato. Briefly, a cDNA library was prepared from tissues that were expressing the genes of interest (in our case floral tissue). Filter replicas were prepared from platings of the library that were representative of the mRNA populations in the tissue or organ of interest. One filter was hybridised with a cDNA probe made from the homologous mRNA from which the library was prepared, and the other filter was hybridised with a probe derived from the tissues not expressing the gene. Clones that hybridised with only the homologous probe and not with the heterologous probe were selected as being specific to the homologous tissue at that developmental stage. These clones were isolated for further characterisation. This method has been used successfully to isolate temporally regulated or tissue-specific genes from a variety of organisms (De Vries *et al.*, 1983; Sargent and Dawid, 1983; Hedrick *et al.*, 1984; Mansson *et al.*, 1985; Nasrallah *et al.*, 1985; Slater *et al.*, 1985).

Since we were interested in developmentally regulated genes with organ specificity, we isolated tissues from flowers sorted into four broad stages that were defined by the length of the sepals; buds of $\leqslant 4$ mm, 6–7 mm, 8–9 mm and mature flowers were isolated. These stages represent, respectively, premeiosis to early meiosis, mid to late meiosis, post-meiosis with tetrads or immature microspores, and flowers at anthesis. Figure 8.1 shows the stages of growth for a typical tomato flower bud from the macroscopic appearance of the bud to anthesis. Under standard conditions, this process requires approximately 24

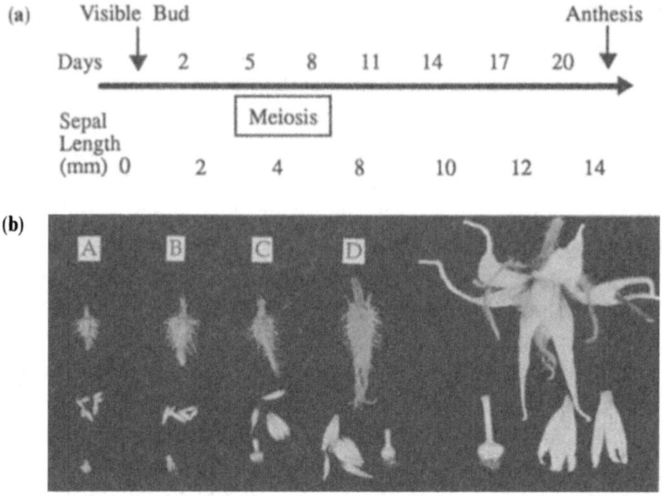

Figure 8.1. Stages in the development of tomato flowers. (a) stages of tomato flower development with sepal length from the first macroscopic appearance of the flower to anthesis. (b) shows the stages of flower development used in the isolation of floral-specific cDNAs. The sepal lengths of these buds are 3 mm (A), 5 mm (B), 8 mm (C), and 14 mm (D). These stages represent premeiosis to early meiosis, mid to late meiosis, post-meiosis with tetrads or immature microspores and flowers 1–2 days before anthesis.

days. If growth conditions are held constant there is a very good correlation between the sepal length and the developmental stage of the flower (McCormick *et al.*, 1987). In the cultivar VF36, meiosis is initiated in buds with sepals approximately 4 mm in length. Some variance occurs among buds when correlating the sepal length to the developmental stage, but this can be attributed to both the physiological state of the plant and the relative position of the bud on the inflorescence. Buds that are initiated first on the inflorescence have a faster growth rate and, therefore, a shorter time from initiation to anthesis.

cDNA libraries were produced from RNA isolated from whole buds less than or equal to 4 mm in length and from pistils and stamens dissected from flowers of the later three stages. Each of these libraries was differentially screened to isolate organ-specific sequences. For example, the stamen library from buds 8–9 mm in length was screened with the homologous mRNA labelled as cDNA and with a cDNA probe prepared from poly-A + RNA from young seedlings. The seedling mRNA was extracted from plants harvested prior to the initiation of floral meristems that included all the vegetative organs of the seedling. The library was first screened at a relatively high density (approximately 300 plaques per cm^2) to include as many plaques as possible in the first round. Plaques showing a strong hybridisation with the homologous probe and none with the seedling probe were selected for further characterisation. Due to the high plating density the first plaque isolates usually

Table 8.1. Identification of stage- and tissue-specific cDNA clones by differential screening of cDNA libraries from young flower buds, pistils and stamens at different stages of development.

		Pistil libraries			Stamen libraries		
	Buds $\leqslant$ 4 mm*	6–7 mm	8–9 mm	Mature	6–7 mm	8–9 mm	Mature[†]
Plaques in primary screen[‡]	20000	20000	20000	25000	20000	40000	40000
Primary screen positive[§]	65	65	16	55	30	180	110
Tertiary screen positive[¶]	10	10	4	9	17	11	20

*Length of the tomato buds from which tissue was collected to prepare cDNA libraries.
†Data on screening of mature anther libraries are from McCormick *et al.* (1987).
‡Total number of plaques from each library used in the differential screening.
§Plaques showing differential hybridisation to the homologous probe and not to the seedling probe after the first set of hybridisations.
¶Total number of cDNA clones which had strong hybridisation with the homologous probe and none with the seedling probe when DNA was isolated from each clone, digested with a restriction endonuclease, electrophoresed and Southern blotted.

contained several independent clones. To purify individual clones and to confirm the results of the first round of screening, the plaque mixtures were plated again at a lower density (approximately three plaques per cm^2). Duplicate filters were prepared from these plates and hybridised with the same probes. The lower density allowed longer exposures so that weak hybridisation signals from the seedling probe could be detected. Individual plaques were picked and DNA was prepared from each clone.

DNA was isolated from selected clones, digested with a restriction endonuclease and electrophoresed on agarose gels. The gels, each of which included several independent cDNA clones, were blotted and hybridised with the same two probes used in the original selection. The larger amount of DNA on these filters relative to the plaque lifts allowed more sensitive detection of hybridisation with the probes. The results of screening seven independent libraries prepared from different developmental stages of tomato flowers are summarised in Table 8.1. These results demonstrate that the differential screening technique worked well for the isolation of sequences that had little or no hybridisation to the seedling probe. Every plaque that hybridised with the homologous probe and had no signal with the seedling probe was isolated in the primary round of screening. Six to twenty-five per cent of the plaques selected in the primary screening showed a differential hybridisation in the tertiary screening. The elimination of clones from primary to tertiary screening was due in part to uneven blotting (failure to attach the lambda DNA to the filter) of the plaques between the first and second lifts that were made from each plate. To reduce false positives during the plaque lifts, the second replica filter was hybridised with the homologous probe, reducing the possibility that the seedling probe would have less DNA to hybridise with because the plates had been lifted previously. The tertiary screening results in Table 8.1 are the sum of only those clones that showed a very strong differential hybridisation between the two probes.

8.3 Characterisation of gene expression in floral organs

We were interested in determining whether the putative organ-specific cDNA clones showed any expression in organs other than those used in the initial screen. The Southern blot filters used for the tertiary screens were, therefore, additionally hybridised with labelled total cDNA probes synthesised from flower stages, or floral organs that were not used in the initial screen. These experiments showed, for example, that transcripts for several of the cDNAs that represent genes expressed in pistils and not in seedlings were also present in stamen RNA. While these experiments were not as sensitive as Northern blotting procedures, they allowed for rapid testing of a large number of clones for the timing and distribution of expression in all floral organs. The Southern blot filters of the cDNA clones were additionally used to test the clones for cross-hybridisation. In several instances individual cDNA clones hybridised to several other clones, indicating that clones had been independently isolated from a single or several closely related genes.

One of the purposes in isolating organ-specific clones from flowers was to identify and study the *cis*-acting sequences responsible for the regulation of the corresponding genes. For these studies, isolation of clones of the genes encoding the cDNAs is necessary. This isolation could be difficult if sequences with strong homology to the cDNA clones were present in high copy number in the tomato genome. Clones deriving from high copy number genes were identified by hybridising the Southern blots containing the cDNA with radiolabelled genomic DNA. The intensity of the signal produced by each clone in these hybridisations allowed for a rough estimation of the copy number of the corresponding gene. We selected clones that gave the weakest signals in these experiments, and showed the strongest floral specificity in the earlier screenings. To facilitate further studies the inserts of the cDNA clones were subcloned from the phage into plasmid vectors.

Northern blotting analysis was used to verify and refine the information gained from the hybridisations of the Southern blots of the cDNA clones. Northern blots containing poly-A + RNA from vegetative and floral organs of tomato plants were hybridised with radioactively labelled inserts from selected cDNA clones. An example of one of these hybridisations is shown in Figure 8.2. The Northern blot was probed with cDNA 108 isolated from the stamen library of buds 8–9 mm in length. The strong hybridisation to RNA

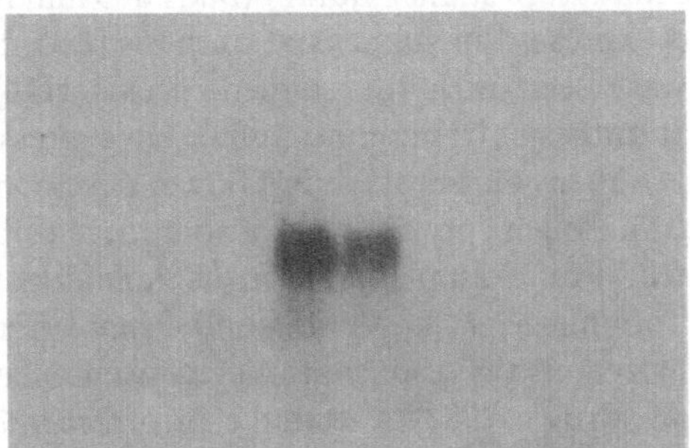

Figure 8.2. Developmental regulation of a stamen-specific mRNA. The Northern blot was hybridised with stamen-specific cDNA clone 108. Each lane contained 2 μg of poly A + RNA from pistils of tomato flowers at anthesis (1), pistils of flowers with sepal lengths of 8–9 mm (2), pistils of flowers with sepal lengths of 6–7 mm (3), stamens of tomato flowers at anthesis (4), stamens of tomato flowers with sepal lengths of 8–9 mm (5), stamens of tomato flowers with sepal lengths of 6–7 mm (6), 14-day-old tomato seedlings (7), tomato leaves (8), tomato fruit at the breaker stage of ripening (9) and green tomato fruit (10). The expression of this gene is limited to stamens during the early stages of flower development.

from stamens of 8–9 mm buds and the complete lack of hybridisation in vegetative organs demonstrated the stringent regulation of this gene. The Northern analysis also showed that transcript accumulation is temporally regulated, with abundant mRNA present from approximately mid-meiosis (6–7 mm) through immature pollen formation, followed by a rapid decline to an undetectable level in stamens at anthesis. Only two of the 11 stamen-specific cDNAs isolated from buds 8–9 mm in length hybridised to RNA isolated from organs other than stamens. Both of these clones (cDNAs 67 and 127c) hybridised to RNA from petal tissues at levels much reduced relative to the levels seen with stamen tissue (A.G. Smith *et al.*, 1990). This result is similar to the findings of other researchers. Stamen-specific cDNAs isolated from a cDNA library prepared from tomato flowers at anthesis (McCormick *et al.*, 1987) show a similar degree of stamen-specific expression (Twell *et al.*, 1989; Ursin *et al.*, 1989). Only one cDNA clone, designated LAT52, hybridised to petal RNA in addition to stamen RNA (Twell *et al.*, 1989). Similarly, pollen-specific clones have been isolated from *Zea mays, Tradescantia paludosa* (Stinson *et al.*, 1987) and *Oenothera* (Brown, 1988).

Tomato cDNAs selected from the staged pistil libraries as flower specific did not show this degree of organ specificity (Gasser *et al.*, 1989). Only the cDNA designated pMON9608 had expression limited to pistils. Over one-half of the clones that were characterised showed expression in the tomato leaf, although this expression was greatly reduced when compared with pistil expression. Several of the clones were also expressed in petals or stamens.

In general, the stamen-specific genes that have been characterised appear to be expressed over a relatively short developmental period. Of the 11 cDNAs isolated from the immature stamen library (buds 8–9 mm in length), only one of the clones was expressed in stamens at anthesis (A.G. Smith *et al.*, 1990). Similar results were seen with the stamen-specific cDNAs isolated from tomato flowers at anthesis (Ursin *et al.*, 1989). No expression of these genes was detectable in stamens during stage 1 of flower development (tomato buds 6–10 mm in length). Pollen-specific genes also appear to be highly developmentally regulated. For example, transcripts homologous to the pollen-specific cDNAs from maize, *Tradescantia*, and *Oenothera* are first detectable after the microspore mitosis and continue to accumulate up to anthesis (Stinson *et al.*, 1987; Brown, 1988). A second class of *Oenothera* pollen mRNAs (identified by cDNA P6) is only detectable during the last 1-2 days before anthesis (Brown, 1988). The results from the analysis of RNA accumulation during stamen and pollen development demonstrate the regulation of gene expression that exists as a bud matures into a flower with mature gametes. Even though this entire process is accomplished in a relatively short time-span, at least three groups of genes with distinct temporal regulation patterns have been identified.

Northern blot analysis demonstrated that the differential screening procedure was sensitive enough to discriminate between clones with a fivefold difference in the expression level between the vegetative and reproductive organs (Gasser *et al.*, 1989). The differential screening method detected both highly expressed transcripts (approximately 1% of the mRNA population) and less abundant ones (approximately 0.1% of the mRNA population). Differential screening was an appropriate approach to generating molecular probes for the study of flower-specific gene expression.

8.4 Tissue and cell specificity of gene expression in flowers

Floral organs such as stamens and pistils are composed of many specialised tissues that can be easily distinguished by their morphology or their differential staining (Esau, 1977). The identification of the specific tissues that are expressing the genes corresponding to the flower-specific cDNAs would aid in determining the possible functions of these genes. Separation of these tissues for RNA extraction and Northern analysis is not possible. Therefore, we used *in situ* hybridisation to analyse gene expression in the tissues and cells that constitute the different floral organs.

Fresh tomato flowers or buds were quick-frozen in a standard mounting medium and used to prepare frozen sections. The sections were fixed and prepared for hybridisation. Radioactively labelled probes representing the sense or antisense strand of the mRNA produced from the flower-specific cDNA clones were hybridised to the sections (A.G. Smith *et al.*, 1987). The sense probes served as controls to determine background hybridisation levels. After hybridisation, the sections were treated with ribonuclease A, washed to remove non-hybridising probe, and covered with a photographic emulsion to localise the hybridising probe.

The hybridisation pattern obtained using the probe prepared from stamen-specific cDNA 108 is shown in Figure 8.3. On Northern blots this clone hybridised only to stamens of buds 6–9 mm in length (Figure 8.2). In Figure 8.3, the silver grains appear as opaque regions over the unstained sections when viewed with bright-field illumination. In these transverse sections of tomato buds at four stages of development, weak hybridisation was first detected in the tapetal layer of the stamens of buds 5 mm in length (Figure 8.3B), and a much stronger signal was seen in the tapetal cells of buds 8 mm in length (Figure 8.3C). Buds 14 and 3 mm in length did not show hybridisation above background (Figure 8.3A and D). The dark appearance of the vascular traces and the mature pollen grains in Figure 8.3 is due to the opaque nature of these cells under bright-field microscopy. No concentration of silver grains above background appeared over these cells. The sense strand probe did not show any differential hybridisation to any floral tissues and was greatly reduced relative to the signals obtained with the antisense probe (data not shown). Thus, the only significant hybridisation signal above background was to tapetal cells of tomato buds from 5 and 8 mm in length. Similar tapetal-

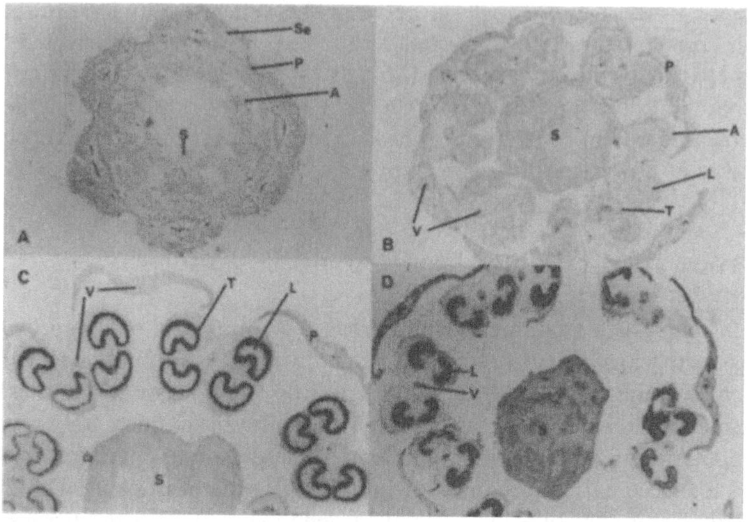

Figure 8.3. Localisation of RNA homologous to the stamen-specific cDNA clone 108 in median cross-sections of buds 3 mm (A), 5 mm (B), 8 mm (C), and 14 mm (D) in length. The sepals (Se), petals (P), style (S), vascular trace (V) and the locule containing the developing gametophytes (L) show no hybridisation above background in any stage of flower development. Weak hybridisation was detected in the tapetum (T) of anthers (A) from buds 5 mm in length, and strong signals were detected in the tapetal tissue of buds 8 mm in length. No homologous transcripts were detected in any organs of buds 3 or 14 mm in length. The cells in the locule of mature flowers and in the vascular traces of all flowers are dark because of their innate refractile properties not from the deposition of silver grains. The sections were exposed for 12 days and photographed using bright-field microscopy.

specific cDNAs have also been isolated from tobacco (Goldberg, 1988). These results show that expression of the gene corresponding to cDNA clone 108 is not only stamen-specific, but is specific to the tapetum of stamens during the period of maximum activity of this tissue.

On the basis of cytochemical and ultrastructural studies, several functions have been proposed for the tapetal tissue. These functions include nutrition for the developing pollen grains (Heslop-Harrison, 1964; Esau, 1977), production of material that is deposited on the pollen exine (Mepham and Lane, 1969; Echlin, 1971; Vithanage and Knox, 1980) and production of material for the dissolution of the tetrads (Frankel *et al.*, 1969; Stieglitz and Stern, 1973). Genes that are only expressed in the tapetum could participate in one of these functions or in the development or metabolism of the tapetal cells themselves. Very little is known concerning the role that individual genes play in these processes or the mechanisms that control their expression. The isolation of tapetal-specific genes may allow the function of the tapetal tissue and the movement of tapetal-specific gene products from this tissue to the male gametophytes to be examined.

Proper tapetal tissue function appears to be required for the production of viable pollen grains. The tapetal tissue is very active physiologically, as determined by the increase in rough endoplasmic reticulum, mitochondria and dictyosomes (Vasil, 1967; Mascarenhas, 1975). The ploidy level of the tapetal tissue of some species has been shown to increase during microsporogenesis and may reach levels of 6C (Heslop-Harrison and Moss, 1968; Liu *et al.*, 1987). The unique nature of the tapetal tissue in combination with its high metabolic rate correlates well with our identification of two additional independent cDNAs with identical expression patterns to cDNA 108. It has been hypothesised that the premature degeneration of the tapetal tissue results in cytoplasmic male sterility, since the first manifestation of the male-sterile phenotype is, in many cases, seen in the tapetal tissue of the developing anthers (Warmke and Lee, 1977; Bino, 1985). These data emphasise the essential nature of the tapetal tissue for normal pollen development.

The expression of the tapetal-specific genes closely parallels the differentiation and degeneration of this tissue. Ultrastructural studies begin to show a difference between the tapetum and the other cell layers derived from the primary parietal cells when the buds are approximately 4–5 mm in length. This correlates with the detection of tapetal-specific transcripts in buds 5 mm in length using *in situ* hybridisations (Figure 8.3). Similarly, the degeneration of the tapetum prior to opening of the flower (buds 12–14 mm in length) correlates closely with the disappearance of mRNA homologous to the tapetal-specific clones.

Several cloned stamen-specific genes expressed in the later stages of flower development have also been localised using similar methods of *in situ* hybridisation. Northern blot experiments had already determined that transcripts to these genes were abundant in mature pollen grains (Ursin, 1989). Localisation experiments revealed that transcripts of these genes were present in both sporophytic and gametophytic tissues. The initial expression of these genes in stage 2 stamens (buds 10–14 mm in length, post-tetrad formation with uninucleate microspores) was confined to the male gametes. However, stage 3 stamens (flowers with binucleate microspores and green petals prior to anthesis) had low levels of expression in the epidermis of the stamen wall, and stage 4 stamens (flowers at anthesis) had high levels of expression in the epidermis and the endothecium in addition to strong expression in the pollen. It was also shown that these genes were expressed in the growing tips of pollen after 18 hours in germination medium (Ursin *et al.*, 1989). The persistence of pollen-specific transcripts has also been examined during *in vitro* germination of maize and *Tradescantia* pollen (Stinson 1987). Unlike the tomato RNAs, the transcripts from maize and *Tradescantia* showed a dramatic reduction in level during the first 60 minutes of germination in culture.

The expression pattern of two pistil-specific cDNAs isolated from tomato flowers at anthesis has been determined (Gasser *et al.*, 1989). The transcripts to these cDNAs accumulate in the mature pistil in dramatically different tissues.

The transcripts to one clone, pMON9617, accumulate in the outer 5–8 cell layers of the transmitting tissue. The expression pattern is uniform from the interface of the stigma and the style to the point where the transmitting tissue meets the ovules. The transmitting tissue cells distinguished by the expression of this gene are not distinguished by staining with saffranin and fast green. The transmitting tissue connects the stigma to the ovaries and is thought to provide nutrients to the pollen tubes as they grow towards the ovules (Esau, 1977). This does not imply a direct correlation between this gene and the nutrition of the pollen tubes. Genes specific to the transmitting tissue could also function in regulating the direction of pollen tube growth or the suppression of pollen tube growth in species that exhibit self-incompatibility. The genes associated with self-incompatibility in *Nicotiana alata* are expressed throughout the transmitting tissue of the style and stigma as well as the epidermal cells of the placenta (Cornish *et al.*, 1987). However, DNA sequence comparisons between pMON9617, the *Nicotiana alata* and other self-incompatibility genes found no detectable homology (Gasser *et al.*, 1989).

Transcripts to a second pistil clone, pMON9608, accumulate exclusively in the ovules, in 1–2 cell layers of the integument surrounding the egg sac. Examination of the temporal regulation of this gene showed that no detectable transcripts are present in buds 6–9 mm in length (flowers in late meiosis to tetrad formation). Strong hybridisation was seen in pistils at fertilisation and in the early stages of fruit development (fruit 10 mm in diameter), and no signal was detected in fruit 25 mm in diameter (Gasser *et al.*, 1989). The cells corresponding to the hybridisation signal may provide nutritive material to the megagametophyte and embryo at early developmental stages (Cooper, 1931). A cDNA isolated from tobacco has also been shown to have ovary-specific expression localised to the cell layer that connects the ovule to the ovary (Goldberg, 1988). Again, the cellular localisation of a transcript suggests a possible group of functions for the flower-specific gene product.

In combination with data from the Northern blot analysis, the *in situ* hybridisation allowed detailed descriptions of the expression patterns of many of the flower-specific genes to be made. Determination of the potential functions for these flower-specific genes will come in part from correlating the spatial and temporal regulation patterns with the known functions and physiology of the cells in which they are expressed. Identification of the locations of the proteins encoded by these genes, using antibodies prepared against peptides from the deduced amino acid sequences of open reading frames, or derived from cDNA sequences expressed in *E. coli*, may also add to our understanding of the role of these genes. The protein products of the flower-specific genes may also be present in cells lacking detectable transcript signals. This would be predicted, for example, for a gene encoding a tapetal-specific mRNA that produces proteins that are exported and eventually deposited in the outer layers of the pollen grains.

In the discussion of the expression patterns of the flower-specific genes, we have postulated functions associated with various tissue or cell types of the flower. Much of the evidence for these putative functions is circumstantial or correlative. The isolation of genes expressed preferentially in differentiated cell types allows specific experiments to be performed that may clarify the role of these cells during flowering. Experiments designed to determine the function and mechanisms regulating the transcription of these genes are described in the next section.

8.5 Discussion

The differential screening of cDNA libraries produced from staged organs has enabled the isolation of clones from flowers of different species that are organ-specific, developmentally regulated and in some cases tissue- or cell-specific. The cDNA clones that have been isolated provide an additional increment in our knowledge of the flowering process. The flower-specific cDNAs will be effective tools in answering important questions regarding the regulation of flower development and the structure and function of the floral organs.

In most cases the regions of a gene that have been found to be most critical in regulating gene expression are 5′ to the start of transcription (Fraley *et al.*, 1986; Chapter 1). Examination of, and experimentation on, sequences 5′ of the start of transcription of the flower-specific genes will aid in elucidating the mechanisms responsible for the tissue- and organ-specific expression of the genes. The DNA sequence 5′ to the start of transcription of the stamen-specific gene, LAT52, has several directly repeated sequences, one of which includes the SV40 core enhancer (Twell *et al.*, 1989). The comparison of sequences 5′ to the start of transcription for two stamen-specific genes expressed in pollen (LAT52 and Zmg13), with two other genes with pollen expression *Adh* 1 and *Waxy* revealed many common sequence elements (Hamilton *et al.*, 1989). The role of the SV40 enhancer, the direct repeats or the homologous pollen sequences in regulating the expression of these genes has not yet been tested.

Experiments are in progress that may functionally define the sequences that control transcriptionally the spatial and temporal regulation of several flower-specific genes. We have isolated genomic clones that correspond to the flower-specific cDNAs and are isolating sequences 5′ to the start of transcription initiation from these clones which may be important in controlling organ-specific and temporal expression patterns. The putative regulatory sequences will be identified in part by fusing these regions to the coding sequences of reporter genes and introducing these chimeric genes into plants. The transgenic plants will be scored for the correlation between reporter gene expression and the expression of the flower-specific gene. When promoters have been identified, deletion and mutagenesis will be used to identify the regions of the promoters that are responsible for the specific regulation (see Chapter 1). Additional experiments can be performed to determine how these controlling sequences interact with other proteins to regulate gene expression. The isolation of these proteins and the genes that encode them will be the next step in determining the mechanisms that control gene expression.

For some clones additional information concerning the functions of the genes has come from sequencing the cDNA clones. Comparisons between the nucleic acid sequence or the deduced amino acid sequence of open reading frames has revealed several interesting homologies. A pollen-specific cDNA from *Oenothera* has strong homology to polygalacturonase from tomato (Brown, 1988). Two of the cDNAs isolated from tomato stamens at anthesis have homology to pectate lyase from *Erwinia* (Wing *et al.*, 1989). The putative activity of these proteins in the digestion of polysaccharides suggests that they may be involved in digestion of the intercellular matrix of the transmitting tissue to facilitate the growth of the pollen tube to the ovary. In other cases knowledge of the enzymatic function of a gene product known to effect flowering may not allow for an obvious proposal of the function of that gene in flowering. For example, a set of male-sterile mutants of *Arabidopsis* are known to lack adenine phosphoribosyltransferase activity (Moffatt and Somerville, 1988). However, even though the phenotype and biochemical lesion of these mutants are known, the mechanism that results in pollen absortion has not been determined.

Having clones of the coding sequences of the flower-specific genes and the regions that control expression will allow additional, more direct experiments to examine the potential functions for these genes. One strategy involves the production of antisense RNA to a flower-specific transcript. The production of antisense RNA has been shown, in other systems, to inhibit or reduce gene expression (Izant and Wientraub 1984; McGarry and Lindquist, 1986; Knecht and Loomis, 1987). Experiments using antisense RNA with *Drosophila* have demonstrated directly that a cloned sequence corresponded to a known mutant phenotype (Rosenberg *et al.*, 1985; Cabrera *et al.*, 1987). Antisense RNA has also been used successfully to reduce gene expression in several plant systems. Antisense constructs were shown to inhibit the activity of plasmid-encoded genes introduced into carrot cells by electroporation (Ecker and Davis, 1986). The expression of nopaline synthase (NOS) in plants transformed with a wild-type T-DNA could be decreased eight- to 10-fold by transforming these plants with a construct that expressed the antisense RNA of the NOS gene under the control of the CAMV 35S promoter (Rothstein *et al.*, 1987). This inhibition was stably inherited in the transgenic tobacco plants. Similarly, transformed plants that expressed the chalcone synthase antisense RNA had decreased flower pigmentation (van der Krol *et al.*, 1988), although the independent transgenic plants with the same construct exhibited very different pigmentation. Antisense RNA to the cDNA clone encoding the tomato enzyme polygalacturonase was expressed under the control of the 35S promoter to inhibit successfully the expression of polygalacturonase in tomato fruits (C.J.S. Smith *et al.*, 1988; Sheehy *et al.*, 1988). The inhibition of polygalacturonase was stably inherited. Even the very highly expressed gene encoding the small subunit of ribulose bisphosphate carboxylase has been effectively inhibited by the production of antisense RNA (Rogermen *et al.*, 1988). Although phenocopy mutants produced by antisense expression may not determine the enzymatic or structural function of the gene, they will allow the identification of genes that are essential for fertility and normal floral development.

The isolation of regions controlling organ- or tissue-specific expression will, in conjunction with plant transformation, allow the targeted expression of genes to the floral organs at specific times in their development. Coding sequences for phytohormone biosynthetic genes are obvious choices for this targeted expression. Phytohormones have been shown to affect many of the steps in the production of flowers. In tomato, auxins have been shown to increase the number of flowers in the first inflorescence (Picken *et al.*, 1985) and induce parthenocarpic fruit development (Mapelli *et al.*, 1978). Cytokinins alone and in combination with gibberellins have been shown to reduce flower abortion during stress periods in tomato (Kinet *et al.*, 1978). These studies involved the exogenous application of the phytohormones to the plants. Exogenous application can involve extended exposure to the phytohormone over long periods and has the unavoidable problem of uneven uptake and translocation leading to unknown concentration gradients within the tissues.

The coding sequences of genes for the synthesis of auxin and cytokinin have been cloned from the T-DNA of *A. tumefaciens* (Chapter 2). The cytokinin biosynthetic gene from the Ti plasmid of *A. tumefaciens* has been used to overproduce cytokinins in petunia and *Arabidopsis* (Klee *et al.*, 1987; Medford *et al.*, 1989). Plants that overexpressed these genes showed phenotypes that would be expected from increased levels of the phytohormone. Similar experiments using the auxin or cytokinin biosynthetic genes in combination with the flower-specific promoters are being performed (Budelier, C.S. Gasser and H.J. Klee, unpublished). Introduction of these chimeric genes into plants should lead to specific production of auxin or cytokinin in floral organs. The targeted synthesis will determine the effect of aberrant production of phytohormone on tissues or cells at specific points of flower development. Translocation of the phytohormone from the site of synthesis will allow the study of hormone gradients within the flowers. These experiments will provide insights on the role phytohormones play in floral functions and morphogenesis.

The methods of molecular biology have allowed the isolation of cDNA clones for flower-specific genes, many of which have now been determined to express in single tissues of flowers. The characterisation of the function and regulation of these genes has just begun and promises to add information that will be invaluable in the attempt to understand completely, define and manipulate the complex and essential process of flowering.

References

Anderson, M.A., Cornish, E.C., Mau, S.-L., Williams, E.G., Hoggart, R., Atkinson, A., Bonig, I., Grego, B., Simpson, R., Roche, P.J., Haley, J.D., Penschow, J.D., Niall, H.D., Tregear, G.W., Coghlan, J.P., Crawford, R.J. and Clarke, A.E. (1986) Cloning of cDNA for a stylar glycoprotein associated with expression of self incompatibility in *Nicotiana alata. Nature* **321**: 38.

Appels, R., Bouchard, R.A. and Stern G. (1982) cDNA clones from meiotic-specific poly(A) + RNA in *Lilium*: Homology with sequences in wheat, rye and maize. *Chromosoma (Berl.)* **85**: 591.

Baker, B., Schell, J., Lörz, G. and Federoff, N. (1986) Transposition of the maize controlling element "Activator" in tobacco. *Proc. Natl Acad. Sci. USA* **83**: 4844.

Bernatzky, R. and Tanksley, S.D. (1986) Toward a saturated linkage map in tomato based on isozymes and random cDNA sequences. *Genetics* **112**: 887.

Bernier, G. and Kinet, J.M. (1985) The control of flower initiation and development. In *Plant Growth Substances*, ed. Bopp, M. Springer Verlag, Berlin, 293.

Bino, R.J. (1985) Histological aspects of microsporogenesis in fertile, cytoplasmic male sterile and restored fertile *Petunia hybrida. Theor. Appl. Genet.* **69**: 425.

Bowman, J.L., Smyth, D.R. and Meyerowitz, E.M. (1989) Genes directing flower development in *Arabidopsis. Plant Cell* **1**: 37.

Brown, S.M.G. (1988) Molecular analysis of gene expression during pollen development in *Oenothera organensis.* PhD thesis, Indiana University.

Burke, D.T., Carle, G.F. and Olson, M.V. (1987) Cloning of large segments of exogenous DNA into yeast by means of artificial chromosome vectors. *Science* **236**: 806.

Cabrera, C.V., Alonso, M.C., Johnston, P., Phillips, R.G. and Lawrence, P.A. (1987) Phenocopies induced with anisense RNA identify the *wingless* gene. *Cell* **50**: 659.

Cooper, D.C. (1931) Macrosporogenesis and the development of the macrogametophyte of (*Lycopersion esculentum*). *Am. J. Bot.* **18**: 739.

Cornish, E.C., Pettitt, J.M., Boning, I. and Clarke, A.E. (1987) Developmentally controlled expression of a gene associated with self-incompatibility in *Nicotiana alata. Nature* **326**: 99.

De Vries, S.C., Harmsen, M.C., Kuiper, M.T.R., Dons, H.J.M. and Wessels, J.G.H. (1983) Molecular cloning of pea mRNAs encoding a shoot-specific polypeptide and light-induced polypeptides. *Plant Mol. Biol.* **2**: 295.

Döring, H.-P. and Starlinger, P. (1986) Molecular genetics of transposable elements in plants. *Ann. Rev. Genet.* **20**: 175.

Echlin, P. (1971) The role of the tapetum during microsporogenesis of angiosperms. In *Pollen: Development and Physiology*, ed. Heslop-Harrison, J. Butterworths, London, 41.

Ecker, J.R. and Davis, R.W. (1986) Inhibition of gene expression in plant cells by expression of antisense RNA. *Proc. Natl Acad. Sci. USA* **83**: 5372.

Esau, K. (1977) *Anatomy of Seed Plants.* John Wiley, New York.

Feldman, K.A., Marks, M.D., Christianson, M.L. and Quantrano, R.S. (1988) A dwarf mutant of *Arabidopsis* generated by T-DNA insertion mutagenesis. *Science* **243**: 1351.

Fraley, R.T., Rogers, S.G. and Horsch, R.B. (1986) Genetic transformation in higher plants. *CRC Critical Rev. Plant Sci.* **4**: 1.

Frankel, R. and Izhar, S. and Nitsan, J. (1969) Timing of callose activity and cytoplasmic male sterility in *Petunia. Biochem. Genet.* **3**: 451.

Gasser, C.S., Smith, A.G., Sachs, K.A., McCormick, S., Hinchee, M.A., Horsch, R., Shah, D.M. and Fraley, R.T. (1988) Isolation of differentially expressed genes from tomato flowers. In *Plant Gene Research,* eds. Verma, D.P.S and Goldberg, R. Springer Verlag, Vienna, 83.

Gasser, C.S., Budelier, K.A., Smith, A.G., Shah, D.M., Fraley, R.T. (1989) Isolation of tissue-specific cDNAs from tomato pistils. *Plant Cell* **1**: 15.

Goldberg, T.B. (1988) Plants: novel developmental process. *Sci* **240**: 1460.

Grierson, D., Tucker, G.A., Keen, J., Ray, J., Bird, J. and Schuch, W. (1986) Sequencing and identification of a cDNA clone for tomato polygalacturonase. *Nucl. Acids Res.* **12**: 8595.

Hamilton, D.A., Bashe, D.M., Stinson, J.R., and Mascarenhas, J.P. (1989) Characterization of a pollen-specific genomic clone form maize. *Sexual Plant Reproduction* **2** (in press).

Haughn, G.W. and Somerville, C.R. (1988) Genetic control of morphogenesis in *Arabidopsis. Developmental Genet.* **9**: 73.

Hedrick, S.M., Cohen, D.I., Nielsen, E.A. and Davis, M.M. (1984) The isolation of cDNA clones encoding T cell-specific membrane-associated proteins. *Nature* **308**: 149.

Heslop-Harrison, J. (1964) Cell walls, cell membranes and protoplasmic connections during meiosis and pollen development. In *Pollen Physiology and Fertilization*, ed. Linskens, H.F. North-Holland, Amsterdam, 39.

Heslop-Harrison, J. and Moss G.I. (1968) Photoperiod and pollen sterility in maize. *Ann. Bot.* **32**: 833.

Huynh, T.V., Young, R.A. and Davis, R.W. (1985) Constructing and screening cDNA libraries in lambda gt10 and gt11. In *DNA Cloning*. Vol. 1, *A Practical Approach*, ed. Glover, D. IRL Press, Oxford, 49.

Izant, J.G. and Wientraub, H. (1984) Inhibition of thymidine kinase gene expression by anti-sense RNA: a molecular approach to gene analysis. *Cell* **36**: 1007.

Kamalay, J.C., and Goldberg, R.B. (1980) Regulation of structural gene expression in tobacco. *Cell* **19**: 935.

Kaul, M.L.H. and Murthy, T.G.K. (1985) Mutant genes affecting higher plant meiosis. *Theor. Appl. Genet.* **70**: 449.

Kinet, J.M., Hurdebise, D., Parmentier, A. and Stainer, R. (1978) Promotion of inflorescence development by growth substance treatment to tomato plants grown in insufficient light conditions. *J. Am. Soc. Hort. Sci.* **103**: 724.

Klee, H.J., Horsch, R.B., Hinchee, M.A., Hein, M.B. and Hoffmann, N. (1987) The effects of overproduction of two *Agrobacterium tumefaciens* T-DNA auxin biosynthetic gene products in transgenic plants. *Genes Develop.* **1**: 86.

Knecht, D.A. and Loomis, W.F. (1987) Anti-sense RNA inactivation of myosin heavy chain gene expression in *Dictyostelium discoideum. Science* **236**: 1081.

Koornneef, M., van Eden, J., Hanhart, C.J., Stam, P., Braaksma, F.J. and Feenstra, W.J. (1983) Linkage map of *Arabidopsis thaliana. J. Hered.* **74**: 265.

van der Krol, A.R., Lenting, P.E., Veenstra, J., van der Meer, I.M., Koes, R.E., Gerats, A.G.M., Mol, J.N.M. and Stuitje, A.R. (1988) An antisense chalcone synthase gene in transgenic plants inhibits flower pigmentation. *Nature* **333**: 866.

Laser, K.D. and Lersten, N.R. (1972) Anatomy and cytology of microsporogenesis in cytoplasmic male sterile angiosperms. *Bot. Rev.* **38**: 425.

Liu, X.C. and Jones, K. and Dickinson, H.G. (1987) DNA synthesis and cytoplasmic differentiation in tapetal cells of normal and cytoplasmically male sterile lines of *Petunia hybrida. Theor. Appl. Genet.* **74**: 846.

McCormick, S., Niedermeyer, J., Fry, J., Barnason, A., Horsch, R. and Fraley, R. (1986) Leaf disc transformation of cultivated tomato (*L. esculentum*) using *Agrobacterium tumefaciens. Plant Cell Rep.* **5**: 81.

McCormick, S., Smith, A., Gasser, C., Sachs, K., Hinchee, M., Horsch, R. and Fraley, R. (1987) Identification of genes specifically expressed in reproductive organs of tomato. In *Tomato Biotechnology*, eds. Nevins, D.J. and Jones, R.A. Alan R. Liss, New York, 255.

McGarry, T.J., and Lindquist, S. (1986) Inhibition of heat shock protein synthesis by heat-inducible antisense RNA. *Proc. Natl Acad. Sci. USA* **83**: 399.

Mapelli, S., Frova, C., Torti, G. and Soressi, G.P. (1978) Relationship between set, development, and activities of growth regulators in tomato fruits. *Plant Cell Physiol.* **19**: 1281.

Mansson, P-E., Hsu, D. and Stalker, D. (1985) Characterization of fruit specific cDNAs from tomato. *Mol. Gen. Genet.* **200**: 356.

Mascarenhas, J.P. (1975) The biochemistry of angiosperm pollen development. *Bot. Rev.* **4**: 259.

Medford, J.I., Horgan, R., EL-Sawi, Z. and Klee, H.J. (1989) Alterations of endogenous cytokinins in transgenic plants using a chimeric isopentenyl transferase gene. *Plant Cell* **1**: 403.

Mepham, R.H. and Lane, G.R. (1969) Formation and development of the tapetal periplasmodium in *Tradescantia bracteata. Protoplasma* **68**: 446.

Moffatt, B. and Somerville, C. (1988) Positive selection for male-sterile mutants of *Arabidopsis* lacking adenine phosphoribosyl transferase activity. *Plant Physiol.* **86**: 1150.

Nam, H-G., Giraudat, J., den Boer, B., Moonam, F., Loos, W.D.B., Hauge, B.M. and Goodman, H.M. (1989) Restriction fragment length polymorphism linkage map of *Arabidopsis thaliana. Plant Cell* **1**: 699.

Nasrallah, J.B., Kao, T.-H., Goldberg, M.L. and Nasrallah, M.E. (1985) A cDNA clone encoding an S-locus-specific glycoprotein from *Brassica oleracea. Nature* **318**: 263.

Picken, A.J.F., Hurd R.G. and Vince-Prue, D. (1985) *Lycopersicon esculentum*. In *CRC Handbook of Flowering*, ed. Halevy, A.H. CRC Press, Boca Raton, FL, 330.

Rogermen, S.R., Abbott, M.S. and Bogorad, L. (1988) Nuclear–organelle interaction: nuclear antisense gene inhibits ribulose bisphosphate carboxylase enzyme levels in transformed tobacco plants. *Cell* **55**: 673.

Rosenberg, U.B., Preiss, A., Seifert, E., Jäckle, H. and Knipple, D.C. (1985) Production of phenocopies by *Krüppel* antisense RNA injection into *Drosophila* embryos. *Nature* **313**: 703.

Rothstein, S.J., DiMaio, J., Strand, M. and Rice, D. (1987) Stable and heritable inhibition of the expression of nopaline synthase in tobacco expressing antisense RNA. *Proc. Natl Acad. USA* **84**: 8439.

Rick, C.M. and Butler, L. (1956) Cytogenetics of the tomato (*Lycopersicon esculentum* Mill). *Adv. Genetics* **8**: 267.

Sargent, T.D. and Dawid, I.B. (1983) Differential gene expression in the gastrula of *Xenopus laevis. Science* **222**: 135.

Sawhney, V.K. and Bhadula, S.K. (1988) Microsporogenesis in the normal and male-sterile stamenless-2 mutant of tomato (*Lycopersicon esculentum*). Can. J. Bot. **66**: 2013.

Sawhney, V.K. and Greyson, R.I. (1972) On the initiation of the inflorescence and floral organs in tomato (*Lycopersicon esculentum*). *Can. J. Bot.* **50**: 1493.

Sawhney, V.K., Chen, K. and Sussex, I.M. (1985) Soluble proteins of the mature floral organs of tomato (*Lycopersicon esculentum* Mill). *J. Plant Physiol.* **121**: 265.

Sheehy, R.E., Pearson, J., Brady, C.J. and Hiatt, W.R. (1987) Molecular characterization of tomato fruit polygalacturonase. *Mol. Gen. Genet.* **208**: 30.

Sheehy, R.E., Kramer, M. and Hiatt, W.R. (1988) Reduction of polygalacturonase activity in tomato fruit by antisense RNA. *Proc. Natl Acad. Sci. USA* **85**: 8805.

Slater, A., Maunders, M.J., Edwards, K., Schuch, W. and Grierson, D. (1985) Isolation and characterization of cDNA clones for polygalacturonase and other ripening-related proteins. *Plant Mol. Biol.* **5**: 137.

Smith, A.G., Hinchee, M. and Horsch, R. (1987) Cell and tissue specific expression localized by in situ RNA hybridization in floral tissues. *Plant Mol. Biol. Rep.* **5**: 237.

Smith, A.G., Gasser, C.S., Budelier, K.A. and Fraley, R. (1990) Isolation and identification of stamen- and tapetal-specific genes from tomato. *Mol. Gen. Genet.* (in press).

Smith, C.J.S., Watson, C.F., Ray, J., Bird, C.R., Morris, P.C., Schuch, W. and Grierson, D. (1988) Antisense RNA inhibition of polygalacturonase gene expression in transgenic tomatoes. *Nature* **334**: 724.

Stieglitz, G. and Stern, H. (1973) Regulation of β-1,3-glucanase activity in developing anthers of *Lilium. Develop. Biol.* **34**: 169.

Stinson, J.R., Eisenberg, A.J., Willing, R.P., Pe, M.E., Hanson, D.D., and Mascarenhas, J.P. (1987) Genes expressed in the male gametophyte of flowering plants and their isolation. *Plant Physiol.* **83**: 442.

Tanksley, S.D., Zamir, D., and Rick, C.M. (1981) Evidence for extensive overlap of sporophytic and gametophytic gene expression in *Lycopersicon esculentum. Science* **213**: 453.

Twell, D., Wing, R., Yamaguchi, J. and McCormick, S. (1989) Isolation and expression of an anther-specific gene from tomato. *Mol. Gen. Genet.* **217**: 240.

Ursin, V.M., Yamaguchi, J. and McCormick, S. (1989) Gametophytic and sporophytic expression of anther-specific genes in developing tomato anthers. *Plant Cell* **1**: 727.

Vasil, I.K. (1967) Physiology and cytology of anther-development. *Biol. Rev.* **42**: 327.

Vithanage, H.I.M. and Knox, R.B. (1980) Periodicity of pollen development and quantitative cytochemistry of exine and intine enzymes in the grasses *Lolium perenne* L. and *Phalaris tuberosa* L. *Ann. Bot.* **45**: 131.

Warmke, H.E. and Lee, S-L. J. (1977) Mitochondrial degeneration in Texas cytoplasmic male-sterile corn anthers. *J. Hered.* **68**: 213.

Willing, R.P., and Mascarenhas, J.P. (1984) Analysis of the complexity and diversity of mRNAs from pollen and shoots of *Tradescantia. Plant Physiol.* **75**: 865.

Wing, R., Yamaguchi, J., Larabell, S., Ursin, V. and McCormick, S. (1989) Floral specific cDNAs from tomato encode proteins homologous to pectate lyase from the plant pathogen *Erwinia. J. Cell Biochem.* **13D**: 294.

Index

GPSR Compliance

The European Union's (EU) General Product Safety Regulation (GPSR) is a set of rules that requires consumer products to be safe and our obligations to ensure this.

If you have any concerns about our products, you can contact us on ProductSafety@springernature.com

In case Publisher is established outside the EU, the EU authorized representative is:

Springer Nature Customer Service Center GmbH
Europaplatz 3
69115 Heidelberg, Germany

Batch number: 10607212

Printed by Printforce, the Netherlands